Springer Series in Design and Innovation

Volume 59

Springer Series in Design and Innovation (SSDI) publishes books on innovation and the latest developments in the fields of Product Design, Interior Design and Communication Design, with particular emphasis on technological and formal innovation, and on the application of digital technologies and new materials. The series explores all aspects of design, e.g. Human-Centered Design/User Experience, Service Design, and Design Thinking, which provide transversal and innovative approaches oriented on the involvement of people throughout the design development process. In addition, it covers emerging areas of research that may represent essential opportunities for economic and social development.

In fields ranging from the humanities to engineering and architecture, design is increasingly being recognized as a key means of bringing ideas to the market by transforming them into user-friendly and appealing products or services. Moreover, it provides a variety of methodologies, tools and techniques that can be used at different stages of the innovation process to enhance the value of new products and services.

The series' scope includes monographs, professional books, advanced textbooks, selected contributions from specialized conferences and workshops, and outstanding Ph.D. theses.

The volumes of the series are single-blind peer-reviewed.

Keywords: Product and System Innovation; Product design; Interior design; Communication Design; Human-Centered Design/User Experience; Service Design; Design Thinking; Digital Innovation; Innovation of Materials.

How to submit proposals

Proposals must include: title, keywords, presentation (max 10,000 characters), table of contents, chapter abstracts, editors'/authors' CV.

In case of proceedings, chairmen/editors are requested to submit the link to conference website (incl. relevant information such as committee members, topics, key dates, keynote speakers, information about the reviewing process, etc.), and approx. number of papers.

Proposals must be sent to publishing editor Mr. Pierpaolo Riva (pierpaolo.riva@springer.com).

Daniel Raposo · João Neves · Ricardo Silva
Editors

Perspectives on Design IV

Research, Education and Practice

Springer

Editors
Daniel Raposo
Polytechnic Institute of Castelo Branco
Castelo Branco, Portugal

CIAUD, Lisbon School of Architecture
Universidade de Lisboa
Lisbon, Portugal

João Neves
Polytechnic Institute of Castelo Branco
Castelo Branco, Portugal

CIAUD, Lisbon School of Architecture
Universidade de Lisboa
Lisbon, Portugal

Ricardo Silva
Polytechnic Institute of Castelo Branco
Castelo Branco, Portugal

ISSN 2661-8184 ISSN 2661-8192 (electronic)
Springer Series in Design and Innovation
ISBN 978-3-032-00732-2 ISBN 978-3-032-00733-9 (eBook)
https://doi.org/10.1007/978-3-032-00733-9

This Springer imprint is published by the registered company Springer Nature Switzerland AG
The registered company address is: Gewerbestrasse 11, 6330 Cham, Switzerland

Preface

In contemporary times, we no longer inhabit an era in which design is regarded solely as an applied field aimed at problem-solving. Rather, it asserts itself as a reflective practice, critically engaged and committed to social transformation, ecological awareness, and a form of practical action rooted in epistemological inquiry.

At the heart of this shift lies a growing awareness that design does not merely intervene in the world but actively participates in its construction: it shapes relationships, instils values, translates ideologies, and projects futures. This transformation challenges designers—and design researchers and educators—to adopt a new role: that of thinkers, mediators, and agents of systemic change.

The fourth volume in the Perspectives on Design series is situated within this context. It is both a mirror of ongoing transformations in design theory and practice and a constructive contribution to the debate surrounding its future. The texts gathered in this book do not merely offer solutions but interrogate the very foundations of what it means today to "research in design", "teach design", and "practice design".

The reflections presented by the authors acknowledge that relevant knowledge for contemporary design often emerges from situated practice, interdisciplinary collaborations, tensions between tradition and innovation, and the demands of a society in constant reconfiguration.

This volume is organized into two parts—Communication Design, Media Design and Design Education as Catalysts for Change, and Fashion Design, Product Design and Innovation—comprising a total of 21 chapters that explore a range of approaches, methodologies, and fields of application.

The first part reflects on the potential of design as a driver of educational, cultural, and social change. It discusses visual and identity strategies, affect-based pedagogical methodologies, and decolonial approaches that propose plural ways of teaching and learning.

Particular attention is given to sustainability, inclusion, regeneration, and participation—whether through critical analyses of hospitality, the revaluation of urban space, or the curatorial engagement with material and immaterial heritage. This part also offers reflections on speculative design and co-design as strategies to address complex problems, rethink belonging, and redesign systems.

The second part focuses on the role of design in material, technological, and ecological innovation. It highlights new pedagogical practices in fashion education, the use of sustainable textile technologies, and the revalorization of local resources through circular models. The chapters further explore emerging forms of design dissemination in digital contexts, mediated by artificial intelligence, promoting more critical and accessible communication. The final reflections introduce innovative spatial and sensorial concepts, explore bioneurodesign as an integrative model connecting cognition, ethics and the environment, and revisit obsolete technologies as renewed sources of aesthetic and critical expression.

This volume, *Perspectives on Design IV*, positions itself as a contribution to an expanding design community—one that is open to experimentation and to questioning its own boundaries. The chapters presented here demonstrate that design can—and should—play a central role in constructing fairer, more conscious and more sustainable futures, not only through form and function, but also through its capacity for listening, imagination, and transformation.

The editors extend their sincere gratitude to all authors and institutions involved in the preparation of this publication. It is our hope that this volume may serve as a platform for dialogue, inspiration, and the collective construction of knowledge that expands the possibilities of design as a practice of research, education, and intervention.

Castelo Branco/Lisbon, Portugal — Daniel Raposo
Castelo Branco/Lisbon, Portugal — João Neves
Castelo Branco, Portugal — Ricardo Silva

Contents

Communication Design, Media Design and Design Education as a Catalyst for Change

Strategic Visuality: Exploring the Role of Imagery in Brand Visual Language—A Case Study of Renault Kwid

Manuela Gastal, Fernando Oliveira, and Daniel Raposo

Abstract Although the concept of image has been examined across multiple disciplinary fields, there remains a notable gap in scholarly literature regarding the role of imagery in constructing a brand's symbolic repertoire, meaning, and communicative strategies. This study aims to investigate the relationship between brand imagery and brand personality, with the objective of proposing parameters that inform the development of photographic aesthetics within a brand's visual language This article focuses on car brands and employs a literature review combined with case study analysis to systematically examine the Imagery (Photographic style and/or Illustration/Iconography) of the Renault Kwid. The study seeks to identify elements within photographic representations that align with the brand's DNA and narrative, in order to define key points of contact that can reinforce these strategic attributes. The findings include the identification of analytical parameters for evaluating brand imagery and demonstrate their applicability in monitoring and assessing the relevance and communicative function of images within brand communication.

Keywords Car brands · Brand personality · Brand visual language · Brand imagery/photography

M. Gastal (✉)
UNIDCOM/IADE, Unidade de Investigação em Design e Comunicação, Lisbon, Portugal
e-mail: manuela.gastal@universidadeeuropeia.pt

F. Oliveira
IADE, Faculdade de Design, Tecnologia e Comunicação, Universidade Europeia, Lisbon, Portugal
e-mail: fernando.oliveira@universidadeeuropeia.pt

D. Raposo
Lisbon School of Architecture, CIAUD, Universidade de Lisboa, Lisbon, Portugal

Centro de Investigação em Design para o Território, Polytechnic Institute of Castelo Branco, Castelo Branco, Portugal

D. Raposo
e-mail: draposo@ipcb.pt

D. Raposo et al. (eds.), *Perspectives on Design IV*, Springer Series in Design and Innovation 59, https://doi.org/10.1007/978-3-032-00733-9_1

1 Introduction

This study offers a comprehensive analysis of key parameters that may influence the selection of a specific typology of Imagery—such as photographic style, illustration, or iconography—for a brand's visual language. It highlights potential correlations between brand personality and the decision-making processes related to imagery. The objective is to identify connections between a brand's visual representations and its personality, thereby proposing parameters that can guide the definition of photographic aesthetics within the brand's distinctive visual identity. Olins (2008) and Raposo (2008) have established connections between Brand Personality or conceptual positioning and core elements of visual identity, including Typography(s), Symbol(s), Color(s), and overall visual style. However, there remains a notable gap in the literature concerning the relationship between Brand Personality and one of the Complementary Elements identified by Wheeler (2018) and Oliveira (2015) as Imagery—referring specifically to the visual characteristics of Photography, Illustration, and Iconography within brand communication. Lupton (2011) acknowledges Imagery as an integral component of a brand's visual language system and attributes significance to it, her work does not provide empirical evidence or in-depth exploration of its associative potential.

From a pragmatic perspective, communicating a specific mood or concept to a photography professional inevitably involves a degree of interpretation. This exchange entails a dynamic interaction between the intended message and the subject being represented, with the resulting image ultimately shaped by the photographer's vision and creative decisions. As Barthes (2018) suggests, the photographer's subjectivity plays a central role in constructing the photograph's meaning, which emerges from both the intentional and the unintended elements captured. Similarly, Joly (2012) emphasizes that the creation of meaning in an image is not fixed but negotiated between the image-maker's choices and the viewer's interpretation.

This is not to diminish the photographer's authorship—their practice is informed by a range of conceptual and technical variables that contribute significantly to the final composition (Shore 2010). Rather, the objective here is to establish a set of guiding parameters that underscore the value of analyzing imagery at its point of origin. These parameters can assist both in articulating visual expectations to professionals and in curating Imagery from platforms or databases designed to offer content aligned with specific communicative or branding contexts (Wheeler 2018).

This study is proposed as a contribution to academic discourse by fostering a dialogue between Imagery and Brand Language, grounded in logical associations shaped by a defined brand personality or conceptual idea. Such associations enable the structuring and teaching of typological units of meaning within a broader visual communication system. From a practical perspective, the research also offers valuable support for the design process in communication and visual identity design within the professional context. The underlying premise is to provide designers with clear and structured requirements for analyzing and articulating the components of visual identity and Imagery so that can be identified as an integral part of the branding system and message strategy.

These requirements not only support the creative process but also contribute to the justification and standardization of visual identity systems. In this regard, beyond the Basic Elements of the Brand System such as name, Typography(s), Symbol(s), and Color(s), (Oliveira 2015), Imagery will be posited as an equally significant component—one that requires distinctive attributes aligned with the personality of the brand system and should be incorporated into its visual guidelines. Furthermore, in sectors such as fashion and automotive, Imagery often plays a central role in communication, serving both as a medium for inspiration and as a direct representation of the product.

In such contexts, Imagery emerges as a crucial component in the construction and expression of brand identity. The integration of photography—or other forms of Imagery—as a core element of the visual identity system is particularly significant within certain typologies of brand identity, such as those found in Place Branding, Technology, and Energy sectors. Van Nes (2012) categorizes these as Dynamic Identities, characterized by flexible and adaptive visual systems that evolve over time or across contexts. Within these identity structures, Imagery frequently operates not as a Complementary Element, but as an integral part of the foundational components—alongside Typography(s), Symbol(s), and Name. It often merges with these elements in a fluid and adaptive manner, reinforcing the dynamic nature of the brand. As such, Imagery assumes a central role in the articulation of identity, contributing to both consistency and differentiation within complex and evolving brand environments.

The objectives guiding the development of this study are threefold: (1) to conduct a literature review that establishes a theoretical framework encompassing visual identity systems and the components of brand personality; (2) to explore the integration of some Elements extracted from the Joly's (2012) Image Analysis Model—comprised of three analytical levels: the Perceptual (formal visual aspects such as composition, color, and framing), the Cultural (symbolic and socio-historical codes), and the Interpretative (meaning-making shaped by viewers' subjective and contextual experiences)—as part of a methodological tool for examining how brand Imagery reflects and reinforces Brand Personality; and (3) to apply this integrated analytical framework in a case study focused on the Renault Kwid, enabling a detailed and systematic investigation of the brand's visual strategies.

Joly's Model work as a theoretical and practical base that together with other theorists are the foundational approaches in the visual communication, such as Barthes' (1984) distinction between denotation and connotation, which reveals how seemingly neutral images carry culturally constructed meanings, and Charles Sanders Peirce's triadic sign classification—*icon, index, and symbol* (Crow 2010), which allows deeper categorization of the representational modes used in branding. These frameworks, together with insights from design theorists such as Wheeler (2018) and Oliveira (2015), support a multidimensional model that connects the understanding of Imagery as a strategic and meaning-making device within brand identity systems with the constructed meanings. By situating brand Imagery within this interdisciplinary theoretical context, the study aims to contribute both to academic research and to design practice, particularly in defining parameters for the creation and analysis of photographic aesthetics in brand communication.

On the framework bellow (Fig. 1) illustrates how foundational approaches in visual communication—drawing from theories of Joly (2012), Barthes (1984) and Pierce (as interprets by Crow 2010), are connect to key concepts extracted from brand identity theories such as Wheeler (2018) and Oliveira (2015), in order to develop an interdisciplinary model that aims to connect the strategic brand meaning with culturally construct meanings, offering a broader perspective on how Imagery can expand the brand system by supporting the visual decision-making on the Imagery Element of the brands.

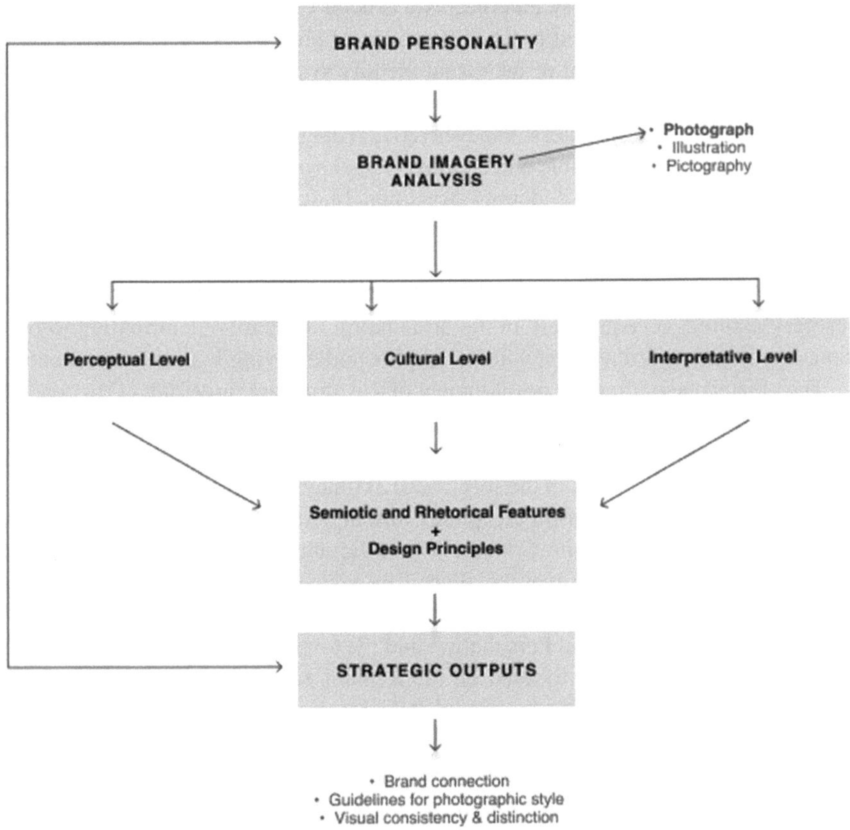

Fig. 1 This framework integrates Joly's (2012) model of image analysis with semiotic theory and principles design to support the analysis of brand imagery. It connects perceptual, cultural, and interpretative levels of image construction to the strategic expression of brand personality, offering tools for improvement on the Imagery field, focusing on the photography imagery in communication design. Author's image 2025

2 Methodology

This study adopts qualitative research (Mason 2018), and adopts an exploratory approach (Kumar 2013) aimed at analyzing the role of brand imagery in the construction and expression of brand personality, and it its supported by a literature review about semiotic theory and design research, the methodology is structured around three interrelated phases: literature review, theoretical framework construction, and case study analysis, aimed at synthesizing information in order to draw connections between references (Martin and Hanington 2012), to generate a theoretical framework on the areas of visual identity systems and their relationship with the personality components of a particular brand.

The literature will serve as support for a case study (Muratovski 2016), where the aim is to verify the possibility of integrating elements of reading and image analysis into the branding process in a pragmatic context, where data triangulation is managed in the transition from theory to cases through analysis processed by researchers (Muratovski 2016).

For the development of the case study, online data collection was employed, focusing specifically on the selected brand's official landing page. This approach follows Creswell's (2014) guidance on qualitative research, which emphasizes the use of multiple sources of evidence in case studies to enhance depth and validity. As Creswell (2014) notes, case study research typically draws on various forms of data, including observations, interviews, documents, and audiovisual materials. In this study, the emphasis was placed on visual analysis, using systematically collected visual content—primarily photographic and promotional imagery—as the primary data source. This choice was aligned with the study's objective to explore the role of imagery in brand expression and identity construction.

The selection of the Renault Kwid as the subject of the case study was guided by its positioning within the automotive sector, specifically in the category of popular or entry-level vehicles in the Brazilian market. The analysis focused on the Brazilian landing page of Renault, reflecting a localized perspective on the brand's communication strategy. This geographic and cultural specificity is essential, as interpretations of visual language are deeply influenced by visual culture, which varies across national and regional contexts (Hall 2012; Mirzoeff 2009).

The concept of visual culture acknowledges that the meanings constructed through images are not universal but are mediated by cultural norms, social history, and collective experience. Therefore, the choice of a region-specific platform reinforces the relevance of conducting a context-sensitive analysis of brand imagery, as the visual strategies employed by Renault in Brazil may differ significantly from those used in other markets.

3 Literature Review

While Wheeler (2018) emphasizes the importance of imagery in the construction and communication of brand identity, she does not offer specific analytical parameters that clarify the relationship between brand personality and the visual characteristics of photography. This absence creates a gap in understanding how visual elements—such as composition, mood, and subject matter—can be systematically aligned with the strategic expression of a brand's identity.

Bergström (2008), in contrast, provides a more nuanced view by defining image typologies according to their communicative function, rhetorical intent, and the message being conveyed. These perspectives align closely with foundational concepts from visual culture studies, which explore how images participate in constructing meaning within specific cultural and ideological contexts (Hall 2012; Mirzoeff 2009). Villas Boas (2010) further contributes to this discussion by highlighting visual aspects such as framing, composition, and the aesthetic treatment of subjects as critical to conveying meaning—making them particularly relevant to branding practices that seek to visually express a brand's core attributes or "DNA".

Given these contributions, this research identifies an opportunity to build upon existing theory by integrating the image analysis model proposed by Joly (2012). Her framework—structured around perceptual, cultural, and interpretative levels—provides a comprehensive approach to examining how meaning is produced through Imagery. Joly's semiotic lens enables a balanced analysis that considers both the formal visual elements and the socio-cultural contexts in which images are interpreted, especially within the domain of advertising. This integration serves as a foundation for the exploratory analysis proposed in this study, which aims to articulate how photographic aesthetics can be systematically aligned with brand personality within visual identity systems.

While brand personality refers to the strategic construction of human-like traits attributed to a brand (Aaker 1997; Mark and Pearson 2003; Martins 1999; Swystun 2022)—formed through communication and public perception—brand positioning ensures that the brand occupies a distinct, strategic space within its competitive landscape (Davis 2006; Elam 2004; Olins 2008). Positioning defines how a brand differentiates itself and establishes a meaningful relationship with its audience.

This strategic brand construction emerges from the alignment of vision, values, and business strategy, grounded in an understanding of consumer needs and expectations (Wheeler 2018). The brand vision serves as both a foundation and a guiding force, and as Wheeler (2018) emphasizes, it must be compelling, expressive, and aligned with the company's long-term objectives. This vision informs the visual and verbal identity system, which communicates the brand's positioning and personality across all touchpoints.

According to Wheeler (2018), the consistency of this system is maintained through a defined set of system elements, including the brand name, graphic mark, typography, color palette, symbols, shapes, tone of voice, and what Oliveira (2015) calls the "fifth element"—a distinctive visual cue that reinforces recognition and differentiation (Raposo and Oliveira 2021). These components, when coherently articulated, form the foundation of a brand identity system.

This notion of a systemized approach to branding is echoed in the work of Olins (2008), who emphasizes that successful brands are not merely visual artifacts but strategic expressions of an organization's culture, values, and promise. Olins argues that coherence across all brand expressions—visual, verbal, and experiential—is essential for building trust and long-term equity.

Similarly, Mollerup (2005) advocates for a systematic, design-led approach to identity development, highlighting the role of structure and rules in ensuring consistency while allowing for flexibility and evolution. He emphasizes that brand systems are not fixed templates but adaptable frameworks that evolve with the brand, market, and culture. In this view, brand systems serve both a communicative and organizational function—supporting recognition, differentiation, and internal alignment.

Together, these perspectives underline the importance of viewing brand identity not as a collection of isolated elements, but as an integrated system—strategically designed and consistently applied to express the brand's positioning and personality across diverse platforms and cultural contexts, and through the proposed research, we aim to integrate the elements of reading and image analysis, as outlined by Joly (2012), to analyze the elements that compose Brand Imagery, relating them to Brand Personality.

Joly (2012) initiates the analysis process by describing the analyzed advertisement, focusing on aspects related to the visual composition of the image. The choice of angle or point of view is also part of the elements analyzed through Joly's model (2012), as it is determinant and can either reinforce or contradict the sense of reality associated with the photograph (Joly 2012). The point of view, in turn, refers to the position from which the scene was captured, as highlighted by Goliot-Leté et al. (2011).

Another point emphasized by Joly (2012) concerns the Composition or interior geography of the visual message, which is one of the fundamental plastic tools. It plays an essential role in hierarchizing the vision and, therefore, in guiding the reading of the image (Joly 2012). In the field of Composition, Bergström (2008) emphasizes the presence of various principles governing the formation of an image. Among them, he highlights the concept of A&O—Attracting and Orienting, which refers to the ability to attract the viewer's attention through a dominant element, followed by orientation to other elements in the image. Within this approach, the photographer is encouraged to establish a sequence of reading the image. Within this framework, the author mentions elements such as symmetry and asymmetry, the rule of thirds, contrasts, and balance.

The interpretation of Shapes is also part of the systematic analysis proposed by Joly (2012), as for the author, as with other plastic elements, it is fundamentally influenced by anthropology and culture. Therefore, to understand the shapes organized in a visual message (and to understand the interpretation they evoke), it is necessary to try to disregard their literal meaning and observe them for themselves, attentively.

The element of Color is interpreted anthropologically in Joly's Model (2012), along with Shapes, which we mentioned earlier, and Light/Illumination, which states that lighting has a psychophysiological effect on the viewer (2012).

4 Case Study

During the execution of this case study, an analysis was conducted on the images extracted from the landing page of the Renault Kwid between February 2024 to March of 2025. The first step is capturing the images of the Renault Brazil website thought print screen. The second step is analyzing the Imagery in isolation from the other Elements of the System—Name, Graphic Mark, Typography (Oliveira 2015) as shown in Fig. 2. Then, aspects of the Brand's DNA/Personality that are described, in this the case, it's analyzed the relationship between the brand DNA with the information systematize from the website of the brand, together with the print screens of the Imagery of the Kwid. The third step refers to the Context. And the fourth involves describing and diagnosing with the Elements and its parameters presented in the brand analyzed, always relating them to the Brand's Personality/DNA.

This process can be carried out using a non-linear approach, and the order of the parameters may differ from the one presented here, in some brand systems may not include all Elements or may be neutral in relation to them, but in the case of the Renault is possible the identify all the Elements and apply the parameters of analysis in all the Elements—Context, Color(s), Light. Form and Object, Signs and Styles, Movement, Angles and Point of View and Composition.

The concept of brand DNA refers to the identity characteristics of the brand that determine its expression and language (Magalhães and Oliveira 2023). In the case of the Renault Kwid, it is possible to perceive the synergy between the elements related to the Mission, Vision, and Values of the Renault in its Imagery. A description found on the brand's website defines: "Our Passion, our Brand, our greatest inspiration is our passion. This feeling is present in every product and service we create" regarding their style, they define themselves as an "easylife", "French style" and as well as "focus on developing new technologies to became more competitive, balance and electrify". The Renault Kwid also features slogan which is "Make your day big" and "Find everything in the SUV of compact cars" which is reflected in the choices of the brand's Imagery focusing on design and unique features on its communication.

The next stage of analysis focuses on the parameter of color, which, as Joly (2012) emphasizes, is deeply shaped by anthropological and cultural factors. To fully interpret the role of color in a visual message, one must move beyond its literal or immediate associations and instead engage in a more attentive and culturally contextualized observation. In the case of the Renault Kwid, the chromatic choices stand out as a distinctive feature that sets the brand apart from its competitors.

The Imagery colors go from orange-yellow tones, goes to blue-sky tones until the grey-concrete tones aligns with the strategic positioning of the vehicle, reinforcing

Fig. 2 Synthesis of the imagery of the Renault Kwid, isolated from other components of the brand system communication. *Note* Original images from Neogama—Communication Campaign for Renault Kwid Brazil 2023—https://www.renault.com.br/. Author's archive, 2025

attributes such as design, balance and technology. In such a context, Renault's color choices operate not only as aesthetic decisions but as semiotic cues, supporting the brand's differentiation strategy positioning as a compact SUV, using a visual communication that reflects its personality and stands out within the category of popular vehicles.

The third Element analyzed is Light, just like Color, the interpretation of Light/Illumination has an anthropological character, but besides that, Illumination has a psychophysiological effect on the viewer (Joly 2012). In the case of the Renault Kwid, the use of natural light is applied due to the outdoors pictures, where its possible to visualize the sky tones as well.

The fourth Element analyzed is Shapes and Depicted Objects. According to Joly (2012), the interpretation of shapes, like other plastic elements, is fundamentally influenced by anthropological and cultural factors. In the imagery of the Renault Kwid, there is a notable emphasis on outdoor environments; however, these spaces cannot be clearly identified as urban, nor can they be defined as natural. Instead, the prominence of architectural elements and structured forms suggests a setting that foregrounds constructed, designed spaces—highlighting a connection to modernity and design-consciousness.

Within the analysis of shapes, the presence of two individuals—a man and a woman—positioned in a stylized manner can also be observed. These figures, who resemble fashion models, introduce a visual representation of gender duality and inclusivity. Moreover, their aesthetic presentation may allude to French cultural influence, particularly the country's longstanding association with fashion and style, which are integral to Renault's brand values. This subtle visual reference strengthens the brand's identity by linking its imagery to French elegance and sophistication, reinforcing its positioning within both automotive and lifestyle markets.

The fifth Element analyzed addresses the presence of Signs, Styles, and Memes and how they are related to the Brand System Elements and the other analyzed parameters. It is possible to perceive that among the signs and rhetorical elements present, there is an Allusion to the influence on fashion, using models on the Imagery of the car. Additionally, there is also Repetition Metaphor identify but the repetitive use of the imagery of the car itself.

The sixth Element analyzed is Movement and its relationship to the brand's DNA and personality. In the imagery associated with the Renault Kwid, movement is subtly suggested through visual cues that imply the vehicle is situated within an urban, dynamic environment. However, movement does not emerge as the primary communicative focus. Instead, the imagery prioritizes a stylized architectural and fashion-oriented context, suggesting that the brand seeks to position the Kwid not only as a vehicle for mobility but as an accessory to a contemporary, design-conscious lifestyle. This choice reinforces a brand personality that values aesthetics, modernity, and cultural alignment over traditional notions of speed or performance.

The seventh Element analyzed is the Angle of Viewpoint, a critical element in shaping the rhetorical impact of an image. As Joly (2012) notes, the choice of angle can either reinforce or challenge the sense of realism typically associated with the photographic medium. In the imagery of the Renault Kwid, the use of diagonal angles introduces a sense of dynamism and sophistication, aligning with rhetorical strategies that aim to suggest movement, modernity, and aesthetic refinement. From the perspective of visual rhetoric, such compositional choices are not merely technical, but persuasive tools that guide the viewer's interpretation and emotional engagement (Hill and Helmers 2004).

Additionally, the frequent use of low-angle or ground-level shots evokes a sense of magnificence, dominance, and power—rhetorical effects commonly associated with establishing authority or elevating the subject's status (Kress and van Leeuwen 2010). These angles contribute to the construction of a brand personality that is confident, bold, and aspirational.

Moreover, the visual strategy includes the use of wide-angle lenses or long-shot perspectives, which allow the viewer to perceive not only the object (the vehicle) but also its spatial context. This aligns with what Barthes (1984) would describe as a rhetoric of the image—where spatial composition and perspective operate as codes that carry connotative meanings. Here, the sense of openness and integration with the environment supports the narrative of the Renault Kwid as a car that is not only functional but also embedded in a lifestyle-oriented, urban visual discourse. Together, these visual strategies function to position the vehicle as more than a product—they frame it as an aspirational object within a larger cultural and symbolic system, reinforcing brand values related to design, freedom, and contemporary urban living.

Finally, the eighth and last Element analyzed, Composition, it is relevant to emphasize that this principle aims to synthesize all the Elements present in the Imagery composition and interpret the elements present in the Imagery composition through the analysis parameters, also relating how the Composition of the Brand Imagery is consolidated and relates to the other Elements of the Brand System.

In the case of the Renault Kwid, the imagery—primarily commercial in nature (Bergström 2008), employs compositions that strive for realism, particularly in photographs intended for product presentation. While certain conceptual images exhibit mild distortion, likely resulting from the use of wide-angle or specialty lenses, this appears to be a deliberate stylistic choice aligned with the brand's visual language. In other images, the use of zoom and cropping serves to highlight specific design details of the vehicle, directing the viewer's attention to key features. This visual strategy reinforces a sense of clarity and pragmatic communication, supporting the primary commercial goal of informing and persuading the consumer in a way that feels natural and accessible car, such as a popular vehicle. When the visual elements are examined in relation to the brand's DNA and the parameters outlined in the broader imagery analysis framework, a clear alignment emerges. The Renault Kwid's Imagery reflects its urban positioning, marked by a minimalist visual approach, subtle references to French cultural influence, and a design language that suggests technological sophistication.

The photographic choices—encompassing composition, focus, lens selection, and spatial framing—play a key role in shaping a cohesive brand narrative that merges functionality with style. These visual decisions reinforce Renault's positioning within the compact SUV segment of the popular vehicle market. Additionally, in another commercial visual application, the predominance of photography within the brand's communication strategy becomes clear, further emphasizing its importance as a core component of the brand's visual identity system as the picture bellow (Fig. 3).

5 Conclusion

The analysis conducted throughout this study demonstrates the feasibility and relevance of integrating image reading and analysis frameworks into the interpretation of brand language, particularly within the automotive sector. The findings highlight the importance of aligning brand Imagery with both the product's characteristics and the brand's personality, reinforcing the strategic role that visual elements play in constructing a coherent and persuasive brand identity. At the same time, it was observed that imagery is often employed to emphasize product attributes, which may not always directly reflect the deeper dimensions of brand personality. Nonetheless, in

Fig. 3 In another visual commercial application of the brand, the predominance of photography in its visual communication is evident, reinforcing its significance within the overall brand system. *Note* Original images from Neogama—Communication Campaign for Renault Kwid Brazil 2023—https://eurorenault.com.br/franca/. Author's archive, 2025

the case of the Renault Kwid, several meaningful connections between brand personality and visual representation were identified, supporting the central argument of this research.

Anchored in a qualitative approach, this study probes to be promising as it seeks to converge information from various sources, following established precepts in the specialized literature. The literature review enabled the consolidation of key concepts related to Visual Identity Systems and their association with the personality components of a particular brand, as well as guiding codes for the Imagery which seems to be a significant element in the brand systems, especially in businesses where visual communication is strongly supported by imagery, such as the automative sector addressed here.

The decision to utilize online systematize data collection—focusing on visual materials from the brand's official Brazilian landing page—proved effective in enabling a comprehensive and context analysis. This approach validated the potential of applying image reading methodologies from the semiotic field in the branding process and confirmed that Imagery can function as a strategic tool, not only for product representation but also for expressing brand values, personality traits, and cultural positioning.

Acknowledgements This study stems from the thesis Design Principles Applied to Semiotics in Brand Imagery: The Cases of Repertory Dance Companies.

The study was supported by UNIDCOM under a grant from the Fundação para a Ciência e Tecnologia (FCT) No. UIDB/00711/2020 attributed to UNIDCOM – Unidade de Investigação em Design e Comunicação, Lisbon, Portugal.

References

Aaker JL (1997) Dimensions of brand personality. J Mark Res 34:347–357. https://ssrn.com/abstract=945432

Barthes R (1984) Image, music, text (trans: Heath S). Flamingo Published by Fontana Paperbacks

Barthes R (2018) A câmara clara. Edições 70

Bergström B (2008) Essentials of visual communication. Laurence King Publ

Campain Renault Kwid (n.d.) Renault Euro. https://eurorenault.com.br/franca/landingpage/acelera-kwid-euro-renault-franca#:~:text=Dias%2020%2C%2021%20e%2022:%20Condi%C3%A7%C3%B5es%20imperd%C3%ADveis,Por%20onde%20for%2C%20v%C3%A1%20de%20Euro%20Renault

Creswell JW (2014) Research design: qualitative, quantitative, and mixed methods approaches, 4ª. SAGE Publications

Crow D (2010) Visible signs [An introduction to semiotics in the visual arts]. AVA Publishing

Davis, S. M. (2006). Brand Asset Management: Driving Profitable Growth through Your Brands. Jossey-Bass

Elam K (2004) Grid systems: principles of organizing type, 1ª. Princeton Architectural Press, New York

Goliot-Leté A, Joly M, Lancien T, Le Mée I-C, Vanoye F (2011) Dicionário da Imagem (trans: Silva V), 2°. Edições 70

Hall S (2012) Representation: cultural representations and signifying practices. SAGE

Hill CA, Helmers MH (eds) (2004) Defining visual rhetorics. Lawrence Erlbaum

Joly M (2012) Introdução à análise da imagem. Edições 70

Kress GR, van Leeuwen T (2010) Reading images: the grammar of visual design, 2nd edn., reprinted edn. Routledge

Kumar V (2013) 101 design methods: a structured approach for driving innovation in your organization. Wiley

Lupton E (ed) (2011) Graphic design thinking: beyond brainstorming, 1st edn. Princeton Architectural Press; Maryland Institute College of Art

Magalhães A, Oliveira F (2023) Visual synthesis languages of the DNA of the Star Michelin gastronomic brands. Converg J Res Arts Educ 13(26):149–165. https://doi.org/10.53681/c1514225187514391s.26.52

Mark M, Pearson CS (2003) O Herói e o fora-da-lei: Como construir marcas extraordinárias usando o poder dos arquétipos (trans: Scoss M), 1ª. Editora Cultrix.

Martin B, Hanington BM (2012) Universal methods of design: 100 ways to research complex problems, develop innovative ideas, and design effective solutions. Rockport Publishers. https://edisciplinas.usp.br/pluginfile.php/5646766/mod_resource/content/1/MARTINHANINGTON_Universal-Methods-of-Design.pdf

Martins JR (1999) A natureza emocional da marca: Como encontrar a imagem que fortalece sua marca, Negócio Ed.
Mason J (2018) Qualitative researching, 3rd edn. SAGE
Mirzoeff, N. (2009). An introduction to visual culture (2nd ed.). Routledge
Mollerup P (2005) Wayshowing: a guide to environmental signage; principles & practices. Lars Müller
Muratovski G (2016) Research for designers: a guide to methods and practice. Sage Publications Ltd
Olins W (2008) The brand handbook. Thames & Hudson
Oliveira FJMS (2015) Diagramas & Marcas: Contributos sobre a Utilização de Diagramas na Conceção e Análise do Discurso visual das Marcas [Tese de Doutoramento, Universidade de Lisboa, Faculdade de Arquitetura]. http://hdl.handle.net/10400.5/13974
Raposo D (2008) Design de Identidade e Imagem Corporativa: Branding, história da marca, gestão de marca, identidade visual corporativa. Edições IPCB. https://www.researchgate.net/publication/41208008_Design_de_Identidade_e_Imagem_Corporativa
Raposo D, Oliveira F (2021) O 5° elemento nas marcas, 1ª. Fless. https://5elementonasmarcas.pt/
Renault Kwid (n.d.) https://www.renault.com.br/
Shore S (ed) (2010) The nature of photographs: a primer, 2nd edn. (rev. exp. redesigned repr.). Phaidon Press
Swystun J (ed) (2022) Glossário De Marcas: Inglês-Português, 1ª. Bookman
Villas Boas A (2010) O que é cultura visual? Multitema
Wheeler A (2018) Designing brand identity: an essential guide for the whole branding team

Teacher Support for Project-Based Learning on Audiovisual and Media Production Degrees

Aldo M. S. Passarinho, Tiago J. C. Nunes, Rui J. P. Travasso, and Teresa S. Barradas

Abstract This article investigates the intersection between Problem-Based Learning (PBL) and the field of audiovisual and media production. The study is grounded in a systematic literature review, through which a curated selection of publications was analysed to extract data relevant to the application of PBL within this domain. The analysis led to the identification of key insights and the formulation of practical recommendations, particularly regarding the often-overlooked need for institutional and pedagogical support for educators implementing PBL. By addressing both theoretical and practical dimensions, this research contributes to the ongoing discourse on active learning methodologies in creative and practice-based educational contexts. Ultimately, this study seeks to serve as a valuable resource to support the effective implementation of PBL in the audiovisual and media production fields.

Keywords Project-based learning (PBL) · Higher education · Audiovisual and media production

A. M. S. Passarinho
CIEQV, Polytechnic Institute of Beja, Beja, Portugal
e-mail: aldo.passarinho@ipbeja.pt

T. J. C. Nunes · T. S. Barradas
Polytechnic Institute of Beja, Beja, Portugal
e-mail: tnunes@ipbeja.pt

T. S. Barradas
e-mail: teresa.penacho@ipbeja.pt

R. J. P. Travasso (✉)
CIAC, Polytechnic Institute of Beja, Beja, Portugal
e-mail: rui.travasso@ipbeja.pt

D. Raposo et al. (eds.), *Perspectives on Design IV*, Springer Series in Design and Innovation 59, https://doi.org/10.1007/978-3-032-00733-9_2

1 Introduction

Higher education is in constant transformation due to the evolution of society and its own internal developments. These challenges manifest in three main ways: (1) changes in the student population; (2) a focus on student competencies; and (3) new technologies that are reshaping education.

Regarding the first aspect, there is a clear trend of increasing student enrolment in higher education. This expansion has led to a broader diversity of student profiles, bringing with it a wide range of projects, expectations, and needs. However, a common characteristic among students is a generally low level of autonomy.

The second challenge arises from the changes introduced by the Bologna Process, which, for example, placed students at the centre of the teaching and learning process. This shift required the abandonment of a traditional, transmission-based model of education in favour of one focused on the development of student competencies. As a result, teaching methods and educators had to adapt, with standardized and one-size-fits-all teaching models becoming obsolete.

Finally, the third challenge relates to the omnipresence of new technologies. Students now have almost constant access to digital tools and resources. The key challenge here lies in guiding students to use that access effectively to acquire knowledge, integrating technology meaningfully into teaching, and supporting educators in adapting to these technological advances (Almeida et al. 2022).

The Project-Based Learning (PBL) methodology emerges as a pedagogical approach capable of addressing these challenges by promoting greater student autonomy, reinforcing the development of competencies, and actively integrating technology into the learning process.

This study aims to explore the application of PBL in higher education within the field of education and training in Audiovisual and Media Production, providing data that may support further research and encourage its implementation in this area. While this approach has already been explored in the field, the present study seeks to deepen the theoretical framework underlying its implementation by identifying best practices. These practices may both facilitate the adoption of PBL by higher education teachers and contribute to ongoing research in the field.

2 Training in the Audiovisual and Media Production Sector

According to the Yearbook of the Audiovisual Production Sector in Portugal (Burnay 2023), entertainment and fiction are the most frequently produced genres by the member companies featured in the study. The 2021 edition of the same yearbook noted that the number of companies dedicated to the production of films, videos, and television programs in Portugal increased between 2019 and 2020. The turnover of these companies also rose every year from 2018 to 2021, as did the number of workers, which has been growing steadily over the past ten years (Burnay and Ribeiro 2021).

In terms of television fiction production hours, Portugal ranks among the countries whose companies produce the most content. In 2017, it was the second-largest producer of television fiction in Europe, with nearly 1400 h, surpassed only by Germany, which produced over 1800 h (Burnay and Ribeiro 2020).

The increase in the number of hours of television fiction production reflects a growing need for well-trained professionals in the field—individuals equipped with solid foundational knowledge, relevant competencies, and cultural sophistication (Youyou and Kit (2025), p. 1148). This requires high-quality training both at the vocational and higher education levels, in line with the European Qualifications Framework in terms of the skills, competencies, and attitudes to be developed. Such quality training must result from strong educational projects guided by a commitment to society and professional development, supported by research and curricular development. It also demands innovation in the implementation of active teaching and learning methodologies, backed by appropriate institutional support.

On the other hand, there is a lack of academic studies in the field of Audiovisual and Media Production, resulting in an underdeveloped epistemological framework when compared to other disciplines, even within the arts and humanities. This situation makes it difficult to establish a connection between theoretical training and practices that truly respond to the needs of companies in the sector, particularly production companies working in audiovisual and multimedia content for television or cinema.

3 Research Context

The present research is grounded in the authors' experience as researchers and educators in the field of education and training in Audiovisual and Media Production. This experience includes the creation and development of course curricula in the field, notably short courses and undergraduate degrees, within which the adoption of Project-Based Learning (PBL) was advocated as a core teaching and learning methodology for the curricular design of training in this area. One such example is the undergraduate degree in Audiovisual and Multimedia at the Polytechnic University of Beja—Portugal, established by Ordinance No. 6539/2019 of July 19, 2019.[1]

For the development of this research, the publications by Thomas (2000) and Kokotsaki et al. (2016) were used—though not exclusively—as foundational references. These publications were chosen due to their relevance and breadth, serving as systematic literature reviews that offer a broad perspective on the adoption of PBL as a project-based teaching and learning methodology, allowing for a transversal application to the area under analysis in this study. Taking this into consideration, the research team found it useful to conduct a systematic literature review focused

[1] https://files.diariodarepublica.pt/gratuitos/2s/2019/07/2S137A0000S00.pdf. Accessed on April 7, 2025.

specifically on the field of Audiovisual and Media Production, covering the period from 2016 onwards and thereby narrowing the scientific scope of the investigation.

4 Problem Based Learning (PBL)

In the context of this research, PBL is understood as an active teaching and learning methodology based on three principles of constructivism: (1) learning is context-specific, meaning it is tied to the context in which it occurs; (2) students should actively engage in the learning process; and (3) learning objectives should be achieved through social interaction and knowledge sharing. This places the student at the centre of the teaching-learning process, characterized by student autonomy, constructive inquiry, group or collaborative work, specific objectives, communication, reflective practices, the presentation of a final product, all framed within a real-world theme and context (Kokotsaki et al. 2016). However, how does PBL differ from group work or assigned tasks, pedagogical aspects that have been part of teaching for decades?

If the project, and its central activity, does not present difficulty for the students, or if they use information or skills already acquired previously to complete it, this is not PBL but rather an exercise (Thomas 2000). Thus, the project must offer transformation and knowledge construction, with no pre-defined final outcome. There must be autonomy and responsibility on the part of the students, and unsupervised work.

It is also important to note that there is evidence that PBL is quite challenging in terms of planning and implementation. There are indications that students may encounter difficulties at certain stages of complex projects, such as at the beginning of reflection, in conducting investigations, in managing time and scheduling, and in effectively using technology. In these cases, guidance or the development of a support system for specific stages that teaches students how to learn independently becomes essential. Many authors suggest that, in such cases, the introduction of an initial scaffold system would be crucial (Kokotsaki et al. 2016). Despite this, various studies show that PBL is an effective learning methodology with results superior to traditional methodologies. For teachers as well, this is a challenging methodology that completely alters the teacher's role. It is a role that contrasts sharply with traditional teaching, where, in executing a project and solving problems, the teacher must step back and refrain from revealing their ideas in order to give students the space and time to explore their own paths and even choose paths different from those the teacher might have taken. This methodology shifts the teacher from being a lecturer to becoming a facilitator and pedagogical guide as students follow an autonomous path (Bender 2012).

5 Project-Based Learning Design and Fundamentals

5.1 Expeditionary Learning

This is a model of PBL that originated in Outward Bound. It involves exploration outside the classroom, and even outside the school, based on intellectual questioning, character development, and community building (Thomas 2000).

5.2 Ill-Structured Problems

This is a model of PBL that originated in the healthcare field, designed to help interns develop their diagnostic skills when facing structured health problems. However, this process has also been adopted in fields such as management, architecture, law, among others, through the incorporation of problem statements, databases, and tutorial processes that enable students to develop their hypothetical-deductive reasoning skills. More recently, the fields of mathematics and social sciences have also incorporated this model.

Regarding this model, it is important to note that the tutorial is a distinguishing element, distinct from task-driven scripted models. It is more closely aligned with models similar to scaffolding, meaning that the role of the tutorial is designed to become less intrusive over time, gradually increasing the user's autonomy.

5.3 Cognition Structure

There are several models based on the cognitive approach, including motivational models—where a goal in the form of a reward discourages public comparison among the students involved—research between experts and novices—simulating an environment and conditions similar to those in which certain experts evolved a particular subject—real-world context—learning is maximized through maximum similarity to real-life environments—and the application of technology—the use of computers and electronic devices provides students with access to data and information, also enhancing external collaboration and connection to the real world, particularly through the production of artifacts.

Regarding the adoption of PBL, it is important to consider what several authors refer to as the implementation stages derived from the principles of project-based learning. Lyu and Ang (2025), for example, argue that project-based learning should be divided into six stages, namely: (i) project design; (ii) plan formulation; (iii) collaborative inquiry; (iv) product creation; (v) product presentation; (vi) summary and evaluation.

6 Audiovisual and Media Production

The field of audiovisual and media production is an emerging area due to its connection to the evolution of technology. While the concept of audiovisual originally emerged in contexts more closely related to cinema and television (Díaz and Rebollo 2013; Greiner 2021), multimedia, which is also an integral part of media production, takes other directions. The term multimedia, although it has existed since the late 1950s, gained a new multidisciplinary focus with the integration of Human-Computer Interaction processes with Media Studies, with the computer becoming a central component. In this way, multimedia is defined as a multidisciplinary field based on art and technology—essentially computational technology (Zagalo et al. 2021). Consequently, the computer now plays at least one of the following three roles: creation, storage and compression, and distribution; highlighting four central characteristics: digital, voluminous, interactive, and real-time or synchronized (Quaranta 2013; Alves da Veiga 2021; Havaldar and Medioni 2009). Software has developed and replaced various media channels that were created throughout the last century, enabling networking, both with other software and with equipment such as digital cameras, audio players, televisions, among others (Manovich 2013).

Finally, we highlight the growing importance of Artificial Intelligence, which is present in various media products. It is enough to consider that Machine Learning is an integral part of many software applications, providing significant support in content production and automation (Zhu et al. 2020).

7 Methodological Approach—Systematic Literature Review

To conduct a systematic literature review, the following steps must be covered: gathering bibliographic sources, strategies for adding knowledge, evaluation and selection of sources to be used, tools to be employed, and presentation of the study. It must have a solid method and planning, and its result should serve to map, discover, critically assess, identify gaps, producing a final result in the form of a report or synthesis. As for the criteria and strategies for conducting a systematic literature review, they should align with the needs posed by the research question (Brizola and Fantin 2016).

8 Research Question

After analysing the bibliographic sources collected by the research team, and in the authors' opinion, the relevance of the recommendations by Kokotsaki et al. (2016) gained extreme importance for the successful implementation of the PBL methodology in the context of audiovisual and media production. Thus, the research team formulated the following question:

How has research in the field of Audiovisual and Media Production on the implementation of PBL, as an active teaching and learning methodology, aligned with the recommendations made by Kokotsaki et al. (2016)?

9 Objectives

This research, in addition to aiming to answer the research question, has inherent objectives that are spread across both the theoretical and practical components. Regarding the theoretical component, the goal is to intertwine the foundational literature on PBL with the development of teaching and learning processes in the field of audiovisuals and media production. As for the more practical objectives, the review and subsequent analysis of the connection between audiovisuals and media production with PBL will allow this document to serve as a resource for teachers and researchers who integrate this connection into their teaching or research activities, thus contributing to the epistemological development of the field as well as its curricular development.

10 Method

To focus on audiovisuals and media production, a review of publications from 2016 onwards was conducted across various academic search engines and specialized journals. This research resulted in the identification of 12 publications that the research team considered most suitable, as they addressed the use of the PBL methodology in education and training. These 12 publications also offer a variety of subfields, thus providing a broad perspective within audiovisuals and multimedia.

With the 12 identified publications, a systematic analysis of their content was carried out, seeking resonance in each of them with the six recommendations from Kokotsaki et al. (2016):

1. *Student support*: Students need to be effectively guided and supported; emphasis should be given on effective time management and student self-management, including making safe and productive use of technological resources.

2. *Teacher support*: Regular support needs to be offered to teachers through regular networking and professional development opportunities. The support from the school senior management is crucial.
3. *Effective group work*: High-quality group work will help ensure that students share equal levels of agency and participation.
4. *Balancing didactic instruction with independent inquiry method* work will ensure that students develop a certain level of knowledge and skills before being comfortably engaged in independent work.
5. *Assessment emphasis on reflection, self and peer evaluation.* Evidence of progress needs to be regularly monitored and recorded.
6. An element of *student choice and autonomy* throughout the PBL process will help students develop a sense of ownership and control over their learning. (p. 274)

With this method, and despite only analysing 12 publications, it was possible to collect information that could answer the research question and meet the proposed objectives.

11 Analysis of Results

From now on, subchapters will be presented, each referring to one of the recommendations by Kokotsaki et al. (2016), in which the data extracted from the analysed publications will be presented.

Systematic literature review based on the recommendations of Kokotsaki et al. (2016)

Author (year)	Student support	Teacher support	Effective group work	Balancing didactic instruction with independent inquiry method	Assessment emphasis on reflection, self and peer evaluation	Student choice and autonomy
McFadyen (2018)	The professor provides support through the different paths chosen by the students, requiring a high level of adaptation	No mention of teacher support	Collective reflection on results and group work from a certain stage of PBL	The professor assumed the role of facilitator, guide, and media production expert, providing support according to students' needs as they pursued their creative directions	The assessment model was co-constructed by the teacher and students, with ongoing feedback sessions and adjustments throughout the PBL process	An initial session to raise preliminary questions and identify topics of interest, after which they began an autonomous exploration of those topics
Martins and Ferreira (2019)	Theoretical foundations and job market orientations, familiarity with the technical aspects. Determine whether students are capable of being critical	No mention of teacher support	Working in pairs of students, who would then review their peers' work, and have a full-group review	Introduction of projects in the classroom enhances students' independent thinking, their problem-solving skills, their creativity, flexibility and adaptability	According to Kuo (2014), no matter how objective the assessment seems to be, there is always a certain degree of subjectivity that cannot be overcome	Introduction of projects increases students' independent thinking, problem-solving skills, creativity, flexibility and adaptability
Untari et al. (2020)	He points out that the scenarios or problems on which the model is based must be well defined	No mention of teacher support	The challenges presented require students to work in groups to solve problems	No mention of balancing	No mention of evaluation	No mention of student choice and autonomy

(continued)

(continued)

Author (year)	Student support	Teacher support	Effective group work	Balancing didactic instruction with independent inquiry method	Assessment emphasis on reflection, self and peer evaluation	Student choice and autonomy
Gasonpan and Temdee (2021)	Simulation activities in order that students can experience problem-solving tasks. Teacher manages groups of students defining the scope of project problems	No mention of teacher support	Collaborative learning in a way to encourage students to exchange ideas with each other during the process	Students learn by themselves, have free time to review lessons and learn from the media and knowledge sources	Peer review, students asked to rate their team members in different aspects	Students take the self-study mode and complete a pre-test before starting a module. After online lectures students are asked to complete a post-test
Demirci et al. (2023)	He mentions the support of the teacher as a facilitator	No mention of teacher support	Refers to the importance of teamwork	An expert professional is invited to balance knowledge and practice	Refers to the fact that teamwork makes it difficult to identify each student's effort	Assumes the importance of autonomous learning
Cuenca-rodríguez et al. (2024)	Prior explanatory sessions on the themes/problems to be addressed	No mention of teacher support	Co-operative work through the sharing of results and processes	Free creative exploration by the students balanced with minimal support from the professor	No mention of evaluation	A free and creative component was present

(continued)

(continued)

Author (year)	Student support	Teacher support	Effective group work	Balancing didactic instruction with independent inquiry method	Assessment emphasis on reflection, self and peer evaluation	Student choice and autonomy
Julio et al. (2024)	Design a plan that mixes PBL with SAFe—Scaled Agile Framework. The latter offered a rigid structure in which PBL functioned	No mention of teacher support	Group work with tasks defined according to the skills of those involved	Autonomous student research on problems aligned with community or industrial needs, followed by a creative process aimed at addressing these challenges	Assessment was both quantitative and qualitative, carried out through feedback sessions and a system for documenting students' learning progress	Autonomy in both the programs being worked on and the direction of the work
Ling et al. (2024)	It is stated that the students may rely on peer knowledge, expert advice, academic articles end similar resources	No mention of teacher support	PBL solutions may facilitate and encourage university students from different cultural backgrounds to work together and accomplish their common goal. Teachers maximise motivation	PBL encompasses the "problem" and "project" methodologies. Using PBL in intercultural scenarios, during creative processes	Noticed systematization: 1—Problem-initiated; 2—Curriculum approach; 3—Learner-centered	Problem selection; Purpose; The nature of educational objectives; Tasks; Presentation of the problem; Format; Process; Resources; Role of the teacher; The demonstration of learning

(continued)

(continued)

Author (year)	Student support	Teacher support	Effective group work	Balancing didactic instruction with independent inquiry method	Assessment emphasis on reflection, self and peer evaluation	Student choice and autonomy
Pérez-Gómez (2024)	Problems complemented by a set of audios and reading suggestions available online	No mention of teacher support	Group work to develop the contents	A balance between students' exploration of foundational content and the prior explanation of core concepts	The author highlights the need for developing a more individualized assessment model aimed at differentiating among students	All questions posed prompted students to draw their own conclusions, and it was through these conclusions that the teacher guided the development of the content
Zhang and Jia (2024)	No mention of student support	No mention of teacher support	Projects can be carried out by an individual or by groups with no size limitation	The project doesn't need to have discipline-related limits. It doesn't need a final product. It centres on identifying issues	It emphasises the importance of public presentation. The experience of this social practice benefits the students' self-realisation	No mention of student choice and autonomy
Yee and Huey Shyh (2024)	Learning the fundamentals of journalism and different formats of distribution (writing, mobile, online). Free tools, such as Google News Initiative	No mention of teacher support	Group work is not clear	Students conduct research and analyze the problem using various sources. Develop solutions, based on research and analysis	Producing a reflective essay to evaluate the effectiveness of the media and information literacy project in enhancing students' abilities	Students, in general, were able to initiate autonomous learning under this PBL approach

(continued)

(continued)

Author (year)	Student support	Teacher support	Effective group work	Balancing didactic instruction with independent inquiry method	Assessment emphasis on reflection, self and peer evaluation	Student choice and autonomy
Youyou and Kit (2025)	Teachers are mentors, offering guidance, support and assistance to management and asking timely questions	No mention of teacher support	Emphasises collaborative learning. Emphasises the importance of group work skills. The group must engage in communication, negotiation and collaboration between members	Teachers act as mentors, at the same time as posing questions that stimulate intellectual exploration on the part of the students	It states that evaluation should be comprehensive and include both the evaluation of results and the evaluation of the process	Students experience autonomy, independently determine project implementation plans, get involved in surveys, present results and reflect on evaluations

11.1 Student Support

The first recommendations, referred to by Kokotsaki et al. (2016) as "Student Support", refer to the support that should be provided to students when PBL is implemented as an active teaching and learning methodology. The authors emphasize that students should be effectively guided and supported. In this regard, emphasis should be placed on effective time management and self-management for students, where the safe and productive use of technological resources should be anticipated.

In the set of articles analysed in the systematic literature review, the emphasis placed on the student, both in terms of the support provided and their role in the development of the project, is recurrent. In the most recent article, Youyou and Kit (2025) underline that project-based learning should prioritize students as the main participants, effectively leveraging their proactive engagement and enthusiasm. The authors highlight that PBL allows students to develop their autonomy, given that they themselves must plan the different stages of project execution, including guiding inquiry processes, presenting results, and reflecting on the evaluation (2025, p. 1154). This occurs in a context where the projects should include practical experience, through which the authors argue that students improve their overall skills.

Demirci et al. describe PBL as a teaching and learning method that allows students to engage in a sustained and collaborative manner in the development of a project, often organized around a guiding question (2023, p. 2) provided by the professor.

On the other hand, Martins and Ferreira emphasize the concern about how teachers can prepare students for the future, both in terms of theoretical foundations and in providing guidance for the job market. The authors emphasize that this should be achieved without neglecting the technical aspects that students must acquire. They also mention the importance of students having the ability to critically assess their own work, so they can identify and overcome their own difficulties.

Some authors, such as Ling et al. (2024) refer to examples of the support that can or should be provided to students, including the use of peer knowledge, expert knowledge, or the recommendation of academic articles.

From another perspective, Gasonpan and Temdee (2021) highlight the importance of teachers proposing simulation activities so that students can experience problem-solving autonomously. The authors also emphasize the teacher's role in managing group members and their responsibility in defining the scope of the project problems.

11.2 Teacher Support

According to the recommendation of Kokotsaki et al. (2016) the support for teachers in implementing PBL as an active teaching and learning methodology should be reflected in the creation of support networks and opportunities that allow for their professional development. The author emphasizes the importance of the support that the school can provide.

Considering this recommendation from Kokotsaki et al. (ibid.), it is noteworthy that in the set of articles analysed, none of the authors referred to the type of support provided to teachers in the various contexts of project implementation. However, the need for support to students, associated with the quality of planning, implementation, management, and evaluation of the projects, is clearly emphasized.

For Youyou and Kit, the teacher plays an important role "when designing project content, it is crucial to incorporate tasks that are highly applicable in real-life situations, embodying the principle of placing practice at the core" (2025, p. 1155). Similarly, they consider that the teacher, mainly functioning as a mentor in providing and assisting with project management, should ask timely questions that allow for greater intellectual exploration (ibid. p. 1154). This implies that teachers should have access to real-world market problems and support when preparing projects, alongside the necessary relationship between theory and practice related to real-world problems from the job market and companies.

In the same vein, Untari et al. (2020) state that the success of this teaching and learning model requires and should be focused "on scenario design that needs to be well designed" also highlighting that the problems presented by the teachers should be applicable in the real world. This requires knowledge and a relationship between the teacher and the real job market, and that they have time for designing these scenarios.

Reinforcing the demanding role of the teacher in implementing PBL as a teaching and learning methodology, Demirci et al. mention, regarding a project involving the creation of brands for social networks, that the teacher "resumed the role of a facilitator and a consultant in helping and guiding students throughout the semester" (2023, pp. 3–6). On the one hand, this task requires theoretical knowledge, but it also demands knowledge of the job market so that the teacher can effectively play the role of both educator and consultant.

Further emphasizing this amplification of the teacher's knowledge, McFadyen (2018) states that the fundamental areas of multimedia include video production, computer science, photography, audio production, and graphic design. However, there is still a demand for a connection with diverse knowledge areas, which later result in products such as films, podcasts, video games, websites, graphic posters, among others. All of these are creative possibilities for PBL, and in certain situations, teachers may face a significant variety of media in PBL projects. This broad scope of knowledge implies an almost superhuman ability for the teacher to master various foundations and processes, highlighting a clear need for support. McFadyen (ibid.) even suggests that the teacher may need to provide individualized instruction for each project and possess a high capacity for adaptation and information retrieval.

11.3 Effective Group Work

Group work is presented in various ways across the analysed publications, highlighting the importance of this aspect in PBL. It plays roles such as improving cooperative work in problem-solving or project development, promoting inclusive group dynamics, enhancing communication and negotiation skills, and encouraging knowledge sharing.

Several methodologies for group work were identified, with the first and perhaps most traditional being the division of the class into groups so that they can develop the project in its entirety or in parts. This grouping can happen at the very beginning of the project or in a more structured way, such as after individual research sessions or theoretical foundations provided by the professor. Julio et al. (2024) go further, planning the grouping of students in detail, with a task distribution previously agreed upon and planned according to the skills of the participants.

Another identified method of group work application is the creation of reflection groups. In this model, while tasks are completed individually, there are intermediate points where participants share their progress and receive feedback from others. As for the size of the work groups, this varies, and there does not seem to be a uniform approach regarding this aspect.

In summary, work groups can take various formats; tasks assigned in their entirety, in parts, or left open, offering students different levels of autonomy; alternation between individual and group work phases to carry out the project; and a purely reflective sharing role as an intermediate or final point for individual tasks.

11.4 Balancing Didactic Instruction with Independent Inquiry Method

At this point, it is expected that students, through self-questioning and the help of the teacher, will acquire new knowledge. The teacher's role is that of a facilitator, guide, and media production expert, rather than a lecturer delivering content.

The introduction of projects, during classroom sessions, with the aim of enhance students' independent thinking, their skills on problem-solving tasks, creativity and adaptability (Martins and Ferreira 2019). Another consideration is the students' ability to learn by themselves, specifically under the modalities of online sessions, when students may also learn from the media and knowledge sources in the self-study mode (Gasonpan and Temdee 2021). Hence, one can include the autonomous research on problems aligned with the community or market needs (Julio et al. 2024).

On the other hand, PBL methodology embodies the application of several disciplines and intercultural scenarios, for example, in the cases where students from different universities and countries work together (Ling et al. 2024). However, authors such as Zhang and Jia (2024) stated that the project doesn't need to have discipline-related limits, nor a final product, just an identification of issues.

11.5 *Assessment Emphasis on Reflection, Self and Peer Evaluation*

Regarding assessment, methods exclusively planned by the teacher, collaborative methods between teacher and students, qualitative assessments, and quantitative assessments were found. Starting with the collaborative methods, we would like to highlight Mcfadyen (2018), who structured assessment into three parameters: questioning, research, and multimedia production. Despite this structure, in the second week of the course, students contributed criteria for building the assessment model. After the multimedia projects began, there was another joint session to optimize the previously established parameters, and at the end, students self-assessed based on the criteria collaboratively agreed upon. There was also an individual meeting with the teacher to discuss the self-assessment and the evaluation. Regarding collaborative assessment, Martins and Ferreira (2019) mention peer evaluation, where students assess each other's work within the group.

Another system used was quantitative and qualitative assessment through feedback sessions and a system for collecting documentation of the learning acquired. This system, in some way, requires a clear and reflective systematization of the initial problem, the curricular approach, and the central learning objectives to be achieved. There should also be assessment related to the evolution of results and the progression of processes.

A problem identified by Demirci et al. (2023) and Pérez-Gómez (2024), is that group work protects the less motivated and committed students from the project's evaluation. This finding leads to the recommendation for reflection on the assessment model to be used and the development of an individualized model based, for example, on what some authors mention about the construction of workgroups where certain tasks are carefully planned individually. It is also mentioned that, regardless of the assessment system, there is always a degree of subjectivity inherent in the process, making it impossible to create a completely fair system.

Zhang and Jia (2024) emphasize the importance of the public presentation of the project generated by PBL, not only from an evaluative aspect but also due to the social experience it generates and the personal achievement. There is an aspect in the public presentation that seems crucial to us because, not only is the goal of PBL the creation of a product, but the audiovisual and multimedia field itself is characterized by transforming ideas, fundamentals, and processes into a sensory product—sonic, visual, or exploring other senses; performative or non-performative; static, dynamic, or autonomous.

11.6 *Student Choice and Autonomy*

According to the references, if there is an element of student choice and autonomy it will help students to develop a sense of ownership regarding the project outcomes.

One of the strategies is to hold initial discussions, in order to establish topics of individual interest, which may encourage autonomous explorations and research (Mcfadyen 2018). The introduction of project topics increases independent thinking and problem-solving skills in general (Martins and Ferreira 2019), at the same time that the students understand the problem purpose, the nature of the tasks involved, its resources and the role of the professor who is leading the process (Ling et al. 2024).

It is fair to reinforce that it is the professor who has the objective of guiding the production of the content, while the students are developing it and reaching their own conclusions (Pérez-Gómez 2024). This process allows students to experience their own autonomy throughout the learning process, with the need of engaging in results presentations and surveys (Youyou and Kit 2025).

12 Discussion of the Results

The discussion presented here is based not only on the analysis of the published works but also on the authors' professional perspectives and experiences (Fig. 1). In general, the recommendations of Kokotsaki et al. (2016) are addressed in the publications reviewed, with the exception of support for the professor.

Firstly, it seems clear that one of the difficulties that may arise in the creation and implementation of PBL in the area of audiovisual and media production stems from the practical characteristics of the field. These characteristics make it difficult to distinguish between what is an exercise and a true PBL system. Thus, PBL seems more effective when it aims at cognitive development as the primary goal, rather than when the main objective is related to practical skills. In other words, if PBL is designed, for example, to acquire skills in sound editing, it will be quite difficult to separate it from the format of a practical exercise and distance the teacher as the central figure, as the teacher will naturally position themselves as a resolver

Fig. 1 Audiovisual, sound and radio production studios of the IPBeja Higher School of Education

of doubts regarding technical processes. However, if the PBL focuses on a real-world problem, such as raising community awareness about animal extinction in a particular ecosystem, there is a focus on a real issue that can take various conceptual and technical directions. The project might involve sound editing, but in the context of creation, thus moving away from the technical exercise. The students' creative pursuit will generate new technical needs, leading to new learning opportunities. In this way, cognitive and technical components will be developed and stimulated across different domains, transcending the mere technical aspect to which the field is often connected or reduced.

Due to the student's autonomy in developing the PBL, another problem that may arise is that certain projects may go beyond the teacher's comfort zone or area of expertise, especially if the project allows the freedom to explore any media. Therefore, the surrounding structure—whether other professors, curricular units, or even the school itself—should have mechanisms in place to provide the necessary support.

However, one of the contributions of this research is a set of recommendations to support the teacher in implementing PBL in this area. These recommendations arise from the fact that a gap in this regard was identified during the research. Considering the second recommendation by Kokotsaki et al. (2016), the main gap identified in the literature review is indeed the reference to the support provided to teachers. If the success of PBL implementation requires working with real-world problems from the job market, it is essential to create institutional conditions that allow for an inquiry into the needs of that job market. On the other hand, since PBL requires a balance between theory and practice, with constant support from the teacher in managing and following up on projects, it is important that teachers are supported by a technical team with both practical and theoretical knowledge, ensuring and promoting the students' work and autonomy. At the same time, it is crucial to stimulate an attitude of innovation among teachers so that the pedagogical issues underlying the introduction of PBL are transformed into inquiry problems, with encouragement for the publication and dissemination of the results.

13 Conclusion

This research, through a systematic literature review, bridges and analyses the connection between PBL and the field of audiovisuals and media production, generating data that result in various recommendations. The findings led to the formulation of practical recommendations aimed at supporting educators and institutions in the effective adoption of this pedagogical approach. These recommendations are particularly relevant in a context where the boundaries between technical exercises and problem-based learning are not always clear, as is often the case in practical fields such as the one addressed during this chapter.

As a future development, it is proposed to conduct an exploratory case study to observe, firsthand, the effects of implementing PBL in real teaching contexts in audiovisuals. Such empirical research could validate, complement, or reformulate the recommendations presented here, contributing to a more solid, reflective, and contemporary practice aligned with the challenges of higher education.

References

Almeida L, Susana Gonçalves S, Ramos do Ó J, Rebola F, Soares S, Vieira F (2022) Inovação Pedagógica no Ensino Superior. Cenários e Caminhos de Transformação. In: A3ES readings n 16. https://www.cnedu.pt/content/noticias/nacional/Inovacao_Pedagogica_no_Ensino_Superior_Cenarios_e_Caminhos_de_Transformacao.pdf

Alves da Veiga P (2021) O Museu de Tudo em Qualquer Parte, arte e cultura digital: inter-ferir e curar. Grácio Editor, CIAC

Bender WN (2012) Project-based learning: differentiating instruction for the 21st century. Corwin

Brizola J, Fantin N (2016) Revisão da Literatura e Revisão Sistemática da Literatura. Revista de Educação do Vale do Arinos 3:23–39

Burnay C (2023) In: Custódio C (ed) Anuário do Sector de Produção Audiovisual em Portugal 2023. Católica, CECC-Centro de Estudos de Comunicação e Cultura

Burnay CD, Ribeiro N (2020) In: Carrão I, Rato IM (eds) Anuário do sector de produção audiovisual em Portugal 2021. Católica, CECC-Centro de Estudos de Comunicação e Cultura

Burnay CD, Ribeiro N (2021) In: Carrão I, Rato IM (eds) Anuário do sector de produção audiovisual em Portugal 2021. Católica, CECC-Centro de Estudos de Comunicação e Cultura

Cuenca-Rodríguez ME, Pascual-Moltó I, Pastor-Prada R (2024) Digital technologies in music education: using digital audio workstations (DAW) with project-based learning (PBL). Revista Latinoamericana de Tecnología Educativa 24(1):101–117. https://doi.org/10.17398/1695-288X.24.1.101

Demirci S, Lee KY, Binamak AM (2023) Marketing education through PBL: a case of social media influencer marketing. Eng Proc 38(1). https://doi.org/10.3390/engproc2023038059

Díaz JM, Rebollo MAP (2013) Historia audiovisual para una sociedad audiovisual. Historia Critica 49:159–183. https://doi.org/10.7440/histcrit49.2013.08

Gasonpan N, Temdee P (2021) Design and deployment of online PBL model for high school students promoting collaborative learning. In: 2021 joint 6th international conference on digital arts, media and technology with 4th ECTI northern section conference on electrical, electronics, computer and telecommunication engineering, ECTI DAMT and NCON 2021, pp 372–375. https://doi.org/10.1109/ECTIDAMTNCON51128.2021.9425752

Greiner R (2021). Cinematic histospheres: on the theory and practice of historical films. In: Cinematic histospheres: on the theory and practice of historical films. https://doi.org/10.1007/978-3-030-70590-9

Havaldar P, Medioni G (2009) Multimedia systems: algorithms, standards, and industry practices. Course technology. Cengage Learning. http://books.google.com/books?id=k83m5inkjXwC&pgis=1

Julio YR, Lascano DMV, Silva MLM, Domínguez CGS (2024) Integration of STEAM competencies in the development of IoT components through social media, project-based learning (PBL), and the SAFe framework. In: ETCM 2024—8th Ecuador technical chapters meeting. https://doi.org/10.1109/ETCM63562.2024.10746132

Kokotsaki D, Menzies V, Wiggins A (2016) Project-based learning: a review of the literature. Improv Sch 19(3):267–277. https://doi.org/10.1177/1365480216659733

Kuo S-Y (2014) Quality in subtitling—theory and professional reality. Unpublished doctoral thesis, Imperial College London, United Kingdom

Ling M, Liu Y, Nechita F (2024) Project-based learning at dracula digital: a comparative perspective from Romania and Brunei. Trends High Educ 3(3):757–778. https://doi.org/10.3390/higheredu3030043

Manovich L (2013) Software takes command. https://doi.org/10.5040/9781472544988
Martins C, Ferreira C (2019) Project-based learning in audiovisual translation: a case study in error analysis. J Audiov Transl 2(1):152–185. https://doi.org/10.47476/jat.v2i1.109
Mcfadyen N (2018) Project-based learning and multimedia production
Pérez-Gómez MÁ (2024) Innovación docente en el aula a través del Aprendizaje Basado en Proyectos (ABP) aplicado en la asignatura Teoría, Producción y Realización de Radio. Eur Public Soc Innov Rev 9:1–19. https://doi.org/10.31637/epsir-2024-1427
Quaranta D (2013) Beyond new media art. LINK Editions
Thomas J (2000) A review of research on project-based learning
Untari RS, Kamdi W, Dardiri A, Hadi S, Nurhadi D (2020) The development and application of interactive multimedia in project-based learning to enhance students' achievement for 2D animation making. Int J Emerg Technol Learn 15(16):17–30. https://doi.org/10.3991/ijet.v15i16.16521
Yee KK, Huey Shyh T (2024) Problem-based learning: media and information literacy project to combat misinformation for future communicators. J Mass Commun Educ 79(3):340–364. https://doi.org/10.1177/10776958241256404
Youyou L, Kit AC (2025) Practical research on project-based learning (PBL) in film and television production in Xiamen vocational education. Int J Comput Exp Sci Eng 11(1):1147–1164. https://doi.org/10.22399/ijcesen.872
Zagalo N (coord), Coelho A, Oliveira L, Branco V, Giannetti C, Contreras-Espinosa RS, de Sousa Dias A, Teixeira PM, Otero N, dos Reis J (2021) Multimédia, novos média e média digitais. http://hdl.handle.net/10773/31419
Zhang C, Jia B (2024) Enriching STEAM education with visual art: education benefits, teaching examples, and trends. Discov Educ 3(1). https://doi.org/10.1007/s44217-024-00354-w
Zhu W, Wang X, Gao W (2020) Multimedia intelligence: when multimedia meets artificial intelligence. IEEE Trans Multimed 22(7):1823–1835. https://doi.org/10.1109/TMM.2020.2969791

Emotionally Adaptive Interaction Techniques for Enhanced Learning Experiences

Amic G. Ho

Abstract This research examines the possibilities of integrating emotional intelligence into educational settings to improve student engagement and results. The study examines the use of interaction modalities such as tactile, auditory, and visual feedback to provide emotionally intelligent learning experiences that correspond with learners' emotional states. The prototype development phase encompasses an interactive learning platform integrating emotional engagement modes to enhance educational quality. The adaptive platform included algorithms in affective computing for the real-time identification of emotions and subsequent adaptation depending on those emotional states. Evaluation of whether the emotionally adaptable learning platform enhances performance relative to conventional techniques is then conducted. Two hundred samples were obtained from participants of various demographic backgrounds. They were randomly allocated to an intervention group using the emotionally adaptive platform and a control group applying conventional learning techniques. Statistical analysis assessed the impact of emotionally adaptable engagement approaches on the learning experience. A qualitative component gathered data via follow-up interviews to better understand participants' feelings and perspectives on learning. This comprehensive method ascertains whether emotional involvement enhances learning outcomes and shapes future educational technology breakthroughs. The results of this study may facilitate the exploration of new avenues for designing with and about emotion and act as a catalyst for the development of more emotionally advanced instructional tools. The findings can assist educators and designers in comprehending how emotion-aware learning activities enhance student engagement in distance education, facilitating the development of computing devices that align with learners' emotional needs while integrating affective technologies with educational technology to enhance learning through interactive experiences.

Keywords Emotional engagement · Affective computing · Educational technology · Interactive learning platforms

A. G. Ho (✉)
Department of Creative Arts, Hong Kong Metropolitan University, Kowloon, Hong Kong SAR
e-mail: amicgh@gmail.com

D. Raposo et al. (eds.), *Perspectives on Design IV*, Springer Series in Design and Innovation 59, https://doi.org/10.1007/978-3-032-00733-9_3

1 Introduction

Embedding emotional intelligence in learning environments may improve student engagement and outcomes. Tech-enhanced learning tools are evolving, and there is a growing demand for a human-centred investigation of mood changes in interactive systems. In this paper, we investigate how interaction modalities such as touch-, voice- and visual feedback can be employed to elicit emotionally intelligent learning experiences that correspond to learners' emotional state. These are all happening simultaneously when they engage with digital educational resources.

Previous literature has been systematically searched for interaction techniques, devices, and modalities used in educational settings that spark emotional engagement among students. The prototype development phase of this research includes an interactive learning platform that combines emotional interaction modalities to improve the quality of education. The adaptive platform will include algorithms in affective computing for real-time recognition of emotions and adaptation based on those emotional states. We will initiate the development process with requirements-specific education objectives and user stories, conducting a detailed needs analysis. A wireframe and user interface using a high level of usability and accessibility will be designed. The prototype incorporates touch interfaces, voice recognition and visual feedback mechanisms that vary with the emotional responses recorded by physiological sensors and self-report surveys. Throughout the development, user testing generates feedback and validates that what is being built continues smoothly, fulfilling both educational needs as well as emotional engagement goals for our target users.

This experimental study assesses whether the emotionally adaptive learning platform improves performance compared to traditional methods. We conducted the study after recruiting 200 diverse participants of various demographic backgrounds, allowing for generalisability. Subjects were allocated randomly to an intervention group, which uses the emotionally adaptive platform, and a control group working with the conventional system of learning. During data collection, emotional responses in participants were monitored through physiological sensors [heart rate and galvanic skin response (GSR)] and self-reporting on emotional engagement and learning outcomes. Motivation, engagement and knowledge retention before and after the intervention were assessed. In addition, ANOVA and regression analysis approaches were employed to evaluate the effects of emotionally adaptive interaction techniques on the learning experience. Participants' emotions and perceptions of learning will be subject to follow-up interviews as part of an additional qualitative strand. This comprehensive method will be employed to determine whether emotional engagement influences learning success and new educational technologies. In this vein, the goal of this study is to identify the types of interaction techniques that may stimulate emotional engagement in learning scenarios. The results of this work can assist in leading new methods of design with and about emotion, as well as inform a starting point for building more emotionally tailored educational technologies. The results will provide educators and designers with insight into emotion-aware learning activities for scaffolding students' engagement in remote education. The results of this study can be useful in designing computing devices to satisfy the emotional demands of learners as well as connecting affective technologies (i.e. devices that detect emotions by reading the physical measure) and educational technologies aiming to enhance learning through interactivity.

2 Emotional Intelligence in Education

Presently, Emotional Intelligence (EI) (Tonguç and Ozkara 2020) is increasingly recognised as an indispensable element of educational institutions. Its influence on student engagement, academic achievement and overall well-being has been demonstrated in many studies. EI is defined as the ability to recognise, comprehend and reason with emotions, both in oneself and in others (Goleman 2021). EI has substantial importance in educational environments (Jordan and Troth 2021). An academic environment encompasses a broad spectrum of competencies: self-awareness, self-regulation, empathy, and social skills, which are essential for overcoming the obstacles presented by learning and life. Educational methodologies that foster EI seek to establish significant connections between students and educators. By fostering EI, educators can create an atmosphere of warmth and reciprocal affirmation, making pupils feel appreciated, understood and inspired to study. This experience of support enhances students' self-confidence, motivation and overall health (e.g. physical well-being) (Ho 2024). When students are emotionally involved, they are more inclined to focus in class, actively participate in discussions with professors and so on.

Furthermore, students are inclined to persist despite challenges. Prior studies have consistently shown that emotional involvement is closely linked to intellectual achievement. Mayer et al. revealed that children with elevated EI scores had superior academic performance, fewer behavioural issues and a greater propensity for prosocial community involvement (Kamiloglu et al. 2020). A meta-analysis conducted by Schutte et al. demonstrated that EI predicted academic standing independent of other personality characteristics, although it was associated with general intelligence (Ho 2024).

Emotional Intelligence (EI) significantly contributes to students' development of empathy, social awareness and the ability to manage relationships. By fostering empathy and respect for other perspectives, educational institutions can cultivate a more compassionate and understanding society. Highlighting the significance of EI in educational contexts fosters an atmosphere that embraces individual variations and is uniformly valued. Consequently, it establishes the foundation for a more equitable and peaceful future.

Emotional Intelligence (EI) can be incorporated into the academic curriculum in several ways. As such, interactive strategies may include structured group discussions that respond to readings, assignments analysing media in which students present their knowledge to an audience or filming staged scenes in a classroom setting. Such efforts are indicators of success in academia and inspire a sense of achievement in fulfilling personal needs. On par with the other factors, socio-emotional learning is instrumental in facilitating the maturation of moral conscience during adolescence.

Experiential learning within the realms of organisational behaviour in vocational/technical or high schools integrates EI to teach practical intelligence and social skills. The educational process is almost completely influenced by educational staff and faculty; therefore, it requires a vision and detailed implementation at the level of the education system. Experiential learning is related to the state of inner being, as good performance plays an important role in creating a stimulating and nurturing learning environment. If we care for the whole of a person, we must accept that he can never give something he does not possess through his educational experience. In summary, EI is an indispensable part of a complete education and plays a vital role in academic performance, development and students' personal growth.

Yet, these effects are seemingly not confined to the self per se; schools develop campuses that embrace diversity and adversity as they foster a community that works together for social empowerment via authentic emotional well-being. With EI becoming increasingly pivotal, so too must the teaching methods. This will not only equip students to lead a comfortable and prosperous life in an increasingly globalised society, but it will also expand the quality of education beyond course content and create a more enriching experience for all members of the school community.

3 Emotional Intelligence in Education

In educational settings, these three types of interactive methods—visual response feedback, communication technology and touch interaction—have progressively gained significance. Educational environments will probably increasingly incorporate a variety of interaction techniques as technology continues to evolve. Converged under this banner, these modalities serve as an essential part of emotional, warm and engaged learning experiences. Developing ways for teachers to create a learning atmosphere that fosters students' emotional engagement gives them real leverage while employing multi-modal interactive techniques. This study aims to explore both the importance of these modalities and what has been said about their effect on emotional engagement in educational contexts.

3.1 Touch Feedback Mechanisms

Touch serves as a method for expressing emotions and facilitating communication. This aspect is very relevant in an educational context; for example, there are haptic gadgets that provide tactile feedback to users. Research indicates that tactile input may evoke emotions and enhance the memorability of learning experiences. Another research indicates that haptic feedback enhances user engagement by rendering interactions more immersive and natural. A study report examining the influence of haptic interfaces on learning outcomes concluded that learners receiving tactile feedback exhibited superior retention of reading material and produced higher-quality essays compared to those exposed just to visual or auditory stimuli. Studies have established that touch can enhance learning and may contribute to better memory recall for students, specifically those with sensory disabilities. In addition, a few studies explored the capacity that touch had to convey emotional nuances. In an experiment on emotion detection by touch—in which participants pressed harder or softer on a force-sensitive screen over several glossy surfaces, leading to different patterns of pressure—researchers found that people consistently recognised disgust, fear and joy, while four other emotions like jealousy were interpreted more erratically. Its expressive capacity through tactile interaction clarifies its importance as a learning technology genre. Touch-sensitive interfaces that are connected to educational programmes can also provide personalised emotional learning environments by detecting individualised states of emotion.

3.2 Voice Feedback Mechanisms

The educational environment is one of the important sectors where verbal communication is used as a mode of interaction. A voice recognition system allows students to communicate with educational organisations seamlessly and intuitively. By using different pitches in the voice or changing the tone or talking style recognisable as specific for a given individual, the engagement is tailored to the learner's level of excitement. Studies show that dialogue with the voice function increases emotional engagement (e.g. students' responsiveness to educational content as articulated and expressed with emotional awareness and compassion). At the end of the day, no matter how cutting-edge a lecture is when it is recorded, one thing becomes very clear: the importance of voice responses within educational systems to create a bond between students and studies.

Additionally, voice integration has led to an increase in emotional engagement with users. In an experiment, participants were asked to rate the tactile stimuli with spoken emotions as regards their emotional valence. The findings suggested that individuals often integrated auditory and tactile stimulus properties to form an experience of emotional cues. This multi-modal approach suggests that voice feedback, when combined with other engagement approaches, could enhance the affective experience of students.

3.3 Visual Feedback Mechanisms

Visual feedback is necessary in multi-modal interactive teaching. From an ethnomethodological point of view, graphical user interfaces are meant to be used in quiet, low-light environments where message boxes pop up unexpectedly. Visual methods, such as images, animations and interactive multimedia, bring the context of words to life for students. Many studies show how visual perception impacts users; for example, students are said to be highly motivated and have a better memory when learning in illustrated classrooms with sunlight passing through stained glass windows. In this way, the visual feedback can be tailored according to how the learner is feeling using adaptive interfaces receiving data from sensors or users feeling emotions.

The use of visual stimuli only magnifies the emotional influence. Research shows that combining visual stimuli with auditory or tactile experiences increases students' empathy and engagement with the material. This unique synergy creates a diverse community among students with an eagerness to learn in a concerted sharing of knowledge and enjoyment towards their studies. Additionally, within these different modes, the modes themselves may change their primary roles; for example, text may be presented in writing on a whiteboard or on-screen and then talked through by the teacher using a microphone supported by visual elements for greater comprehension.

3.4 Multimodal Interaction Techniques

Based on a review of the related literature, there seems to be an increasing number of studies about multi-modal interaction and its effect on emotional involvement in education. By combining the strengths of multiple modalities (e.g. tactile, auditory, visual cues), these approaches offer a holistic learning experience that fits the needs of students. Students who learned under these practices ultimately scored significantly higher than their contestant graduate students. Contemporary educators assert that innovative methods of engaging children may foster deeper emotional ties, ultimately resulting in enhanced academic performance. An interdisciplinary approach used psychology, neurology and computer science to examine the interplay of various modalities in human-computer interactions. The current study emphasises that without an emotional connection to the information, students find it challenging to attain success, indicating that effective multi-modal designs must include individual variations in mode choice and methods of integration.

Furthermore, the study highlights the potential of emotional computing technologies to enhance multi-modal engagement in educational settings. Algorithms enable these technologies to identify learners' emotional states using physiological data (e.g. variations in heart rate) or behavioural indicators (e.g. facial expressions) as they engage with instructional tools. By customising material to real-time emotional assessments, educators may provide personalised learning experiences that enhance engagement and increase motivation. The literature concludes that touch, speech and visual feedback are not just modalities in teaching. By using multi-modal engagement tactics, instructors may enhance students' receptiveness to educational messages, resulting in improved academic performance. The current research substantiates that the integration of diverse feedback modalities enhances the learning experience.

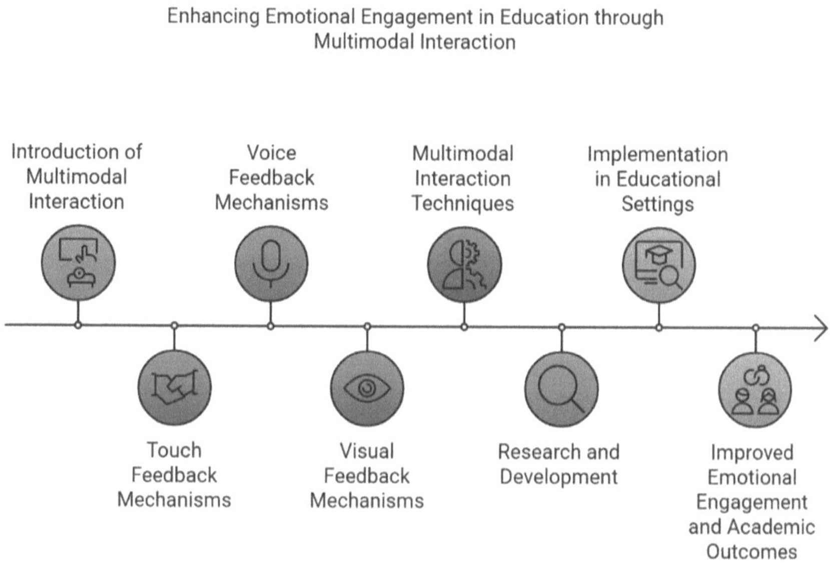

Fig. 1 An illustration of enhancing emotional engagement in education through multi-modal interaction

Research must persist as technology evolves. Future creative educational tools must focus on EI and adaptable learning settings (Fig. 1).

4 Affective Computing in Emotion Recognition

Affective computing is fundamentally a novel field within human-computer interaction that specifically emphasises the medical recognition of emotions in real-time. This emerging discipline amalgamates computer science, psychology and neuroscience. The objective is to create systems that can perceive, understand and act in harmony with human emotions. Affective computing uses sophisticated algorithms and physiological sensors to create emotionally intelligent systems capable of adapting to users' emotional states, resulting in enhanced overall interaction experiences.

4.1 Real-Time Emotion Recognition Algorithms

The pinnacle of emotion identification technology—crucial to the field of affective computing—is recognised as the capacity to identify moods. To achieve this objective, researchers have devised several algorithms and methodologies, leveraging diverse disciplines such as machine learning, signal processing and pattern recognition. A prevalent method involves the application of machine learning algorithms to physiological data obtained from sensors. These algorithms are trained on extensive datasets of annotated emotional patterns, allowing them to discern the regularities and associations included in physiological signals and their various emotional states. Prevalent machine learning methodologies used in emotional computing include support vector machines (SVMs), artificial neural networks and decision trees (Stock-Homburg 2022; Schutte et al. 2001; Mayer et al. 2001). Support vector machines (Rani and Dhenakaren 2016) were used to differentiate emotional states based on physiological indicators, including heart rate, breathing, and skin conductance, achieving over 80% accuracy in real-time identification (Nourani et al. 2019; Singh et al. 2021).

Deep learning techniques provide another possible way to achieve real-time emotion recognition. For example, deep neural networks have shown promising results in this field because they can extract features from raw data automatically. Researchers have also begun using deep learning methods for other aspects of physiological computing. They are looking into how to process signals into emotional states from the raw data we have collected. For example, in a study of emotion recognition from electroencephalography (EEG) (Rios and Burke 2020), researchers utilised a model based on a convolution neural network, showing that deep learning techniques can be successfully applied to affective computing.

4.2 Physiological Sensors and Self-Report Surveys

Real-time detection of emotions carried out by affective computing is closely connected to the evaluation of physiological sensors and self-report questionnaires. It forms a strong basis for generally accepting emotions. To an extent, the sensors reveal different sides of physical measurement, which are tied to a person's moods (Zheng et al. 2018; Huang 2021). It is also a consistent stream of data, which could be applied in the course of artificial EI applications. The most popular measurements are heart rate and heart rate variability. The sensor measures the electrical activity of the heart, presenting a reliable and valid technique for checking the arousal and valence of emotion. For instance, an increase in heart rate is usually related to terror or profound enjoyment, whereas its decrease might be accepted as tranquillity.

The GSR sensor (Bamonte et al. 2023), or skin conductance sensor, measures the skin's electrical conductivity, which changes with sweat gland activity. GSR is incredibly adept at indicating emotional arousal. As the activity of our sweat glands increases, so does the conductivity level, which is a trait widely intertwined with the intensity of experience.

EEG sensors (Huang 2021) are needed to achieve emotional computing. Researchers have discovered that specific patterns of an EEG signal correspond to a type of emotional state, thus enabling the detection of sentiment through changes in brain waves. Furthermore, facial electromyography sensors can be useful for determining the level of interest in facial muscles through the electrical signals each is transmitting. These facts link the data, allowing affective computing systems to tell how you are feeling (Stock-Homburg 2022).

EEG data and other sources of physiological sensors are combined to provide further confirmation of what you are feeling. However, physiological sensors are not the only ones used by researchers in emotional computing. Inter-subject variability means that sole reliance on physiological sensors is not advisable. Self-report surveys, which include questionnaires for users to complete writing about how they feel, are important because they validate emotion-detection algorithms. The purpose of a survey is to help humans determine their feelings about certain predetermined experiences or specific films and then rate how much they agreed with certain statements. In addition, surveys help to ensure identification using different emotions by ensuring that a method reflects actual sentiments. The Self-Assessment Manikin, developed by Lang in 1980, was one of the first self-report tools in emotional computing (Ho 2024). This tool uses puppets to represent three components of emotions: valence, affective arousal and dominance. The puppet chosen by a participant most accurately describes their current emotions, which also allows emotional reactions to be evaluated quickly and more intuitively.

Physiological sensors use supplemental self-report surveys to create a solid foundation for the detection of emotions in real-time. The combination of the two substantially increases the accuracy and reliability of sensing and analysing the systems of identification used. By combining objective data from sensors on the body with the subjective experiences recorded en masse, researchers may create more accurate algorithms that suit our situations and our responses to them. It has been shown that using data from multiple physiological sensors (e.g. EEG, heart rate and GSR) is significantly more precise in identifying various emotional states than using a single sensor. The use of both physiological and self-report measures helps the understanding of the relationship between sensory responses and reporting of emotions.

However, there are significant weaknesses in the physiological sensors and self-report scales that the aides use. There are wide variations in the physiological responses of people to certain stimuli because of differences in basic physiology, variations in social settings and variations in culture (Bamonte et al. 2023). These variations make it extremely difficult to identify universal patterns tying certain physiological signals to certain emotions. In addition, self-report surveys may differ from the respondents' actual feelings depending on social desirability and the ability to accurately delve into thoughts about whether one is correctly judging some issues.

To summarise, a reliance on physiological sensors and self-report surveys in the field of emotional computing allows for the detection of emotions in real-time. In addition to using objective experiences and the determination of subject experiences, researchers can create algorithms with more accurate detection. More powerful EI systems that can cater to the situations different learners find themselves in, including feelings, are now being developed by affective technology. However, this diverse data cannot adequately encompass the extensive emotional spectra of living entities.

Consequently, enhancing the terminologies will only be feasible by considering inter-subject variability and subjective reporting aspects.

4.3 Challenges and Future Directions

Affective computing has made substantial progress in real-time emotional recognition. However, there are still many issues that need to be researched. The chief issue lies in individual variations when it comes to physiological responses to emotional stimuli. People can exhibit different patterns of physiological activity even when they are experiencing the same emotion, making it difficult to develop universal emotion recognition algorithms. The nature of emotions is another concern in real-life contexts. The same physiological response can mean different things depending on the situation or the task at hand. Affective computing systems must be general enough to work in many contexts and account for the rich interplay of emotions, cognition and movement. Despite these difficulties, the future of effective computing is bright. Given technological advancements, it is not surprising that researchers are increasingly exploring new approaches to the recognition of emotions. For instance, recognising emotions through wearable sensors, as well as multi-modal data fusion and the modelling of emotions at the individual level, are some of the technologies that researchers are testing.

It should be noted that affective computing systems are more accurate and adaptive to the customer if they combine the overall data from multiple physiological sensors and consider each system user's experience (e.g. by utilising user-specific or personalised experiences). The application of affective computing across various domains such as education, healthcare and entertainment may prove beneficial. Specifically, in educational settings, affective computing can help fashion adaptive learning conditions that change with the student's emotional state, providing pupils with incentives and engagement. In healthcare, affective computing could be of service in diagnosing and treating mental health disorders by offering objective measurements of mental well-being. Finally, in the world of entertainment, affective computing can create truly holistic experiences that are also emotionally arousing and personalised to each user's own preferences and time-scale responses.

In summary, affective computing is instrumental in building systems that might be emotionally intelligent and offer the ability to recognise real-time emotions. From advanced algorithms to biosensors, affective computing technology makes it possible to build adaptive interfaces to meet people's emotional needs. The specifics of the development of a plethora of these interfaces are discussed in this study. As the field continues to grow, placing effective computing within the context of a multitude of settings has the potential to change them in meaningful ways—ways that both revolutionise technology and alter how at ease we feel with it.

5 Methodology and Participants

The methodology for developing an interactive learning platform that incorporates emotional interaction modalities is structured into three main components: prototype development, user interface design and experimental study design. This experiment included 200 people in a fully controlled setting, aimed at evaluating the efficacy of sophisticated emotionally adaptable learning systems compared to classic simple versions. The experimental group is a random assortment of individuals, demographically representing various generations, socioeconomic levels and occupations within society. The diversity of its population enhances the validity of this series beyond what was previously conceivable, allowing a broader spectrum of study identifications via the integration of statistics and market analysis. The gender distribution among participants was balanced. Approximately fifty per cent were male, and fifty per cent were female. The participants were between 18 and 35 years old, including graduate students from many fields, including psychology and computer science, as well as postgraduates in mechanical engineering. The research aimed to provide a comprehensive profile of engagement with Hokou's emotionally adaptable platform by uniting learners from diverse cultural and educational backgrounds where such chances previously lacked. To guarantee successful learning and emotional engagement in the experiment, students were pre-screened for their familiarity with various digital learning tools.

This study investigated a learning system with adaptive feedback that no prior self-identified technology enthusiast had ever encountered, reducing the impact of preconceived biases on these specific technological issues' results. Participants needed a stable internet connection and access to a camera. This was an online experiment aiming to collect physiological data. Additionally, our study team employed university-based environments to find suitable candidates. They directly sent the invitation for participation or the corresponding course-related advertisement and asked participants to fill out a pre-study questionnaire to express their intention to participate. We used their responses to analyse their initial emotional state and preferred learning system, ensuring that all subjects started on an equal foundation for both aspects of academic performance: motivation and engagement. Upon recruitment, the subjects were randomly allocated to either the treatment preparation employing an established emotionally adaptable platform or traditional learning methodologies. The random allocation of participants sought to minimise the risk of permanent systemic mistakes and biases when applying emerging actions for the subjective cognitive appraisal system. This allocation has also bypassed judgements that could be influenced by factors other than chance. Random assignment is an efficient way to distribute a single study programme across many candidates. It did not consider any possible source of variance, even if their scales were minor before the random selection began. For the current study, we used a collection of biological sensors during all procedures, including a heart rate monitor and a galvanic resistance sensor, to measure the student's emotional response. The questionnaires were later used to measure the participant's subjective feelings after processing the study's participants.

This analytical method was more comprehensive than previous historical models or early Chinese influences. It analysed the involvement of emotional states in various academic units of individuals from different backgrounds and walks of life. For this research, we chose a sample of 200 respondents, and none were academically diverse. All 200 participants lived widely across the province, both geographically and culturally. Using these stratified participants seeks to generalise findings about students' experiences in learning systems that adapt to their characteristics and may be practically and theoretically significant. This study employed rigid recruitment criteria and required demographic equity to understand technology's role as an enabler.

5.1 Prototype Development

Development of the prototype must start with the creation of an interactive learning platform where a different mode of emotional engagement plays a significant role. As an example, this may include tactile input or voice communication, which are becoming increasingly popular in teaching. The learning platform should automatically adapt the feedback, considering the emotional changes of students in real-time, which should improve the quality of the educational experience. One of the benefits of this approach is that it reduces the likelihood of subjectivity in assessment. Another requirement is conducting a detailed analysis of requirements using educational goals and user stories as a foundation, ensuring that the result of the work would indeed correspond to the needs of students and teachers. This research collects user feedback, which indicates that there are specific features that can improve emotional engagement and learning in general. The development team must create a platform that contains a solid educational ethos merged with an emotionally supportive environment, as it will enhance the overall quality of the product.

5.2 User Interface Design

During user interface design, various design concepts emphasise usability and accessibility, which are considered priorities for a smooth experience when using an application. The proposed steps imply that before the interface is finally launched, usability testing should be initiated. Stakeholders, in turn, can preview the design of a specific product and how 'manageable' it could be, as wireframes show the structure and flow of a program and represent the first visual integrity of the developed user interface. It is very important to understand, however, that the visual design of the wireframes should first aim to elicit some emotional responses from users when different interface elements are encountered. The latter may involve colour schemes that may be expected to have specific feelings associated with them and some interactive elements that are emotionally responsive and can react to users.

5.3 User Interface Design

The experimental study design involves obtaining a mixed sample of 200 participants to determine whether the emotionally adaptable platform is superior to the standard American educational system. In this context, an interventional group responsible for employing the emotionally adaptable platform and a control group of individuals who apply the standard facilitators are chosen randomly by lots. The selection prevents any possible biases and ensures that all differences in the learning outcomes are a direct result of utilised interaction modalities by the platform. To achieve the results, the study applies a combination of measures, including sensorial recordings of heart rate and GSR and self-report of emotional state responses during interactions to determine a level of emotional involvement and its influence on motivation, as well as knowledge retention. Thus, it is possible to infer that the study aims to establish whether the use of an emotionally engaging environment leads to a significant improvement in student's response and academic performance in comparison to the standard methods. As a result, the developed research will pave the way for a noticeable breakthrough in understanding how interaction modalities can be effectively and efficiently used in education technology to the greatest advantage. The method lays out a systematic approach to creating an interactive platform that would facilitate and foster students, teachers and educational institutions in the most efficient and effective ways of implementing emotionally intelligent behaviour patterns. In conclusion, the goal of all the lessons is to provide the best learning opportunities for the students to ensure their emotional engagement and high academic performance.

6 Data Collection

Data collection is crucial in this investigation. It encompasses two main sectors: the examination of emotional responses and assessments based on performance defined by credit points or other quantitative metrics.

6.1 Monitoring Emotional Response

The Monitoring Emotional Responses course utilises physiological sensors and self-reporting instruments to evaluate students' feelings reliably. These two types of equipment allow for measuring physiological signals that correspond to emotions and students' subjective experiences during their interaction with an ecological teaching platform. Depending on the case, students are assessed via heart rate monitors and GSR devices to determine their real-time physical signals. The heart rate sensors measure the electrical activity of the heart and are located on students' chests with an electrocardiogram. Changes in students' heart rate corresponding to their emotional states are recorded for analysis. For example, an elevated heart rate indicates excitement or anxiety, and a decreased rate corresponds to calm. GSR sensors record variations in the electrical conductivity of the skin (Bamonte et al. 2023) due to sweat gland changes and are a good indicator of students' emotional changes. As a rule, an elevated GSR reading is considered to be high arousal.

Moreover, researchers may also objectively determine the arousal level of participants by observing them as they complete a range of learning exercises. Apart from these tools, self-reporting instruments are typically used to assess students' subjective emotional experiences and learning outcomes. Usually, these tests contain several questions that a student has to address and assess their emotions using standardised scales or labels. The Self-Assessment Manikin (SAM) is a widely used instrument and is a non-verbal pictorial assessment technique that evaluates three dimensions of emotional experience (Kamiloglu et al. 2020). By relying on both self-reported and physiological responses, researchers can evaluate the efficacy of emotion-detection algorithms and obtain more accurate information on how various cognitive states affect learning through them rather than relying solely on human judgements.

6.2 Assessment Metric

Another crucial aspect of data collection for the current study is the employment of the assessment metrics for motivation, engagement and knowledge retention, both pre-and post-intervention. Several validated tools are used for these constructs. To measure motivation, the Academic Motivation Scale (Zheng et al. 2023) is used, which measures the levels of intrinsic and extrinsic motivation among the learners. Knowledge retention and engagement are measured using an observation checklist or a self-report survey. The first one is designed to measure the attentiveness and participation level of the learners involved in the learning process. The latter includes pre- and post-intervention tests that are conducted among the learners. Both data types require the employment of a statistical test for interpretation. Indeed, Analysis of variance and regression analyses (Sutter et al. 2024; Immordino-Yang and Gotlieb 2020) were used to compare the performance of an intervention group interacting with an emotionally adaptive platform and a control group learning through traditional methods. ANOVA is a widely used tool to analyse the differences between two groups on the same dependent variable, facilitating the identification of any statistically significant differences between the two groups regarding motivation, engagement and knowledge retention. Simultaneously, regression analysis can ascertain the relationship between the intervals of monitoring emotional engagement and learning outcomes. By using these tools, a researcher can determine the impact of using emotionally adaptive interaction techniques and learning interventions. In the long run, these results can be used in the development of context-sensitive adaptive systems for education. The unique contribution of this paper, in turn, is a multi-modal approach that incorporates both quantitative data obtained through the use of physiological sensors and qualitative insights from participants' reports (Pekrul et al.

2020). This kind of approach is intended to make the results of the research more precise and comprehensive.

6.3 *Findings*

To visualise the collected data from the emotional engagement surveys of the 200 participants in this research study, data is presented in various tables that visualise data in terms of the emotional engagement metrics presented in the form of the SAM. The tables below detail the participants' emotional valences and their arousal and dominance scores, which represent the users' emotional states before and after the interaction with the emotionally adaptive learning platform.

Emotional Engagement Survey The data summarises information on the results of an emotional engagement survey of 200 participants and reflects the emotional experiences and states observed before and after using the emotionally adaptive learning platform. Table 1 presents the demographic characteristics of the research participants, indicating a fair and diversified representation across genders, ages and academic majors. This information is crucial for the generalisability of the results obtained for various groups of learners.

Pre-intervention emotional engagement and the resulting scores are presented in Table 2. The pre-intervention valence scores were limited within the range from 3 to 7, and the average equalled about 5. The possible interpretation of the average score is the neutrality or slightly positive emotional state before the intervention. Arousal scores ranged from 2 to 6, with an average of 4, indicating the participants differed in their excitement or anxiety levels. The dominance scores ranged from 4 to 8, with an average of about 6, reflecting the level of the participants' control over their mood or emotions at this moment, which ranged from 'most people' to 'no one'. The results of the intervention and the scores from the post-intervention emotion engagement are presented in Table 3.

In Table 3, the data concerning the post-intervention valence scores shows that they increased. Post-intervention valence scores ranged from 5 to 9. The average post-intervention valence score was about 7. This could be seen as an indication that the experience with the platform led participants to a shift in emotions towards better feelings. Post-intervention arousal scores ranged from 3 to 7, and the mean value was slightly under 5. This variance showed that regarding the use of the platform, the subjects exhibited a heightened degree of engagement as well as excitement. Post-intervention dominance scores ranged from 5 to 9, and their average value was about 7. It might be inferred that participants had different degrees of control (i.e. helped themselves) regarding their use of the platform in some cases. Table 4 shows the significant data related to the change in the emotional engagement scores. The valence scores for ID 1 increased by 2 and for ID 200 by + 3. Arousal for the same examples varied by + 1 and 0. Thus, the three tables together show that the

Table 1 Table captions should be placed above the tables

Participant ID	Gender	Age	Major
1	Male	18	Computer Science
2	Female	19	Psychology
3	Male	20	Engineering
4	Female	21	Education
5	Male	22	Business
6	Female	23	Mathematics
7	Male	24	Biology
8	Female	25	Art
9	Male	18	Physics
10	Female	19	Chemistry
…	…	…	…
200	Female	25	Chemistry

Table 2 Pre-intervention emotional engagement scores

Participant ID	Valence (1–9)	Arousal (1–9)	Dominance (1–9)
1	3	2	4
2	4	3	5
3	5	4	6
4	6	5	7
5	7	6	8
…	…	…	…
200	4	6	8

Table 3 Post-intervention emotional engagement scores

Participant ID	Valence (1–9)	Arousal (1–9)	Dominance (1–9)
1	5	3	5
2	6	4	6
3	7	5	7
…	…	…	…
200	9	6	8

emotionally adaptive learning platform caused a significant change in the emotional engagement scores. The most crucial changes were observed in the increase in the valence and dominance scores and the improvement in the arousal score. This finding leads to the assumption that the use of emotional interaction features as part of educational technology should be implemented to ensure that the technologies used in the learning process create an emotionally interactive and supportive environment.

Table 4 Post-intervention emotional engagement scores

Participant ID	Pre-valence (1–9)	Post-valence (1–9)	Change in valence (Post–Pre)	Pre-arousal (1–9)	Post-arousal (1–9)	Change in arousal (Post–Pre)	Pre-dominance (1–9)	Post-dominance (1–9)	Change in dominance (Post–Pre)
1	3	5	+ 2	2	3	+ 1	4	5	+ 1
2	4	6	+ 2	3	4	+ 1	5	6	+ 1
3	5	7	+ 2	4	5	+ 1	6	7	+ 1
4	6	8	+ 2	5	6	+ 1	7	8	+ 1
5	7	9	+ 2	6	7	+ 1	8	9	+ 1
…	…	…	…	…	…	…	…	…	…
200	4	7	+ 3	6	6	+ 0	8	8	+ 0

Engagement Surveys Motivation and Engagement Metrics The collected data (from the emotional engagement surveys focusing on motivation and engagement metrics for 200 participants) is organised into several tables: pre-intervention scores, post-intervention scores and changes in motivation and engagement metrics.

The motivation and engagement data shown in these tables give a comprehensive view of the participants' metrics before and after using the emotionally adaptive learning platform. Table 1 summarises the demographic background of participants to maintain a balance in demographics in terms of gender, age and academic majors. This variety is vital to improve the study's generalizability results. Based on Table 5, the scores for pre-intervention motivation and engagement were presented between 0 and 50, and intrinsic motivation scores from approximately 15 to 25 with an average score of about 20. Intrinsic motivation equalled 12, extrinsic motivation scored 10–20 (mean of $\pm$ 15), and engagement scored 3–7 (moderate pre-experience-transition scores). Table 6 demonstrates marked improvement in all these indicators after intervention. Intrinsic motivation level post-intervention rose to a score of 20–30, with an average score of approximately 25. There was a subsequent improvement in extrinsic motivation scores as well, varying from 15 to 25, with an average of about 20. Participants had increased engagement scores, with most participants scoring between 5 and 9 post-platform use. The change in motivation and engagement scores presented person-level responses to the intervention, reflecting a substantial degree of inter-individual variability. For example, Participant ID 1 showed an improvement of 11 points in intrinsic motivation; Participant ID 200 showed an improvement of + 8. The changes in these factors indicate that the emotionally adaptive learning platform is effective in increasing both intrinsic and extrinsic motivation among users. In the end, these results add to the growing body of evidence for ways that implementing emotional interaction modalities in educational technologies may help to deepen learner engagement and motivation.

Table 5 Pre-intervention emotional engagement scores

Participant ID	Intrinsic motivation score (1–30)	Extrinsic motivation score (1–30)	Engagement score (1–9)
1	18	16	5
2	24	15	4
3	18	14	6
4	24	15	6
5	24	14	5
...	...	...	...
200	16	20	3

Table 6 Post-intervention emotional engagement scores

Participant ID	Intrinsic motivation score (1–30)	Extrinsic motivation score (1–30)	Engagement score (1–9)
1	29	22	7
2	28	16	8
3	23	15	6
…	…	…	…
200	24	21	5

7 Qualitative Research Study

In the first study, after analysing the data sources and understanding that the emotionally adaptive learning platform had had a significant impact on the learner's adjustment of emotional states, a follow-up qualitative study was conducted through interviews with a portion of the target sample that chose to participate: thirty participants from the main sample of two hundred according to different genders, age ranges and main courses of study they undertake, applying a single-case sampling technique. A semi-structured interview was conducted to promote freedom in the conversation through an interview guide and facilitate an in-depth discussion of the issue concerning the participants' emotions during the learning process. Each interview took about thirty-five to forty-five minutes and was held in physical presence or through video conference based on the interviewee's preferences. The responses were audio-recorded and shared with the respondents for final confirmation.

7.1 Follow-Up Interviews

To ensure the comprehensiveness and depth of the information obtained, the interview included questions such as 'Can you describe how you felt when you were using the emotionally adaptive learning platform?', 'What features of the learning platform did you find engaging?' and 'When you felt that you were succeeding in learning, can you describe how you felt?'. The main questions were extended by follow-up questions, depending on the respondents' replies. For example, if they explained that they were excited, the follow-up question we raised was 'What made you so excited during that activity?' or 'How did you feel inside when you were that excited?'

7.2 Data Collection and Analysis

All interviews were recorded on tape with the permission of the participants and then transcribed verbatim to analyse the data. We then conducted an inductive thematic analysis of the transcripts, which allowed us to identify and analyse the commonalities and themes related to emotional engagement and learning within the sample of our participants. By applying inductive thematic analysis, we gained an understanding of participants' views on how different modalities of emotional interaction allowed them to be better motivated. Also, the ability to analyse a smaller number of materials allowed for a more detailed and complex analysis of the interaction data and learning experiences of the participants. From the data collected, the following themes emerged, as illustrated in Fig. 2.

Enhanced Emotional Engagement A number of the participants implied in their interviews that emotionally adaptive features used throughout the course significantly enhanced their emotional engagement with the material. According to the transcripts, the required interaction with content that, in turn, interacts with the participants and gives them feedback caused such responses as excitement and curiosity.

Sense of Control The participants in the course often mentioned in their interviews that they felt that they were 'in control' of their learning experience, referring to the interactive nature of the material used in the course. In other words, they felt that they were moving the learning process forward. According to the findings, higher levels of self-efficacy and the ability to better control their learning were part of the cause of higher motivation and overall higher positive experience regarding the course.

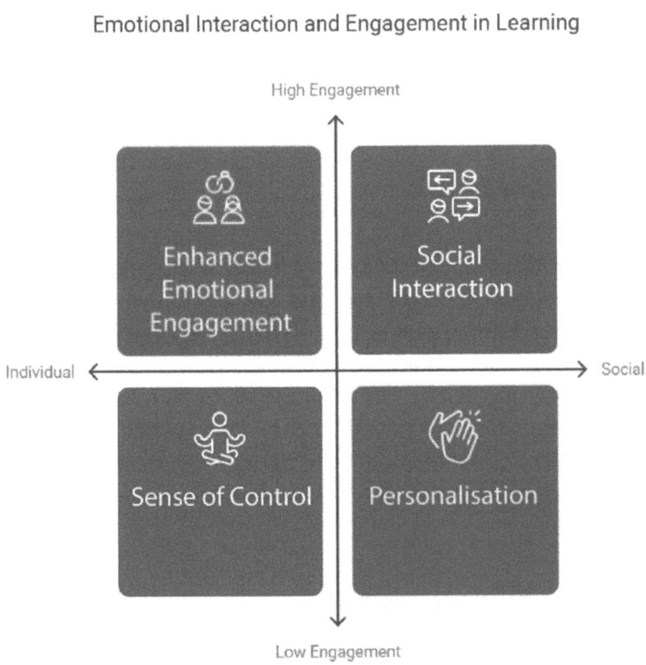

Fig. 2 An illustration of the abilities involved in emotional interaction and engagement in learning

Social Interaction Some of the interviewees also mentioned that within the course, they felt a sense of the social dimension, i.e. interdisciplinary interaction and collaboration between professionals. Features that allowed them to receive positive feedback and support from colleagues were an integral part of their response to learning interactive material.

Personalisation The participants stressed that they enjoyed receiving personalised feedback within the course and not being treated as a faceless group of attendees. A majority of the interviewees specifically mentioned the latter factor as being a positive contributor to their interaction with the material.

7.3 Implications for Future Research

The follow-up interviews added to the insights of the participants' negative emotions and shed new light on the further improvements that can be made in the emotionally adaptive learning system. While the current findings of the study provided a general understanding of the concepts of positive and negative affect, the qualitative data revealed additional information that needs to be known. Interviews of this kind could be incorporated into further studies to make sense of how emotional interaction modalities influence learner engagement and outcomes. Such findings, in turn, were crucial to the development of better educational tools in the future. Overall, this qualitative research showed that the exploration of learners' subjective experiences should go hand-in-hand with assessments that rely on numbers.

8 Discussion

Despite the multifaceted nature of the influence of emotional engagement, emotions can directly and indirectly shape student motivation, learning experience, academic achievements and overall satisfaction in educational settings. Emotional engagement is the affective responses that students experience when engaging with learning materials and environments, which might include how interested or enthusiastic a student is when participating if a student feels like they belong, and so on. Data from the emotional engagement surveys reveals evidence of how emotional engagement factors into learning outcomes when users engage on an emotionally adaptive learning platform.

The quantitative results showed that interacting with the platform caused participants to have much greater intrinsic motivation and engagement scores than before they began. Specifically, intrinsic motivation scores increased from an average of 20 pre-intervention to approximately 25 post-intervention. In other words, there was a robust improvement in students' internal drive to study the material. This can be supported by existing literature, which indicates the critical role of positive emotion in cultivating the intrinsic motivation needed to learn (Nourani et al. 2019; Singh et al. 2021). In addition, after the intervention, participants scored higher for engagement in diving into more specific learning tasks through the emotional, adaptive features of the platform.

Interview data that followed up on these findings qualitatively explained the affective experience of participants as they interacted with the platform. The participants from our study who enjoyed the content they read, saw or heard using some form of adaptive media expressed surprise and joy when other emotionally relevant pieces were shown as well. 'Because of that, I could relate to the content more because it was reacting to how I felt at that moment, and eating made learning fun'. This is in line with the idea that an emotionally appealing environment can create a relationship between students and content, which, ideally, increases long-term emotional responses to the material (Rios and Burke 2020).

The interviews demonstrated the significance of emotional engagement experiences in attaining a sense of control over the learning process. Respondents mentioned more than once that the responsiveness of the platform enabled them, in a literal sense, to have control over their learning experience. In turn, this feeling of agency is fundamentally important as a factor contributing to resilience and persistence in learning environments. In their own words, participants commented, 'I used to feel now for me it became more similar to my way of thinking, and I became triggered when I moved on my way, giving great motivation to keep going'. This perspective emphasises the interaction between emotional engagement and self-regulation, implying that emotionally well-supported students are better able to engage in adaptative learning strategies (Zheng et al. 2018). This is also seen in the gains from Part I to Part II. Part I establishes initial emotional engagement (Ho 2020), so increased recall supports these students later in their academic learning (Ho 2017). Greater amounts of emotional engagement during platform interactivity were associated with increased knowledge retention scores from both pre- and post-intervention assessments. This finding aligns with previous research indicating a positive correlation between emotional engagement and academic achievement when adherence to behavioural norms is elevated (Zheng et al. 2018; Huang 2021). The qualitative data supported this finding; participants who enjoyed a feeling of emotional engagement often reported greater memory recall at assessment.

Emotional engagement also predicts cognitive outcomes and social aspects of learning. Any feature that allows students to peer collaborate is said to increase their emotional tie-ins within the experimental study (Ho 2017). One trainee revealed, 'To be working with like-minded people made me feel part of a community: it pushed me to give more'. It also underscores the importance of social engagement in promoting emotional involvement and indicates that collaborative learning experiences support not only individual motivation but collective success as well (Huang 2021). Participants also experience issues with engaging emotions. Technical concerns and the adoption of new forms of interaction occasionally hindered the ease at which they could experience these moments. This points to the need for educational technologies to continually evolve so that they better serve emotional engagements without creating obstacles. This study was an examination demonstrating how emotional engagement contributes to the development of quality learning; EE-related emotions are excellent predictors of motivation, student success and social relations among learners. This is the first time a combination of quantitative data showing increased intrinsic motivation and qualitative insights revealing personal narratives has been used to bring deep context to how emotions might impact students' educational journeys (Bamonte et al. 2023). It is becoming increasingly important that with educators and designers working to create more engaging and healthful learning experiences, emotional engagement must make its entry into education if these well-rounded learners are forged. By understanding and incorporating these affective qualities that contribute to the diversity of a holistic student success story, we can better support the whole child in educational contexts, helping students succeed in the future.

9 Conclusion

The results from this study are indicative of the importance of emotional engagement in increasing educational practice utilising emotionally adaptive interaction strategies. The study shows that when EI is interwoven in learning environments, learners develop motivation from within to learn the course materials due to increased retention and engagement of knowledge. The quantitative data obtained from the emotional engagement surveys showed significant growth in intrinsic and extrinsic motivation scores before and after the intervention with p values, indicating that an emotional, adaptive learning platform could be effective for a better educational experience. This increase from 20 to 25 is an average five-point jump in participants' intrinsic motivation scores, demonstrating value in helping learners want to interact with the material. Moreover, themes derived from follow-up interviews in the qualitative analysis revealed vibrant stories of participants who found that emotionally responsive features of the platform had a positive effect on their emotional state contextually, with an enhanced sense of content immersion and a feeling of control over learning experiences.

In addition to academic benefits, the results highlight how emotionally responsive communication practices can contribute to a more inclusive and supportive learning environment. As content adapted to the emotional states of the participants, they shared feelings of excitement and intrigue about new formats that linked questions on feelings, illustrating that elevating the emotional content allows a learning experience that is not only informative but also engaging and rewarding. The qualitative data underscored the need for social interaction in the learning platform, as many participants mentioned that collaboration tools helped support a feeling of community and presence. The emotional engagement part is what we need to focus on, but we neglect to develop a complete education enterprise that corrals human beings not just around cognitive outcomes but towards well-being.

Along with the findings that participants who experienced a higher level of emotions showed better knowledge maintenance, these results attest to the belief that emotions and other psychological events are related to cognitive processes. This result is consistent with the existing literature, which shows that academic achievement increases when we can emotionally engage learners. Therefore, educators should use emotionally adaptive technologies that adapt to the student's emotions through their interaction with the system but in typically real-time responses, helping create a more tailored educational learning experience based on students' requirements and preferences. These technologies can adjust depending on physiological feedback and perhaps self-reported emotions, both providing a rich seam for enhancing student engagement and improving outcomes from learning efforts of all kinds. This study serves not only to understand how emotional engagement can contribute to learning experiences but also to offer a glimpse of the various findings that still need further investigation. The study necessitates further exploration of which technologies may effectively thrive in various educational and research contexts. Future studies must further examine the long-term impacts on academic trajectories at all levels and also influence student well-being, yielding significant benefits for student support systems. Future studies should focus on how these tools might be customised for students with varying learning styles, cultural backgrounds and individual inclinations. Future research studies, therefore, should emphasise the development of technology integration best practices in established curricula and must include training for teachers to use them effectively. Overall, the current research illustrates how emotion-aware interaction methods can play a transformative role in educational practices. Educators can create more dynamic and responsive learning environments that enhance student well-being by prioritising emotional engagement through innovative technologies. The implications support a revision of the education paradigm in which emotion is given more than passive recognition but is designed into activities to ensure improved outcomes for everyone. The research on the applications of effective technologies in education continues to be an active area for developing future practices as we move forward and to help our schools address the increasingly diverse needs of students in a complex world.

References

Bamonte MF, Risk M, Herrero V (2023) Emotion recognition based on galvanic skin response and photoplethysmography signals using artificial intelligence algorithms. In: Congreso Argentino de Bioingeniería. Springer Nature, Switzerland, pp 23–35

Goleman D (2021) Leadership: the power of emotional intelligence. More Than Sound LLC

Ho AG (2017) Emotional typography: the effectiveness of current typographic design categories in eliciting emotion. Int J Vis Des 11(2):37–44. https://doi.org/10.18848/2325-1581/CGP/v11i02/37-44

Ho AG (2020) Exploring cultural recognition: enhancing creativity in interactive communications. New media spectacles and multimodal creativity in a globalised Asia. In: Digital culture and humanities (Challenges and developments in a globalized Asia), vol 3. Springer, Singapore, pp 141–158. https://doi.org/10.1007/978-981-15-7341-5_7

Ho AG (2024) Emotion guideline workbook: managing the design process. CRC Press

Huang C (2021) Recognition of psychological emotion by EEG features. Netw Model Anal Health Inform Bioinform 10(1):12

Immordino-Yang MH, Gotlieb RJM (2020) Understanding emotional thought can transform educators' understanding of how students learn. In: Educational neuroscience. Routledge, pp 244–269

Jordan PJ, Troth AC (2021) Managing emotions during team problem solving: emotional intelligence and conflict resolution. In: Emotion and performance. CRC Press, pp 195–218

Kamiloglu RG, Fischer AH, Sauter DA (2020) Good vibrations: a review of vocal expressions of positive emotions. Psychon Bull Rev 27:237–265

Mayer JD, Salovey P, Caruso DR, Sitarenios G (2001) Emotional intelligence as a standard intelligence. Emotion 1(3):232–242. https://doi.org/10.1037/1528-3542.1.3.232

Nourani V, Molajou A, Najafi H, Danandeh Mehr A (2019) Emotional ANN (EANN): a new generation of neural networks for hydrological modeling in IoT. In: Al-Turjman F (ed) Artificial intelligence in IoT. Transactions on computational science and computational intelligence. Springer, Cham. https://doi.org/10.1007/978-3-030-04110-6_3

Pekrul SR, Flis T, Schloesser R (eds) (2020) Autism spectrum disorder across the lifespan, Part I. An issue of psychiatric clinics of North America, vol 43, no 4. Elsevier Health Sciences

Rani VMK, Dhenakaren SS (2016) Tumor intensity ratio model using support vector machine. In: 2016 international conference on advanced communication control and computing technologies (ICACCCT). IEEE, pp 188–191.

Rios K, Burke MM (2020) The effectiveness of special education training programs for parents of children with disabilities: a systematic literature review. Exceptionality 29(3):215–231. https://doi.org/10.1080/09362835.2020.1850455

Schutte NS, Malouff JM, Bobik C, Coston TD, Greeson C, Jedlicka C, Rhodes E, Wendorf G (2001) Emotional intelligence and interpersonal relations. J Soc Psychol 141(4):523–536. https://doi.org/10.1080/00224540109600569

Singh S, Sharma C, Sharma S, Verma NK (2021) Re-learning emotional intelligence through artificial intelligence. In: 2021 9th international conference on reliability, infocom technologies and optimisation (Trends and future directions) (ICRITO). IEEE, pp 1–5. https://doi.org/10.1109/ICRITO51393.2021.9596091

Stock-Homburg R (2022) Survey of emotions in human–robot interactions: perspectives from robotic psychology on 20 years of research. Int J Soc Robot 14(2):389–411. https://doi.org/10.1007/s12369-021-00778-6

Sutter CC, Totonchi DA, DeCoster J, Barron KE, Hulleman CS (2024) How does expectancy-value-cost motivation vary during a semester? An intensive longitudinal study to explore individual and situational sources of variation in statistics motivation. Learn Individ Differ 113:102484. https://doi.org/10.1016/j.lindif.2024.102484

Tonguç G, Ozkara O (2020) Automatic recognition of student emotions from facial expressions during a lecture. Comput Educ 148:103797. https://doi.org/10.1016/j.compedu.2019.103797

Zheng W-L, Liu W, Lu Y, Lu B-L, Cichocki A (2018) Emotionmeter: a multi-modal framework for recognising human emotions. IEEE Trans Cybern 49(3):1110–1122. https://doi.org/10.1109/TCYB.2018.2797176

Zheng Y, Janiszewski C, Schreier M (2023) Exploring the origins of intrinsic motivation. Motiv Emot 47(1):28–45. https://doi.org/10.1007/s11031-022-09969-8

De-academize Design: Learning from the Pluriverse

Veronica Magno de Moraes and Carla Paoliello

Abstract This article explores the de-academization role in reshaping design education through a pluriversal lens, promoting a more inclusive and expansive approach to design learning. By challenging conventional frameworks, it emphasizes the importance of diversifying design history and moving away from established terminology rooted in traditional structures, specifically by introducing nature-centered design. The study investigates what academic and design education can learn from a pluriversal perspective, emphasizing models of horizontal learning and other design pedagogies that foster mutual knowledge exchange. Through these methods, the study emphasizes the value of vernacular and community-driven design practices, calling for a more cultural connection between design and society. Additionally, it critically examines whether these alternative expressions should be classified as "design" or if such categorization imposes an external framework. This raises essential ethical considerations, questioning whether defining these practices within institutional paradigms risks recolonizing what is meant to be decolonized. Finally, this research invokes a pedagogic turn that respects and includes diverse ways of knowing to enable a more equitable and reflective design discourse that moves beyond Western-centric narratives.

Keywords Pluriversal design · De-academization · Design education · Alternative design pedagogies

1 Introduction

In recent years, critical discussions in design education have increasingly highlighted the limitations of traditional academic frameworks and the need to embrace diverse forms of knowledge. The concept of de-academization has emerged as a radical concept that seeks to bridge the gap between formal academic knowledge and

V. M. de Moraes (✉) · C. Paoliello
Faculdade de Belas-Artes, Centro de Investigação e Estudos em Belas-Artes, Universidade de Lisboa Largo da Academia Nacional de Belas-Artes, Lisbon, Portugal
e-mail: Veronicamdemoraes@gmail.com

D. Raposo et al. (eds.), *Perspectives on Design IV*, Springer Series in Design and Innovation 59, https://doi.org/10.1007/978-3-032-00733-9_4

real, lived experience. Scholars such as Catherine Walsh (2012) and Linda Tuhiwai Smith (1999) demonstrate how it challenges the dominance of institutional knowledge by valuing community-based and embodied ways of knowing. Similarly, de la Cadena and Blaser (2018) highlight how Indigenous and relational worldviews offer alternative ways of understanding the world, inviting us to think with different knowledges instead of imposing academic frameworks. De-academization does not mean rejecting academic knowledge but expanding the discourse to include alternative epistemologies and methodologies that recognize the rich plurality of design practices, like design, worldwide (Kothari and Garg 2019).

A pluriversal approach to design education, as proposed by Escobar (2018), calls for a shift beyond Western academic norms, moving forward from universalist principles and industrial production models and toward a more inclusive and contextually rich curriculum that values non-academic knowledge. This perspective agrees with the broader discourse on decoloniality, highlighting that design must engage with multiple worldviews rather than enforcing a singular, dominant paradigm (Schultz et al. 2021). Scholars advocating for a pluriversal approach argue that design education must integrate epistemologies from outside academia, including community-based practices, Indigenous methodologies, and craft traditions (Van Amstel 2023).

If, as Meggs (1992) suggests, design has been an intrinsic part of human expression since prehistory, we must also ask: where does design begin, and where does it end? However, this is not an easy line to draw. For example, the distinction between craft and design often feels like an artificial boundary. Both are intrinsically connected to cultural expression, problem-solving, and material innovation. By embracing the dialogic nature of design, we create opportunities for craft to inform contemporary practices, challenging the rigid hierarchies that have traditionally shaped the field.

As we explore the concept of de-academizing design, this paper examines how learning from the pluriverse—diverse, lived spaces of design practice—can inform and innovate design education. We advocate for more equitable, culturally attuned frameworks that move beyond traditional academic constraints by integrating ontological design, vernacular practices, and other kinds of perspectives.

The main intention of this paper is to critically examine design practices that occur outside formal classrooms, exploring the discourse around de-academization in the design field and how other artistic expressions can influence it. The aim is to enrich design education by debating how these practices contribute to a broader, more inclusive understanding of design. To achieve this, the paper adopts an exploratory methodology, beginning with a comprehensive literature review that analyses both academic and non-academic perspectives on what constitutes the concept of de-academization and what should be recognized as a plural design. It draws comparisons with other artistic fields of knowledge.

Additionally, this research emphasizes the importance of alternative design practices and calls for a deeper cultural connection with design's role in society. Not only questioning whether these forms of expression should be considered part of design, but also whether they seek such classification. If doing so, it would constitute an act of colonizing what we aim to decolonize. It critiques institutionalized knowledge systems, suggesting that academia can significantly benefit from engaging with and critically examining non-academic design practices, both embracing and challenging this connection, to enrich the field and ultimately broaden its epistemological horizons.

2 Literature Review

The practice and teaching of design have long been dominated by academic institutions that define what constitutes "good" design, who can be a designer, and what methods are deemed legitimate. However, an increasing collection of research questions the established paradigm, arguing that design does not, and should not, belong exclusively to academia. The concept of de-academization in design education proposes a shift away from rigid frameworks and toward more inclusive, pluriversal approaches that recognize design as a practice outside formal institutions. Researchers such as Manzini (2015), Fry (2017), and Ansari (2020) argue that design education must break free from its academic confines to acknowledge and integrate the diverse ways in which people design in everyday life.

Academia can be understood as a constructed space of ideas shaped by a community engaged in scientific inquiry and education, often perceived as exclusive by those outside of it (Kurylo and Yu 2016, p. 76; Lavoie and Roth 2002, p. 83). It is characterized by largely self-imposed ethical norms and principles of mutual recognition. The terms "Academizing" and "De-Academizing" refer to the processes of situating knowledge or scientific claims within an academic context or extracting them from it, respectively.

The de-academization of knowledge represents a neoliberal process that fundamentally reorganizes the internal structures and logic of the university as an institution. It redefines academia from a narrowly focused, market-driven entity and extends its influence to a Different University. This transformation alters its location and its societal and politico-economic frameworks. The effects are felt by all who contribute to the university community daily, including students, faculty, administrative staff, cleaners, and the surrounding community. These conceptions align with what Ronald Barnett refers to as the ideological imaginary (the neoliberal university) and the utopian imaginary (an alternative university) (Barnett 2013).

Risager and Thorup (2016, p. 19) have coined the concept of de-academization of knowledge as the "processes by which academic knowledge production has become deskilled as just any other knowledge production located in or outside the university. This has also been called a shift from *modus 1* to *modus 2* knowledge production (Gibbons et al. 1994) or a shift to post-academic science (Ziman 2000)." For Saavedra (2011), it correlates with Derrida's concept of deconstruction to *destrenzando*, a Chicana/Latina epistemology to exhume and expose González's (2001) concept of *trenzas y mestizaje*. It brings a sociopolitical and historical consciousness that allows for *nuevas posibilidades* (New possibilities). Or, as Smith (1999) contended, "research 'through imperial eyes' describes an approach which assumes that Western ideas about the most fundamental things are the only ideas possible to hold" (p. 56).

However, de-academization differs from a non-academic or everyday approach; knowledge that has never been situated within an academic framework cannot undergo the process of de-academization. De-academizing does not necessarily reject academic knowledge entirely; instead, it seeks to balance it with other forms of learning and understanding. This strategy intends to create more holistic, inclusive, and flexible educational and knowledge-production systems that can better serve diverse learners and communities. It involves constantly questioning fundamental assumptions to inspire more radical explorations of regionally associated theories and philosophies that remain unexamined. Why de-academization? Why should we recognize diverse forms of knowledge and learning? Or promote more inclusive and accessible educational experiences and encourage creativity and innovation outside traditional academic frameworks?

Maybe the answer is in ontological design, a philosophical approach to design that explores the fundamental nature of existence and how design shapes our reality. For Willis (2006, pp. 69–70), "Ontological designing implies a radically different understanding of design as practice and object than those generally available; it also implies different ways of understanding how we, as modern subjects 'are' and how we come to be who/what we are in the modern world." It is a concept that emphasizes the significant influence that designed objects, systems, and environments have on human beings and their lifeworlds.

Ontological design suggests that design is not merely a tool for creating objects, but a force that shapes how we perceive and engage with the world. This concept is rooted in the philosophy of Escobar (2018), Fry (1994, 1999), Latour (1992), and Heidegger (1962). It posits that design is inherently tied to cultural, political, and social structures. It's both shaped by and actively constructs the realities we inhabit. For these authors, every design decision contributes to constructing specific ways of being-in-the-world, influencing material realities and social and cultural structures. This perspective shifts design from being merely about objects or functionality to being about creating possibilities for life itself. Escobar goes further and pledges that there is no separation between mind and body, humans and nonhumans, and subjects and objects. He argues that this dualistic framework has led to extractive practices and ecological crises. Therefore, the ontological design seeks to move beyond this by embracing rationality and acknowledging the interconnectedness of all entities and systems, as de-academization also does.

In many non-Western traditions, design education could be inspired by cosmovisions—worldviews that integrate spirituality, ecology, and social responsibility (Gudynas 2011). For instance, *Buen Vivir* or *Sumak Kawsay*, originally from the Quechua language, is a concept emphasizing harmonious living with nature and others. This can be related to proposals for sustainable design that focus on eco-efficiency, the use of renewable materials, the idea of product sharing, and the concepts of social innovation and sufficient consumption (Acosta 2013). Unlike industrial design's emphasis on efficiency and consumption, these approaches prioritize rationality, sustainability, and holistic thinking. By integrating ontological design principles, design education can shift from being a technical skill set toward a transformative practice that critically engages with the ways design shapes social and ecological systems (Willis 2006).

Vernacular design can also be a reference when we are talking about analysing design in the pluriverse spaces. It refers to localized, context-driven design solutions that emerge organically within communities, often outside formal academic training (Oliver 2003). These practices appear across various cultures and locations while exemplifying extensive understanding of material culture along with environmental adaptation and social unity. The incorporation of vernacular design into modern design discussions disrupts Eurocentric dominance while promoting other valuable knowledge systems essential for shaping design education (de Moraes et al. 2025). For example, Ankara fabric in Nigeria and Ghana is produced using traditional wax-resist dyeing methods, embedding cultural symbols and historical narratives into patterns that convey social messages. These textiles function beyond decoration, as they act as societal communicators, transmitting messages about social status and identity as well as political statements. Local communities continuously reinterpret and develop designs that evolve naturally according to regional preferences and requirements. These examples emphasize how vernacular design is an essential part of the pluriverse, cultivating diverse knowledge systems that challenge the dominance of Western-centric design paradigms. Also, it differs significantly from the traditional industrialized construction methods taught in formal design programs, which frequently prioritize globalized materials and standardized aesthetics over local wisdom (Rudofsky 1964).

The integration of vernacular design principles into educational settings teaches students to value localized knowledge and participate in it, as opposed to applying unrelated solutions (Rapoport 1969). Pedagogical approaches such as field-based learning and apprenticeships with artisans, along with participatory design methodologies, emerge as vital tools for developing critical empathy and contextual sensitivity.

On the other hand, the definition of design has been widely debated in academia, yet a vast field of creative production remains outside this formal scope. Practices such as street graffiti, handicrafts, DIY (do-it-yourself) culture, vernacular signage, and indigenous traditions embody visual and functional solutions that share principles of design but do not always recognize themselves as such or are classified as such. This introduces a key question: does categorization matter to those who practice these forms of creation, or is it only relevant to those who hold the power to name?

Naming is an act of power, as emphasized by Nêgo Bispo in his discussions on Afro-Indigenous epistemologies. For Bispo (dos Santos 2015, 2023), knowledge is not merely an abstract academic exercise, but a living practice embedded in the land, community, and ancestry. He critiques the imposed hierarchies of colonial modernity, which prioritize written, institutionalized knowledge over oral traditions, embodied experiences, and intergenerational transmission. His discussions on "*escrevivência*" (writing as lived experience) and "*tecnologia dos encantados*" (spiritual and ecological technologies) emphasizes how Afro-Indigenous peoples have continuously developed sophisticated ways of thinking, being, and creating despite colonial erasure.

The displacement of Indigenous and Afro-descendant communities represents not merely a material loss but a profound disruption of their ways of understanding and interacting with the world. In this context, design transcends aesthetics and function; it reflects how communities construct and maintain their lives in harmony with their environment. When academia seeks to recognize certain practices as design, is it genuinely validating these knowledge systems, or is it merely incorporating them into an institutionalized discourse, perpetuating colonial dynamics? The risk of reclassifying vernacular and Indigenous productions as "design" may result in a form of appropriation, stripping these practices of their conceptual and epistemological autonomy. As Latour's (2008) asserts about modernity, his suggestion of breaking away from established dichotomies such as nature and culture could also be useful in this discussion. Design education could move beyond fixed definitions and foster a more fluid understanding of creative processes. This perspective can be fruitful, allowing us to view these practices as fluid knowledge systems rather than forcing them into fixed categories. If we follow this line of thought, academia may need to learn from these alternative ways of "doing design" without attempting to dominate or translate them into its logic.

Best and Kellner (2008) attribute to Nietzsche a construct that serves as a framework for this perspective: Perspective is akin to a lens through which we see. A perspectival approach involves interpreting the uniqueness of concrete phenomena, granting us access to "a complex form of specificity" (Nietzsche 1968, p. 340) and enabling a more nuanced and comprehensive understanding of the particularities of these phenomena. Moreover, viewing issues from conflicting perspectives allows individuals to appreciate otherness and difference, as well as to recognize the uncertain, provisional, hypothetical, and "experimental" nature of all knowledge (Best and Kellner 2008, p. 5).

So it is the opposite of simplifying. Reducing text and information lightens the cognitive load, minimizing what needs to be stored, explained, or remembered. At first glance, this appears beneficial, enhancing accessibility and efficiency. Nevertheless, simplification does not mean stripping knowledge of its academic rigour. Instead, it often creates the illusion that depth is diminishing, leading to increased claims that lack substantial grounding. When knowledge is condensed excessively, reliance shifts toward immediate perception—visual analogies, sensory impressions, and emotional responses—rather than critical analysis. This shift distorts comprehension, replacing nuanced understanding with surface-level associations that may be mistaken for knowledge itself. More concerning is how this process facilitates manipulation, allowing information to be framed without historical, contextual, or functional accuracy. By prioritizing immediacy over depth, simplification risks eroding the integrity of knowledge, leaving it vulnerable to misinterpretation and misuse.

How institutions could incorporate these perspectives into design curricula means rethinking not only what is, but how knowledge is produced and shared (Tlostanova

2017). This may involve: Diversifying design history courses to include non-Western contributions to arts fields; Decentering English and Eurocentric design terminology in favour of multilingual, context-specific frameworks; Engaging in horizontal learning models where students, educators, and community members collaborate as equal knowledge producers; Alternative design pedagogies, like speculative and critical design workshops, that challenge mainstream product-driven approaches and imagine radically different futures.

2.1 *Diversifying Design History*

Expanding cultural perspectives becomes attainable when design actively incorporates the study of practices from diverse cultures and regions. This includes exploring other civilizations' arts, crafts, and philosophy. Such an approach acknowledges that design innovation and creativity are not confined to European or American contexts but thrive in a multitude of non-Western traditions and practices.

Initiatives like the Diversity in Design (DID) collaborative, led by Herman Miller, aim to address this imbalance by creating pathways for underrepresented communities in design education and careers. Some educational institutions (University of Sydney, University College London, Parsons School of Design, Ontario College of Art and Design University, and University of New Mexico, to name a few) also work to decolonize design curricula by introducing various voices and perspectives. They included efforts to achieve gender balance, increase racial diversity, and incorporate ideas and viewpoints in design education and discourse (Kaufmann-Buhler et al. 2019; Clark and Brody 2010; Sparke 2013; Margolin 1997). They have also involved faculty and students in coursework that combines Indigenous theory with design studio-based education, privileging ancestral ways of knowing.

Another critical aspect involves challenging dominant narratives in design history. This shift reflects a growing recognition of the need to transcend the prevailing "heroic" and aestheticized accounts that have traditionally centred on a select group of celebrated—predominantly Western—designers. Instead, contemporary discourse advocates for a more inclusive and pluralistic historical framework that acknowledges the contributions of diverse cultural traditions, marginalized communities, and non-Western epistemologies. This reorientation enriches the field and fosters a more comprehensive understanding of design as a global, multifaceted practice shaped by various sociopolitical, economic, and technological contexts.

Therefore, diversifying design history is a continuous effort to expand the scope and perspectives of traditional design narratives, recognizing the contributions of underrepresented groups and challenging the longstanding Eurocentric focus that has dominated the field. It endeavours to foster a more inclusive and comprehensive understanding of design's global influence and evolution. As Khanchandani (2020) stated: "Change happens when people from all sorts of backgrounds, including those representing the dominant culture, try to address it."

2.2 Decentering Design Terminology

Decentering design terminology is an emerging approach that seeks to challenge and move beyond the entrenched frameworks of traditional human-centred design practices (Nicenboim et al. 2023). Nature-centered design (Paoliello et al. 2024) is at the heart of this shift, which extends design thinking beyond anthropocentric paradigms and incorporates non-human elements into the design process. This new direction questions human-centric assumptions and opens up design possibilities that acknowledge the interconnectedness of all life forms—humans, non-humans, and the environment.

The goal of decentering design terminology is to broaden the scope to include diverse perspectives, such as multispecies considerations, which encourage designers to engage with and consider various nonhuman species' needs, rights, and experiences. "By decentering humans from design, I do not mean removing humans from the design process, or the goals of design, but rather opening up design to include nonhumans and their lifeworlds, as well as our own, in design theory and practice" (Smith 2019, p. 12). By moving toward a more inclusive, multispecies approach, design becomes a more holistic process that recognizes the complex interdependence of all living beings.

This approach also intersects with feminist technoscience (Haraway 1991), new materialism (Coole and Frost 2010), and object-oriented ontology (Harman 2018), which offer alternative frameworks for understanding the relationships between humans, nonhumans, and material objects. Feminist technoscience, for example, critiques the hierarchical distinctions between nature and culture, proposing more egalitarian relations between human and nonhuman entities in scientific and design practices. New materialism emphasizes the agency of materials, suggesting that nonhuman elements are not passive but have their roles and contributions to the design process. Meanwhile, object-oriented ontology challenges the traditional subject-object divide, offering a philosophical lens that treats all entities—human or nonhuman—equally important and capable of influencing the world in unique ways.

By incorporating these ideas, decentering design terminology pushes the boundaries of traditional design thinking, cultivating the creation of methods and tactics that prioritize ecological balance, empathy, and cohabitation with nonhuman life forms. This shift transforms how designers approach problem-solving and aligns design practices with broader environmental and ethical considerations, paving the way for a more inclusive and sustainable future.

2.3 Engaging a Horizontal Learning Model

Engaging a horizontal learning model emphasizes peer-to-peer interactions, experiential learning, and the development of key competencies essential for navigating the complexities of modern society. Gawinek-Dagargulia and Tymoshchuk (n.d.) explore the concept of horizontal learning, emphasizing its five main features and its role in contemporary education: (1) it focuses on developing knowledge, skills, and behaviours that strengthen key competencies; (2) it creates an environment where learners can learn from each other through participatory methods on equal footing; (3) it centres on existing knowledge of learners and builds on that by introducing new themes; (4) it changes the role of educators from professors to facilitators; (5) It is based on a holistic learning approach that emphasizes the active components of learning, such as discovery, reflective observation, trial and error, and growth with challenges or collaboration.

Learners require a safe space for active participation to act as autonomous and responsible individuals. This safe space fosters trust, critical thinking, and personal agency, enabling students to engage meaningfully with the material and one another. The horizontal learning model requires creating environments that support diverse learning styles and paces, encourage cross-sectional exchanges, and value individual experiences (Freire 2000). These diverse methods ensure that learners feel acknowledged and respected, regardless of their background or expertise (Brookfield 2015).

Moreover, an education model that addresses practical, real-world issues (Lave and Wenger 1991) is essential for enhancing relevance, engagement, individual voice, and self-confidence. By engaging with tangible, context-based challenges, learners are empowered to connect theory with practice, preparing them for active citizenship and personal empowerment. This approach also encourages innovation and adaptability, which are increasingly important in today's rapidly changing world.

This learning model, therefore, not only fosters academic growth but also strengthens social bonds by cultivating empathy and collaboration across diverse perspectives (Kolb 2015). In embracing a horizontal approach to education, institutions can contribute to developing a more inclusive, resilient, and dynamic society, where designers are prepared not only to respond to challenges but also to shape their environments through cooperative action and continuous learning proactively.

2.4 Alternative Design Pedagogies

Alternative design pedagogies are distinguished by their emphasis on nontraditional approaches to teaching and learning. Experiential learning is a cornerstone of alternative design pedagogies, which emphasize learning through direct engagement with project-based approaches. For instance, the Case Problem (Experimental) Model (CPM), developed by Martin Symes and Alexi Marmot in the 1980s, encourages students to immerse themselves in authentic design challenges rather than relying solely on theoretical knowledge. This hands-on approach develops practical skills and nurtures critical problem-solving abilities, as students are required to think creatively and adapt to the constraints and ambiguities inherent in real-world scenarios (Kolb 2015). By bridging the gap between theory and practice, experiential learning enables designers to see the direct application of their ideas and decisions, providing a deeper understanding of the consequences of their design choices.

Nontraditional pedagogies also embrace an interdisciplinary approach, which dissolves the boundaries between distinct academic disciplines. Such pedagogies encourage students to explore design from multiple perspectives and integrate insights from various fields, thus enhancing their ability to think holistically. For example, the Community-based Design Learning Model (C-bDM) encourages students to address design problems rooted in social, environmental, and cultural contexts. This model integrates knowledge from sociology, environmental studies, and community development, encouraging students to understand design as a social practice that impacts and is shaped by the community (Lave and Wenger 1991). The ability to draw on multiple disciplines fosters a deeper appreciation for the broader implications of design decisions, making the learning process more relevant to complex, interdisciplinary challenges faced by designers in the real world.

Another significant feature is the emphasis on collaborative learning. In contrast to traditional models that focus on individual achievement, alternative pedagogies prioritize interactions, collective knowledge building, and the co-creation of ideas. As said in the previous item, horizontal learning models are particularly notable in this regard, as they foster an egalitarian approach where all participants—students, educators, and even community members—contribute equally to the learning process. It is to value diversity, including different perspectives, backgrounds, and experiences, which enriches the learning environment (Freire 2000). By promoting collaboration, these pedagogies cultivate communication skills, teamwork, and the ability to engage in constructive critique, all essential for designers in multidisciplinary and diverse teams.

These pedagogies also place a strong emphasis on the design process itself, rather than focusing solely on the end product. For example, Design Thinking encourages students to engage in iterative exploration, testing, and refinement of their design ideas throughout the learning experience. By prioritizing process over product, students learn to embrace experimentation, failure, and revision as integral components of the design journey (Brown 2009). This iterative approach fosters creative thinking and encourages students to develop resilience and flexibility—skills critical for success in an ever-evolving design landscape. This process-oriented model focuses on developing critical thinking, reflective practice, and continuous learning rather than pursuing a single, final solution. Ultimately, alternative design pedagogies are redefining design education by cultivating a more holistic, inclusive, and process-oriented approach to learning.

3 Conclusion

As the design field continues to change in the face of globalization and technological advancement, it becomes necessary to move beyond rigid academic structures and develop a practice that prioritizes collaboration, reciprocity, and inclusivity. To achieve this, design education must experience a paradigmatic shift, one that not only includes marginalized voices but also actively disrupts hierarchical modes of knowledge production. This change demands that teachers, designers, and institutions reimagine their roles, not as knowledge guardians but as facilitators of collective learning, co-creation, and mutual exchange. Rather than replicating dominant frameworks, de-academization calls for embracing the fluidity of design, making space for vernacular practices, and recognizing that knowledge exists in multiple, intersecting forms beyond traditional academia.

De-academization of design education does not mean abandoning academic rigour, but opening its horizons to embrace plural ways of knowing and doing (Escobar 2018). Learning from the pluriverse enriches design practice by making it more contextually responsive, socially responsible, and ecologically aware. This involves an intentional attempt to decenter institutionalized design narratives, which have historically dictated what is "valid" knowledge. Otherwise, embracing non-Western perspectives, such as Indigenous, Afrocentric, and local community-driven traditions, allows for a more holistic, positioned, and ethical approach to making and thinking.

A pluriversal approach to design does not simply try to increase representation, but it actually reshapes how we understand and engage with the world. It challenges hegemonic power structures and questions the way that the design field is imposed upon communities rather than emerging organically from within them. This shift destabilizes the conventional definitions of innovation and expertise, as we recognize that design knowledge is not exclusive to academia but exists in grassroots movements, everyday practices, and intergenerational knowledge systems.

The challenge, however, is in the way in which these perspectives are being conveyed. Are they really being integrated into the design discourse, or are they rather being categorized and reinterpreted through an academic lens that holds the authority to define what constitutes 'design'? Moreover, are we genuinely collaborating with these knowledge systems, or are we, in fact, colonizing their knowledge, imposing our perspectives onto their traditions under the guise of inclusion? The question is whether the synthesis of these diverse perspectives reflects a balanced exchange, or merely replicates existing power relations within design education and practice.

Therefore, the central question is not whether vernacular, Indigenous, or community-based practices are "design" or not, but whether such classification is even relevant to them. The educational systems should acknowledge the value of diverse creative expressions, without appropriating them into current academic structures. The challenge could be in cultivating dialogue that respects the autonomy of knowledge systems, so that design education could become a more fluid, participatory, and responsive discipline without imposing rigid traditions as a mode of validation.

This shift also requires rethinking the terminology and narratives that dominate design discourse. Terms such as "human-centered design" reveal a Western tendency to separate the designer from the designed, often positioning humans as external agents of change rather than integral participants in a living system. Additionally, the move towards horizontal learning models and pedagogies are also essential for dismantling hierarchical education structures that privilege certain forms of knowledge over others. Conventional design education tends to be founded on a top-down model, where knowledge is transmitted from expert to student. Horizontal learning method, by contrast, promotes reciprocity, collective discovery, and mutual learning between individuals. Within this framework, lived experience is also seen as a legitimate form of competence, creating space for intergenerational exchange and engagement with local material cultures.

Last but not least, the future of design lies in its ability to listen, adapt, and co-create with the various worlds that shape it. A pluriversal approach does not simply add diversity to the design lens, but instead transforms design into a practice rooted in mutual respect, sustainability, and equity. By embracing this shift, we move toward a more inclusive design approach, where knowledge is shared rather than owned, and where the voices of many, rather than a privileged few.

Acknowledgements FCT—Fundação para a Ciência e a Tecnologia, I.P.—finances this project within the scope of Strategic Projects with reference UIDB/04042/2020.

References

Acosta A (2013) El Buen Vivir: Sumak Kawsay, una oportunidad para imaginar otros mundos. Icaria Editorial

Ansari A (2020) Decolonizing design: towards a pluriversal framework. Design philosophy papers

Barnett R (2013) Imagining the university. Routledge

Best S, Kellner D (2008) The postmodern turn in philosophy: theoretical provocations and normative deficits. Retrieved 25 Mar 2025 from Kellner's website: http://www.gseis.ucla.edu/faculty/kellner/essays/postmodernturninphilosophy.pdf

Brookfield SD (2015) The skillful teacher: on technique, trust, and responsiveness in the classroom. Jossey-Bass

Brown T (2009) Change by design: how design thinking creates new alternatives for business and society. Harper Business

Clark H, Brody D (eds) (2010) Design studies: a reader. Berg Publishers

Coole D, Frost S (eds) (2010) new materialisms: ontology, agency, and politics. Duke University Press

de Moraes VM, Paoliello C, Mourato JM (2025) Reshaping design education: curriculum diversification for a pluriversal pedagogy. In: Raposo D, Neves J, Silva R, Correia Castilho L, Dias R (eds) Advances in design, music and arts III. EIMAD 2024. Springer series in design and innovation, vol 48. Springer, Cham. https://doi.org/10.1007/978-3-031-74975-9_37

dos Santos AB (2015) Colonização, Quilombos: Modos e Significações. INCTI/UnB

dos Santos AB (2023) A Terra Dá, A Terra Quer. Ubu Editora

Escobar A (2018) Designs for the pluriverse: radical interdependence, autonomy, and the making of worlds. Duke University Press

Freire P (2000) Pedagogy of the oppressed. Continuum

Fry T (1994) Remakings: ecology, design, philosophy. Envirobook

Fry T (1999) A new design philosophy: an introduction to defuturing. UNSW Press

Fry T (2017) Design for/by the real world: rethinking design education. Bloomsbury

Gawinek-Dagargulia M, Tymoshchuk M (n.d.) What is horizontal learning? Horizontal School. Retrieved 25 Mar 2025 from https://horizontal.school/wp-content/publications/rethinking-citizenship-education-03-what-is-horizontal-learning.pdf
Gibbons M, Limoges C, Nowotny H, Schwartzman S, Scott P, Trow M (1994) The new production of knowledge: the dynamics of science and research in contemporary societies. SAGE Publications
González FE (2001) Haciendo que hacer—cultivating a mestizaje worldview and academic achievement: braiding cultural knowledge into educational research, policy, and practice. Qual Stud Educ 14:641–656
Gudynas E (2011) Buen Vivir: today's tomorrow. Development 54(4):441–447
Haraway DJ (1991) Simians, cyborgs, and women: the reinvention of nature. Routledge
Harman G (2018) Object-oriented ontology: a new theory of everything. Pelican Books
Heidegger M (1962) Being and time. Basil Blackwell
Kaufmann-Buhler J, Pass VR, Wilson C (eds) (2019) Design history beyond the canon. Bloomsbury Academic
Khanchandani P (2020). Decolonising and diversifying design history. Royal College of Art. Retrieved from https://www.rca.ac.uk/news-and-events/news/decolonising-and-diversifying-design-history
Kolb DA (2015) Experiential learning: experience as the source of learning and development. Pearson Education
Kothari CR, Garg G (2019) Research methodology: methods and techniques, 4th edn. New Age International Publishers
Kurylo A, Yu Y (2016) Inter/cultural communication: representation and construction of culture. SAGE Publications
Latour B (1992) Where are the missing masses? Sociology of a door. In: Bijker W, Law J (eds) Shaping technology/building society. Cambridge University Press, pp 225–257
Lave J, Wenger E (1991) Situated learning: legitimate peripheral participation. Cambridge University Press
Lavoie D, Roth J (2002) Critical thinking in business. Thomson Learning
Manzini E (2015) Design, when everybody designs: an introduction to design for social innovation. MIT Press
Margolin V (1997) Getting to know the user. Des Stud 18(3):227–236
Meggs PB (1992) A history of graphic design, 2nd edn. Van Nostrand Reinhold
Nicenboim I, Oogjes D, Biggs H, Nam S (2023) Decentering through design: bridging posthuman theory with more-than-human design practices. Human Comput Interact 40(1–4):195–220
Oliver P (2003) Dwellings: the vernacular house worldwide, 2nd edn. Phaidon Press
Paoliello C, Li B, Metzdorf E (2024) Toward a nature-centered design. Diid—Disegno Ind Ind Des (83):12
Rapoport A (1969) House form and culture. Prentice-Hall
Risager K, Thorup M (2016) The politics of everyday life: constructing and representing the democrat. Transcript Verlag
Rudofsky B (1964) Architecture without architects: a short introduction to non-pedigreed architecture. Museum of Modern Art
Saavedra C (2011) De-academizing early childhood research: wanderings of a Chicana/Latina feminist researcher. J Latinos Educ 10(4):286–298
Schultz PW et al (2021) The psychology of sustainable behavior: tips for empowering people to take environmentally positive action. Minnesota Pollution Control Agency
Smith LT (1999) Decolonizing methodologies: research and indigenous peoples. Zed Books
Smith NE (2019) Decentering design: bees, biomimicry, and the sixth mass extinction. Indiana University
Sparke P (2013) An introduction to design and culture: 1900 to the present, 3rd edn. Routledge
Tlostanova M (2017) Postcolonialism and postsocialism in fiction and art: resistance and re-existence. Palgrave Macmillan
Van Amstel F (2023) Pluriversal futures for design education

Willis A-M (2006) Ontological designing. Des Philos Pap 4(2):69–92
Ziman J (2000) Real science: what it is and what it means. Cambridge University Press, Cambridge
De la Cadena M, Blaser M, (eds.) (2018) A world of many worlds. Duke University Press
Walsh C (2012) Interculturality and decoloniality. Cult Stud. 26(2¬3): 341–370. https://doi.org/10.1080/09502386.2012.651803

Design Practices for Regenerative Approaches to Hospitality in the Tourism Field

Elena Elgani and **Umberto Monchiero**

Abstract This chapter explores how design can facilitate the transformation of hospitality-related systems of spaces, products and services into increasingly more sustainable models. By adopting a regenerative perspective to address the complexity of these systems, it is possible to outline processes that enhance biodiversity, mitigate the damage inflicted by human activity, and restore ecological balance, as the tourism sector exerts a considerable impact on natural ecosystems (U.N. Environmental Program, 2023). This approach aims not only at mitigating its effects, but rather to generate a net positive one, reinstating conditions that enable the harmonious coexistence of flora, fauna, and human communities. The initial section of this chapter introduces a research-driven approach that establishes a conceptual foundation for embedding design-focused strategies into the planning and evolution of hospitality spaces within the tourism industry. The second section focuses instead on the analysis of relevant case studies for designers and stakeholders: on the one hand, research projects developed during research and didactical activities at Politecnico di Milano. On the other hand, case studies actively implemented and operating at high levels will also be featured in order to observe the potential practical and applicative outcomes and feedback of regenerative design methodologies. The featured projects deal with a range of topics, from soundscape ecology to water clean-up, from multispecies coexistence to the creative reuse of waste, but are unified in the belief that progress in the tourism industry lies in introducing and enhancing design strategies that implement regenerative approaches in hospitality business opportunities.

Keywords Design for hospitality · Regenerative spaces · Sustainable tourism

---

E. Elgani (✉)
Department of Design, Politecnico di Milano, Milan, Italy
e-mail: elena.elgani@polimi.it

U. Monchiero
School of Design, Politecnico di Milano, Milan, Italy
e-mail: umberto.monchiero@polimi.it

D. Raposo et al. (eds.), *Perspectives on Design IV*, Springer Series in Design and Innovation 59, https://doi.org/10.1007/978-3-032-00733-9_5

1 Introduction

1.1 The Unsustainability of Tourism

Tourism is a very important sector for local economies, as it allows the economic growth of countries and for some specific destinations it represents the only source of livelihood for communities. The 2014 Resolution of the United Nations General Assembly (UN, A/RES/69/234) emphasizes the role of tourism in eradicating poverty, developing communities and protecting biodiversity. At the same time, tourism experiences allow for cultural exchanges and opportunities for personal growth as well as integration between peoples. In 2018, the UNWTO defined cultural tourism as tourism activities aimed at learning about, discovering, experiencing and consuming the cultural products and attractions of a destination (UNWTO 2018). There are also forms of passive and active tourism specifically related to the discovery of natural ecosystems and sports (Tarfanelli 2010). Overall, these experiences support the pursuit of a dimension of well-being for individuals (Bustreo and Micheletto 2021).

The exponential growth of international tourism is a constant and it is anticipated that the current forecast of reaching 1.8 billion international travelers by 2030 will most likely be exceeded, reaching at least two billion (UNWTO 2011). The large number of travelers makes tourism a sector that has a significant impact on the environment and on social communities all over the world. According to the UN Environment's Green Economy Report, "in a 'business-as-usual' scenario, by 2050, tourism is expected to generate an increase of 154% in energy consumption, 131% in greenhouse gas emissions, 152% in water consumption, and 251% in solid waste disposal" (UNWTO 2024). Furthermore, in the post-pandemic era, there has been an increase in negative phenomena related to mass tourism, namely overtourism. The increased flow of visitors, in relation to the limited existing infrastructure, determines the excessive use of environmental and social resources connected to the overdevelopment of destinations, creating negative impact on the environment, such as depletion of natural resources and local communities, as well as social and economic inequality, and difficulties in accessing basic services (Vagena 2021).

At the same time, there are sustainable tourism experiences that aim to support environmental, social and cultural value; however, these experiences are often developed within a global tourism system that is currently unsustainable (Becken 2019).

Considering data and forecasts, tourism is a field that requires continuous multidisciplinary research and experimentation aimed at implementing sustainability, reducing environmental impact through mitigation and adaptation strategies, and integrating local communities into tourism practices.

For this reason, the Glasgow Declaration on Climate Action in Tourism was drawn up in 2021 at the UN Climate Change Conference in Glasgow in November 2021 (COP26), with a commitment to act immediately and accelerate climate action to cut global tourism emissions by at least a half over the next decade and reach Net Zero emissions as soon as possible before 2050.

The declaration promotes five pathways of the Declaration: Measure, Decarbonise, Regenerate, Collaborate and Finance to accelerate and co-ordinate climate action in tourism. In this document, the theme of regeneration, which is the key theme of this chapter, emerges as one of the most significant to be developed for a radical transformation of the tourism sector.

1.2 *The Environmental Impact of Hospitality Spaces in Touristic Territories*

Within the tourism sector, a significant role is played by tourist accommodation facilities as part of the hospitality industry, a heterogeneous sector that includes all the economic activities that directly or indirectly contribute to, or depend upon, travel, tourism and hospitality. This type of commercial and business-oriented interpretation of hospitality places hospitality spaces, mainly hotels, and all spaces related to the provision of specific services, at the center of the experience.

Hospitality spaces often replicate spatial layout models, room types, services offered, which despite the continuous search for hyper-personalization of experiences by guests are essentially spatial models coded and replicated in different geographical contexts, personalized only as regards the interior (FF&E), to satisfy the growing number of people accessing the tourist experience. Therefore, as tourist experiences grow, there is an increase in hospitality facilities that are newly built or refurbished to meet the growing demand for hospitality. According to The Business Research Company (2024), in 2024 the global hospitality market expanded to approximately $4.9 trillion, driven by increased leisure and "bleisure" travel, rising international tourism, and higher profits due to demand recovery in key regions like the Middle East and Europe. escalating from $4673.63 billion in 2023 to $4993.71 billion in 2024 at a compound annual growth rate (CAGR) of 6.8%.

While the increase in tourist experiences and the spread of hospitality facilities has a positive impact on local economies, as well as an increase in employment for the architecture sector and in particular interior design, called upon to develop increasingly innovative and creative projects for the hospitality industry, on the other hand, there is a significant environmental impact of accommodation facilities on both urban and natural ecosystems.

Furthermore, in hotels, the consumption of spaces, but also products and objects, that quickly become obsolete stimulates processes of 'soft refurbishments', partial or complete, as well as fit-out on FF&E (furnishing, finishes and equipment). Fit-outs and refurbishments are scheduled at short intervals, every 5–7 years, and for food and beverage spaces even more frequently (Perkins and Will 2021) both because of the necessary maintenance required by the frequent use of the spaces and the need for aesthetic communicative updating.

For this reason, it is necessary to study in depth how design-led strategies can concretely support the reduction of the environmental impact of hospitality structures in the contexts in which they are built, through a rethinking of the design and construction processes of hospitality spaces to participate in the decarbonization of the tourism supply chain. At the same time, it is essential to apply design-led theories and practices to broaden the horizon of action and consider how hospitality structures can be the vector of transformation of entire tourism systems, which are very complex and multidisciplinary. Through a holistic approach aimed at regeneration, this approach could allow hospitality spaces to make a concrete contribution to the regeneration processes of the local area.

2 Regenerative Approaches to Tourism Systems

2.1 Tourism Transition Through the Regeneration Promoted by Design

From the need to abandon an extractive logic and models, based on the acquisition of raw materials, products, experiences, spaces to be acquired or purchased and consumed for the satisfaction of one's own needs comes the regenerative approach, which intends to promote opposite, contributive logics. These new models promote the ability to take care of contexts, ties, and people to generate positive impacts or values, not only economic, but also social, cultural, relational, and innovative, increased with respect to how much value is consumed. Starting from a relational idea, these experiences reflect on interdependence, reciprocity, coexistence and harmony to delineate an idea of growth that does not correspond to a purely quantitative increase, in size as in numerical results, but in the proliferation of solutions that improve the human dimension, as well as a positive impact, in relation to an environmental and multi-species context that must be taken into account (Elgani and Monchiero 2024).

In tourism, regeneration is is «a transformational approach that aims to fulfil the potential of tourism places to flourish and create net positive effects through increasing the regenerative capacity of human societies and ecosystems [...]» (Bellato et al. 2023, p. 9).

This definition ties in with the definition of sustainable tourism defined as 'Any form of tourism development, planning or activity that respects and preserves natural, cultural and social resources in the long term and contributes equitably and positively to the economic development and full realisation of the people who live, work or stay in protected areas' in the European Charter on Sustainable Tourism.

In fact, the objectives of sustainable tourism are aimed at promoting climate adaptation, ecosystems restoration, climate justice, and regenerative projects.

At the same time, the concept of regeneration relates to the concept of social tourism, which is usually referred to as that set of activities that aims above all at deepening human relations, the need for sociability accompanied by the need of the organisations that organise it to create opportunities for cultural enrichment and the promotion of the area's resources. Tourism, therefore, aimed more at the human aspect than at the economic prestige of the place visited. Founded on the values of socialisation, personal growth and respect for the environment, it is a factor of social cohesion and cultural enrichment. It is a form of tourism that encourages encounters and socialisation. Moreover, it is made up of that set of tourism activities capable of responding to a widespread need for relationality; it is not, therefore, a tourism aimed only at specific social categories of disadvantaged persons (Musella and D'Isanto 2010).

In addition, regeneration applied to tourism takes into account new categories of users, such as climate-sensitive travellers, travellers and tourists who are increasingly aware of the changes and risks of climate emergencies and are therefore able to change their holiday times, routes and places in order to feel comfortable and safe. They are climate-sensitive travellers who pay attention to data and information on the adaptability of products, services and destinations that protect people and also contribute to tackling the climate emergency. Climate-sensitive travellers want to play their part and be seen as co-leaders in the fight against climate emergencies by the destinations to which they choose to travel (Fondazione Santagata 2024).

The ability of regenerative processes to combine the promotion of a positive impact on ecosystems with processes of connection between travellers and residents and the involvement of local communities, and thus with the social dimension of tourism, thus promoting a holistic approach that addresses the complexity of systems without forgetting all their components, makes regeneration a concept to be explored, also through the tools of design. Indeed, in 2023 was written the implementation of the Glasgow Declaration on Climate Action in Tourism, led by UNWTO (2021) within the framework of the Sustainable Tourism Programme of the One Planet network, where UNWTO collaborates closely with the UN Environment Programme (UNEP). In this document it emerged that «the regenerate pathway is less addressed, despite it potentially offers the greatest opportunity for tourism to play a leading, transformative role. At present, regeneration is too often seen as being just about providing financial support to others through offsetting, carbon removal or nature-based solutions. There is less focus on how to redesign core tourism products and experiences to become regenerative» (UNEP 2025).

It thus clearly emerges how design is the discipline capable of structuring sets of strategies and tools, to be applied through experimental and prototyping processes to different tourism contexts. These experimental processes capable of connecting skills in a multidisciplinary manner will allow the definition of guidelines and practices that can be applicable, with the correct adjustments, to different contexts. Experimental design research permits to investigate the relationship between design and positive environmental impact from a regenerative perspective, starting from the rethinking of hospitality spaces. This makes it possible to approach hospitality spaces with a multi-sensorial approach that proposes overcoming the dynamics of the experience economy in favour of experience ecology.

2.2 *Expanding Tourism from Experience Economy to Experience Ecologies*

The contemporary hospitality industry has fully embraced the paradigms of the Experience Economy (Gilmore and Pine 1999). In this economic model, the value of a service does not lie solely in its utilitarian function, but also in the quality of the experience it offers the customer; for the purposes of hospitality, this translates into the design of space and services aimed at evoking involvement and creating memorable experiences: from luxury hotels to innovative hostels, to the immersive experiences offered by boutique hotels and resorts, hospitality is increasingly configured as a complex system in which the guest is no longer just a user of a service, but the protagonist of an experience built specifically for them.

In the era of growing environmental and social awareness, the traditional experiential approach is evolving towards a new perspective, which we could define as the Experience Ecology (De Kerckhove 2022). If the Experience Economy focuses commodities, goods and services on creating memorable experiences to attract and retain customers, the Experience Ecology introduces a critical dimension, focusing on the systemic context in which the experience is lived. The guest, in fact, interprets the quality of the experience not only in terms of comfort and aesthetic pleasure, but also through perceptive filters that allow them to evaluate the degree of short and long-term sustainability of the experience offered. The exploitation of local resources and ingredients, the use of disposable products in the rooms, waste management, energy efficiency applied to the comfort of life in the hotel spaces become fundamental factors in determining the quality of hospitality, going beyond the mere spectacularization of the experience to instead embrace a regenerative logic.

Another key step in this evolution is to overcome the predominance of the visual component in the hospitality experience. Historically, the design of spaces has been based on design paradigms that favor the visual and (to a lesser extent) the tactile aspects, the main formal languages of the architectural disciplines. However, the experiences lived in a particular place and within its accommodation facilities are never limited to visual perception alone, but involve the entire sensory system. This perspective is based on the ecological theory of visual perception by Gibson (2014), according to whom perception is not a passive process, but a dynamic interaction between individual, environment and time. In this context, the experience of hospitality is co-constructed by the guest through their multisensory involvement with the space, the materials, the sounds, the scents and the atmosphere.

In the experience of regenerative hospitality it is particularly important to also emphasize the importance of the "soundscape", a term brought to prominence by the Canadian composer Schafer (1977); a contemporary and fitting definition of the term may be the following: "the collection of biological, geophysical and anthropogenic sounds that emanate from a landscape and which vary over space and time reflecting important ecosystem processes and human activities" (Pijanowski et al. 2011). For this component of reflecting ecosystem changes through sound, the study of the soundscape also takes on an important task of ecological supervision; anthropic activities and sound pollution can modify the relationships between living organisms and their environment, and the study of sounds and their variation over time in a given setting can help us understand the relationship between human societies and life, as well as sustainable coexistence with flora and fauna (Schafer 1977, p. 205). This led to the birth of a discipline that can be defined as *Soundscape ecology*, whose aim is to preserve natural landscapes at risk of extinction due to invasive human activities, starting from the study of sound to actively bring people together with the environments around them and encourage them to take care of the environment, to preserve it and to enhance it (Pijanowski et al. 2011). The contemporary hospitality project cannot therefore avoid evaluating the importance of even immaterial factors in the construction of the identity of a place and in the quality of the experience lived by customers, staff and the local community; the careful management of both the visual and sound landscape becomes a determining factor in increasing the well-being of each person who experiences the hospitality project.

In this perspective, it is interesting to note that sustainable multisensory experiences in the hospitality sector may arise more easily in small to medium-sized contexts, where the management of the accommodation facility is entrusted to local communities or families that operate with a different logic and on a different scale compared to hotel chains: while large international groups often have to focus on a standardized experience that can be reproduced on a large scale, independent businesses have the opportunity to forego some immediate profit in order to promote other values, such as the regeneration of the local area, environmental protection and the strengthening of local social networks. These innovative structures, which can be agritourism centers, eco-lodges, diffused hotels or more traditional ones, offer the possibility of approaching a hospitality model in which the regenerative approach is not a simple marketing element, but a guiding principle that permeates every aspect of the proposed experience in a circular perspective (Caruso 2023).

The transition from Experience Economy to Experience Ecology therefore represents a necessary evolution for a sector that can no longer avoid responding to contemporary environmental and social challenges: the guest is no longer just a consumer of experiences, but a conscious actor who creates and evaluates their stay based on its consistency with principles of sustainability and multisensory well-being, and this process must be implemented by combining a memorable experience with positive contributions to both the material and immaterial features of the local area.

3 Regenerative Approaches Applied to Spaces for Hospitality: A Research Methodology

Taking into account the potential role of design in regenerative processes applied to tourism field this research has been conducted with the aim of identifying answers to the following key questions:

- How can regenerative tourism experiences be designed and implemented?

- How can integrated systems of spaces, products, and services be structured to effectively enable regenerative tourism experiences?
- What is the role of design within the tourism industry, particularly in relation to hospitality spaces?

These inquiries are grounded in the theoretical premise that the future of tourism can only be truly sustainable if behavioral transformations occur. This involves fostering awareness, facilitating knowledge dissemination among individuals, and encouraging responsible actions from tourists, ultimately leading to a shift in the way tourism experiences are conceived and lived. Indeed, from a design point of view, applying the regenerative approach to the hospitality industry means designing with a holistic view systems of spaces-products-services that not only try to reduce the impact of hospitality spaces on ecosystems while supporting tourism practices that are responsible for the environment and local communities, but above all design systems that participate in the broader local tourism system by trying to have a positive impact.

The research shows through the analysis of significant processes and case studies that the prerequisites for starting to apply a regenerative approach are:

- change in the mindset for planners, destination managers, economic entrepreneurs, stakeholders, local communities who must participate in networks of relationships to establish productive comparisons.
- A shift of focus from the individual tourist who must continuously consume (and therefore spend) new and engaging experiences to the overall territory-spaces-community-people system that must generate growth, not just economic growth, from which all the entities involved can benefit.
- The cross-fertilisation between disciplines that are usually deputed to the design of tourist experiences and hospitality spaces with environmental, biological and engineering sciences that can bring nature-based or nature-inspired solutions capable of transforming processes within hospitality. The fertilisation and exchange of competences can also take place with the humanities in order to understand more precisely the needs of guests and local communities, also through participatory design practices.
- The definition of benchmarks that allow the concrete evaluation of the impacts of individual hospitality spaces as well as of the internal tourism systems in which they participate in certain contexts.

Within this framework, in this research the perspective of interior design is adopted, as «interiors can assemble meanings and give value and consequences to a situation» (Brooker and Stone 2010). By its very nature, interior design is an interdisciplinary field and design itself can function through a multi-layered approach, playing a key role in shifting professionals' mindsets toward a broader adoption of ecological perspectives in the built environment. Simultaneously, design can redefine processes of contextual analysis, ideation, and conceptual development concerning spaces, services, and products within the hospitality sector, encouraging the integration of sustainability and regeneration principles throughout the industry.

The research methodology employed in this study is primarily based on desk research, establishing a conceptual framework for incorporating design-driven approaches into the development of hospitality spaces within the tourism sector. It specifically focuses on the practice-led paradigm of regenerative development (Mang and Reed 2011; Mang and Haggard 2016), applying it to hospitality environments to define innovative regenerative spaces that reconnect with natural ecosystems and foster new models of regenerative tourism. The strategies identified in theoretical research are then tested through an analysis of case studies, both already realized and still conceptual, serving as reference points for stakeholders committed to developing regenerative approaches in hospitality services and spaces. These case studies are partly derived from the authors' research activities at the Dept. of Design and also teaching activities at the School of Design, both in the Politecnico di Milano. The projects in question are set in diverse locations, they originate from different design assumptions trying to answer to different needs, such as improvement of well-being and sensorial experiences as well as restoring of marine ecosystems through the collection of microplastics and the improving of host awareness; however, they share a common recognition of the need on the part of the designers to innovate the approach to tourism experiences and above all to spaces for hospitality, shaping them through a systemic vision where the tourist-guest engages with a network of hospitality spaces conceived as interconnected elements within broader natural and social ecosystems.

Within this context, solutions are identified not only at the architectural and interior design scale but also in terms of products—both physical and digital—and services that can facilitate immersive and participatory experiences. These solutions are devised to actively involve local communities and administrations, ensuring tangible sustainability across environmental, social, and economic dimensions.

4 Regenerative-Based Design Practices

4.1 Soundscape Hospitality: Designing Hotels in Harmony with Silence

Among the many disciplines of the design macrocosm, the one interacting with sounds is usually defined as "sound design", an often vague and undefined term that ranges from the world of cinema and theater to that of design in the traditional sense. Due to the intangible nature of sound, sound designers operate at the intersection of communication and digital product design, and they can be defined as those «who carry out a sound project to achieve a dimensional and functional definition, to the qualification of the procedures necessary for the production of a sound event-object» (Marchetta 2010, p. 12).

The primary goal of this discipline is to create new sonic scenarios, sometimes in an invasive manner. However, in the late 1970s, the concept of a designer dedicated to preserving existing sound environments emerged. This role, termed "acoustic designer", focuses on auditory respect, symbolic sound values, and strategies to balance compromised soundscapes (Schafer 1977, p. 327). Raymond M. Schafer emphasized of the role of this hypothetical professional as being that of raising people's awareness of the need to take care of the surrounding landscape as a main components for the development of a sustainable environment:

«It devolves on us to invent a discipline we might call 'acoustic design,' an inter-discipline in which musicians, acousticians, psychologists, sociologists, and others would study the world soundscape together in order to make intelligent recommendations for its improvement. [...] New methods of educating the public to the importance of environmental sound must be devised. The final question will be: is the soundscape of the world an indeterminate composition over which we have no control, or are we its composers and performers, responsible for giving it form and beauty?» (Schafer 1977, p. 15).

A group of Students from the MetaDesign I2 Studio in the Interior Design Bachelor's program at the Politecnico di Milano explored these questions by integrating soundscape methodologies into a sustainable adaptive re-use project. The course challenges students to apply a meta-design methodology (Collina 2005) to repurpose the historic San Lorenzo Convent in Sondrio (IT), a site with origins dating back to 1100 (Bustaffa 2006). Previously tied to female religious orders, the structure was recently acquired by Mamete Prevostini, a wine producer aiming to create in it an innovative and regenerative hospitality opportunity. The student group composed by Alice Centenaro, Costanza Nardon, Ester Preosti, Cinzia Rizzi analyzed in the 2024/2025 edition of the Studio the site's acoustic relationship between the reverberant vaulted corridors interiors in contrast with the external mountain silence. This inspired their concept centered on silence and its many possible interpretations.

Identifying the epicenter of their concept in the three key words "listening, pause, silence", the students identified the relationship with sound (or the absence of it) as the common thread between a monastic life of prayer, contemplation and a strict modus vivendi, and a rebirth of the place where regenerative experiences for the user's body and mind can be offered, creating spaces of calm, reflection and inviting connection with oneself to rediscover a rhythm of life far from the contemporary hectic pace. Their project, *Adagio* (Latin for "with ease", but also a musical term meaning to play slowly), interprets silence broadly: acoustically, it enhances natural soundscapes by controlling intrusive human noise, while visually it is organized into simple and orderly spaces free from distractions, which also allow one to focus on bodily silence, understood as an invitation to stop and reconnect with oneself in order to release physical and mental tensions.

Adagio offers a refuge from technology, fostering reconnection with nature, by reconfiguring research methods into guest experiences: instead of traditional hiking, visitors partake in "soundwalks", guided explorations of environmental sounds and ambiances in a given area (Schafer 1977, p. 212). Initially a research tool, in the *Adagio* concept this activity becomes a meditative experience, heightening awareness of the surrounding landscape and sounds inhabiting it; guests are also asked to document their observations in a questionnaire, noting unexpected sounds and personal impressions. Another activity, "Sound Searching," equips visitors with portable recorders and noise-isolating headphones, allowing them to focus on and capture compelling ambient sounds. These recordings are collected by hotel personnel and transformed into personalized soundscape compositions, serving both as souvenirs for guests but also and contributing to an evolving digital archive of the Valtellina soundscape for future ecological studies.

This concept offers valuable insights for design strategies emerging from the hybridization of soundscape ecology with hospitality. Key takeaways may include:

- Low-Impact Acoustic Design: Structures should minimize noise pollution by limiting vehicular sounds, optimizing building systems (e.g., HVAC) to reduce operational noise, and encouraging guests to appreciate tranquility without contributing to create noise.
- Soundscape Integration: Instead of eliminating sounds, design disciplines should foster conscious listening, encouraging guests to engage with the existing acoustic environment through curated experiences
- Research as a communitary experience: Data collection, through engaging activities proposed to customers, is of fundamental importance for monitoring changes in the local landscape. Customers thus actively participate in soundscape research and preservation, growing the knowledge and understanding of potential issues for local professionals, while also deepening their connection with the environment.
- Ecological Responsibility: The preservation of both auditory and visual landscapes should be a guiding principle, ensuring that architectural interventions are in harmony with the site's environmental and cultural identity.

Thus, at *Adagio*, the guest simultaneously becomes audience, composer, and designer of their own spatial and auditory experience (Schafer 1977, p. 205). Moreover, they play an active role in research and preservation efforts, contributing to the protection of both the acoustic and visual landscapes of the region.

4.2 Water and Marine Ecosystems Restoration Through Hospitality

Restoration involves individual actions aimed at restoring environmental conditions to a state that supports life. These efforts often intersect with preservation and protection initiatives, which, while essential, can sometimes impact local economies by restricting access to natural areas. However, restoration combined with responsible human approaches offers a way to maintain tourism and hospitality activities in natural settings while achieving a positive environmental impact.

Water pollution is among the most pressing environmental issues today. According to WWF (Tekman et al. 2022), plastic waste, primarily from major corporations, continues to degrade water quality. Once in marine environments, plastic decomposes into microplastics, contaminating entire ecosystems. Land-based activities near coastlines and rivers serve as the primary contributors, carrying vast amounts of plastic into the ocean. Europe alone releases between 307 and 925 million pieces of waste annually, with plastic comprising 82% of the total (González-Fernández et al. 2021). Even if plastic production ceased immediately, microplastic levels in the ocean would double by 2050 and increase fiftyfold by 2100 (Tekman et al. 2022).

Beyond harming biodiversity, water pollution also affects tourism, particularly activities like scuba diving. The DTMGT Report 2023 notes that the diving sector is growing at an annual rate of 10.4%, driven by more certification opportunities and emerging destinations. Younger generations prioritize eco-friendly experiences, pushing the design community to create tourism models that support environmental regeneration. Notably, microplastic contamination accelerates marine ecosystem degradation, including coral bleaching (Lai 2025).

Addressing this issue, a team from Politecnico di Milano, composed by students Rosanna Caldarella, Giulia Ettori, Davide Grasso and Rosanna Caldarella and led by researcher Elena Elgani, developed *Hòstraka*, a regenerative floating resort for diving enthusiasts. Designed within the Sustainable Hospitality Challenge by Hotel School The Hague, *Hòstraka* aligns with the UN's 14th Sustainable Development Goal: «Conserve and sustainably use the oceans, seas, and marine resources».

Inspired by nature, *Hòstraka*'s name merges "host" and "ostrica" (oyster), referencing oysters' ability to filter water and transform impurities into pearls. Similarly, the resort actively cleans the sea while offering a unique hospitality experience. Its structure includes communal areas anchored to the seabed and mobile housing units called *Stràka*—small boats serving as both private suites and eco-friendly transport.

A key innovation is its integrated water filtration system. Each *Stràka* collects microplastics as guests travel between resort areas and dive sites, actively contributing to ocean cleanup. The system functions by drawing seawater into the engine, where it is used for cooling. Before being expelled, the water passes through a filter that captures microplastics, preventing further dispersion.

Hòstraka also fosters ecological awareness by engaging guests in immersive experiences. Expert-led excursions educate visitors on marine ecosystems and filtration technologies, deepening their understanding of plastic pollution and possible solutions. This approach transforms tourism into an active force for environmental conservation.

The resort follows sustainable design and circular economy principles. Its floating structures, made from low-impact materials, are entirely removable to avoid long-term environmental damage. The modular housing units, designed as catamarans, accommodate two to five people and are linked by modular blocks, allowing flexible expansion. Oceanite anchors, developed by IntelliReefs, stabilize the resort while promoting coral growth and marine life repopulation.

Hòstraka exemplifies the principle of "reduce and regenerate." Instead of merely minimizing tourism's environmental impact, it turns tourism into an active force for marine ecosystem protection. If widely adopted, this model could make tourism a solution to pollution rather than a contributor. Over time, stays at *Hòstraka* could lead to measurable improvements in ocean water quality.

Its life cycle design (LCD) enhances sustainability by applying circular economy strategies. Guided by Life Cycle Assessment (LCA) principles, every stage—from raw material selection to final disposal—is optimized for minimal environmental impact.

In pre-production, materials are sustainably sourced, preferably locally. Manufacturing occurs near the resorts to reduce transport emissions, with prefabricated structural components shipped for on-site assembly using dry-joint techniques to enable reuse and recycling.

For distribution, *Hòstraka* prioritizes low-consumption vessels, optimized transport routes, and biodegradable packaging. Solar energy powers the resort, which also features advanced waste and wastewater management systems. Each unit collects up to 1080 kg of microplastics annually, actively reducing pollution.

At the end of its lifespan, *Hòstraka*'s modular design allows for extensive reuse and recycling, extending material durability and minimizing waste. Floating structures can be repurposed for local communities, while mobile suites have an estimated 30-year lifespan. The system is thus designed to support circular and regenerative hospitality.

More than a resort, *Hòstraka* offers an immersive eco-conscious tourism experience, blending luxury, adventure, and sustainability. By allowing guests to explore the ocean while actively contributing to its conservation, it introduces a new hospitality model that could transform tourism into a tool for marine pollution management and environmental restoration.

Hòstraka exemplifies the principle of "reduce and regenerate." Instead of merely minimizing tourism's environmental impact, it turns tourism into an active force for marine ecosystem protection. If widely adopted, this model could make tourism a solution to pollution rather than a contributor. Over time, stays at Hòstraka could lead to measurable improvements in ocean water quality.

Its life cycle design (LCD) enhances sustainability by applying circular economy strategies. Guided by Life Cycle Assessment (LCA) principles, every stage—from raw material selection to final disposal—is optimized for minimal environmental impact.

- In pre-production, materials are sustainably sourced, preferably locally. Manufacturing occurs near the resorts to reduce transport emissions, with prefabricated structural components shipped for on-site assembly using dry-joint techniques to enable reuse and recycling.
- For distribution, Hòstraka prioritizes low-consumption vessels, optimized transport routes, and biodegradable packaging. Solar energy powers the resort, which also features advanced waste and wastewater management systems. Each unit collects up to 1080 kg of microplastics annually, actively reducing pollution.
- At the end of its lifespan, Hòstraka's modular design allows for extensive reuse and recycling, extending material durability and minimizing waste. Floating structures can be repurposed for local communities, while mobile suites have an estimated 30-year lifespan.

The system is thus designed to support circular and regenerative hospitality. More than a resort, Hòstraka offers an immersive eco-conscious tourism experience, blending luxury, adventure, and sustainability. By allowing guests to explore the ocean while actively contributing to its conservation, it introduces a new hospitality model that could transform tourism into a tool for marine pollution management and environmental restoration.

4.3 Dealing with Waste Through Luxury Experiences

Contemporary regenerative hospitality spaces are evolving beyond mere accommodations, becoming vital community hubs that foster social sustainability. By nurturing relationships between people, place, and local ecosystems, regenerative hotels can serve as platforms for cultural exchange and environmental stewardship. Among the projects that demonstrate how thoughtful design strengthens both community ties and ecological resilience in the tourism Industry, Desa Potato Head Studios and Suites perfectly exemplifies this approach by transforming a resort into a nexus of regenerative practices that tangibly proves that tourism can actively repair social and environmental fabric while maintaining luxury—setting a precedent for hospitality as a force for systemic renewal.

Desa Potato Head, located in Seminyak, Bali, is a resort that merges luxury hospitality with a deep commitment to environmental and social regeneration. Founded in 2010 by Ronald Akili with the goal of creating an eco-village (Teo 2022), it exemplifies how architecture, design, art, and gastronomy can converge into an experience that transcends traditional tourism.

The resort tackles environmental challenges—particularly waste management and resource consumption—through innovative solutions that minimize ecological impact while fostering Bali's cultural and social revival. Its carbon-neutral achievement marks it as the first Asian hospitality brand to reach this milestone (World's 50 Best 2024). Guided by the motto «Good Times, Do Good», Potato Head practices regeneration not only in its architecture but also through community engagement and responsible waste systems (Schalkx 2024).

The architecture itself is a manifesto: the beach club, the resort's first structure, drew immediate attention for its Colosseum-inspired façade built from 6600 reclaimed wooden shutters sourced from disused Indonesian buildings. Designed by Indonesian architect Andra Matin in collaboration with OMA's Rem Koolhaas (Teo 2022), the use of salvaged materials is both aesthetic and strategic—demolished wood, handmade Balinese bricks, and furniture incorporating recycled plastics and industrial waste. Inside, sustainability extends to innovative interiors, where spaces feature functional furniture crafted from upcycled plastic and wood by designers like Max Lamb and Faye Toogood (Brady 2025). This approach reduces reliance on virgin resources while raising awareness about material life cycles, ensuring efficiency without compromising comfort (Teo 2022).

Bali's waste crisis, which sees over 3000 kg of trash collected daily from rivers and illegal dumps (Teo 2022), is addressed through Potato Head's closed-loop system, which prioritizes reduction, reuse, and recycling. Upcycling is central—97% of the resort's waste is repurposed, up from 50% in 2017. A striking example is the sculpture Pointman—River Warrior, created entirely from waste retrieved from local waterways to highlight pollution (Brady 2025). The resort also educates guests through partnerships with organizations like Sungai Watch, which removes plastic from Bali's waters. Visitors can tour the waste management center to observe how discarded materials are sorted, shredded, and transformed into new products, fostering environmental awareness (Schalkx 2024).

Potato Head's impact extends beyond its boundaries, regenerating local social and economic fabric. It collaborates with Balinese artisans to produce furnishings and textiles, reviving traditional low-impact construction methods. Community programs include recycling workshops for children and support for NGOs cleaning rivers and beaches; the resort also strengthens ties to local culture by hosting events featuring Balinese and international artists, while sourcing most food and goods from nearby Indonesian suppliers to bolster the regional economy (Teo 2022).

Desa Potato Head Studios and Suites provides key design insights for architects and designers focused on sustainable hospitality:

- Local and reclaimed materials: Using autochthonous recycled or waste materials reduces environmental impact while celebrating local cultural identity.
- Closed-loop waste systems: Upcycling and sustainable waste management strategies significantly lower a structure's ecological footprint.
- Adaptive spatial design: Areas and spaces can take on different functions across seasons and times of day, serving guests and locals alike responding to their evolving needs.
- Community engagement: Partnering with local artisans and NGOs boosts economic development and environmental awareness, facilitated through spaces that encourage guest-community interaction.

The resort exemplifies hospitality that actively regenerates its environment rather than merely minimizing harm. Its architectural, design, and waste solutions offer valuable lessons for creating responsible yet luxurious accommodations.

5 Conclusion

Considering hospitality spaces, which are strictly interconnected to the tourism experiences, the theoretical critical research conducted, and the analysis of selected case studies, allow us to conclude that for the design of hospitality spaces participating in a regenerative process is necessary:

- the evolution of design process for the hospitality space, in terms of conception and concept development. In particular, it is important that from the conceptual phase, the design of the host space sets impact objectives, in addition to the return on investment, capable of generating positive environmental and community value.
- the implementation process for the development of a system of hospitality spaces, in terms of executive realisation (executive design, material procurement, implementation, maintenance) and management of these facilities until the end of their life cycle when they should be decommissioned without leaving a trace in the existing context.

The implementation of new development processes and models is a slow process that requires confronting cultural and processual resistance from planners, stakeholders and designers, a global legislative gap, and the slow evolution of certain processes (e.g. building management processes).

In addition to these transformations that mainly concern the design of spaces, from a design point of view promoting a regenerative approach means designing not only hospitable spaces, but requires the application of a creative process capable of connecting the design of a system of spaces with technologies and innovative processes or innovation from other disciplines within a product-service system framework. Design can be the discipline capable of connecting the design of individual spaces and equipment (FF&E) with access to services, and thus with multisensory experiences, understood as an opportunity for growth with a value that is not only economic.

The outputs to be achieved by these processes are:

- concrete reduction of the environmental impact of hospitality facilities that apply regenerative logic compared to those that do not yet apply it. Resumption of the preservation of certain natural contexts that have been severely damaged by human action, as well as mitigation of some environmental impacts related to climate change in order to preserve biodiversity.
- tangible positive impact in contexts where forms of regenerative tourism are applied that place newly designed structures at the center. This means restoration of natural ecosystems, proliferation of local flora and fauna life, better living conditions for inhabitants.
- feedback on the social impact that should promote equity and well-being not only for economic operators active in the tourism sector, but for the entire local communities through a fair distribution of resources.

The case studies analysed show how the regenerative approach to the design of hospitality spaces is still experimental and not very well established in the hospitality industry. However, the theoretical approach and design projects demonstrates a high potential that if applied widely would allow the implementation of hospitable spaces as well as the possibility to load them with many meanings in addition to the appealing aesthetic lynchpin and to register concrete positive impacts.

References

Becken S (2019) Decarbonising tourism: mission impossible? Tour Recreat Res 44(4):419–433. https://doi.org/10.1080/02508281.2019.1598042

Bellato L, Frantzeskaki N, Nygaard CA (2023) Regenerative tourism: a conceptual framework leveraging theory and practice. Tour Geogr 25(4):1026–1046. https://doi.org/10.1080/14616688.2022.2044376

Brady K (2025) Bali brutalism—art, design, and sustainability at the Desa Potato Head. Pin–Up Magazine. https://www.pinupmagazine.org/articles/desa-potato-head-bali-review. Last accessed 2025/03/24

Brooker G, Stone S (2010) What is interior design? RotoVision Books, Mies

Business Research Company (2024) Global hospitality market analysis: size, drivers, trends, opportunities and strategies. https://blog.tbrc.info/2024/10/hospitality-market-analysis/. Last accessed 2025/03/24

Bustaffa F (2006) Monastero di San Lorenzo, sec. XI – 1805. Lombardia Beni Culturali. https://www.lombardiabeniculturali.it/istituzioni/schede/11500669/. Last accessed 2025/03/23

Bustreo M, Micheletto V (2021) Turismo culturale, relazioni con il benessere e la qualità della vita. Turismo e Psicologia 14(1):84–90

Caruso L (2023) Toward regenerative hospitality business models: the case of "Hortel". Tour Hosp 4(4):618–641. https://doi.org/10.3390/tourhosp4040038

Collina L (2005) Design e metaprogetto: Teorie, strumenti, pratiche. Poli.Design, Milano, Italy

De Kerckhove D (2022) Dall'economia dell'esperienza all'ecologia dell'esperienza. https://www.diculther.it/rivista/dalleconomia-dellesperienza-allecologia-dellesperienza/. Last accessed 2025/03/21

Elgani E, Monchiero U (2024) Interiors in the age of uncertainty. Future-proofing design practices. Design International. FrancoAngeli, Milano

Fondazione Santagata ETS, Studio Giaccardi & Associati (eds) (2024) Il turismo diventa climate-sensitive. Cosa cambia e come affrontare il cambiamento. Edizioni Fondazione Santagata per L'Economia della Cultura, Torino

Gibson JJ (2014) The ecological approach to visual perception, 1st edn., 1979. Psychology Press, Taylor & Francis, New York. https://doi.org/10.4324/9781315740218

Gilmore JH, Pine BJ (1999) The experience economy. Harvard Business Press, Cambridge, MA

González-Fernández D, Cózar A, Hanke G, Viejo J, Morales-Caselles C, Bakiu R, Barcelo D, Bessa F, Bruge A, Cabrera M, Castro Jimenez J, Constant M, Crosti R, Galletti Y, Kideys AE, Machitadze N, Pereira De Brito J, Pogojeva M, Ratola N, Rigueira J, Rojo-Nieto E, Savenko O, Schöneich-Argent R, Siedlewicz G, Suaria G, Tourgeli M (2021) Floating macrolitter leaked from Europe into the ocean. Nat Sustain 4:474–483. https://doi.org/10.1038/s41893-021-00722-6

Lai O (2025) What is coral bleaching? https://earth.org/what-is-coral-bleaching. Last accessed 2025/03/22

Mang P, Haggard B (2016) Regenerative development: a framework for evolving sustainability. Wiley, Hoboken, US

Mang P, Reed B (2011) Designing from place: a regenerative framework and methodology. Build Res Inf 40(1):23–38. https://doi.org/10.1080/09613218.2012.621341

Marchetta V (2010) Passaggi di Sound Design. Riflessioni, Competenze, Oggetti-Eventi. Design della Comunicazione, vol 2. Franco Angeli, Milano

Musella, M., D'Isanto, F. (2010) Turismo sociale. Primi passi verso una teoria economica del turismo sociale. Rass Econ 73(1):7–21

Perkins&Will (2021) Net-zero now. Hospitality. https://issuu.com/perkinswill_london/docs/net-zero_now_hospitality_issuu. Last accessed 2025/03/22

Pijanowski BC, Farina A, Gage SH, Dumyahn SL, & Krause BL (2011) What is soundscape ecology? An introduction and overview of an emerging new science. Landsc Ecol 26(1):1213–1232. https://doi.org/10.1007/s10980-011-9600-8

Schafer RM (1977) The soundscape: our sonic environment and the tuning of the world, 1st edn. Alfred Knopf, New York. Italian Translation: Nemesio Ala, Ricordi & LIM editrice (1985)
Schalkx, C. (2024) Potato head suites & studios. Condé Nast Traveler. https://www.cntraveler.com/hotels/bali/potato-head-suites-and-studios. Last accessed 2025/03/21
Tarfanelli, E. (2010) Sport e Turismo. Come fare business col turismo sportive. FrancoAngeli, Milano
Tekman MB, Walther BA, Peter C, Gutow L, Bergmann M (2022) Impacts of plastic pollution in the oceans on marine species, biodiversity and ecosystems. WWF Germany, Berlin. https://doi.org/10.5281/zenodo.5898684
Teo M (2022) Suite life: Desa Potato Head is making Bali a better place. The Straits Times. https://www.straitstimes.com/life/travel/suite-life-desa-potato-head-is-making-bali-a-better-place. Last accessed 2025/03/23
U.N. Environmental Program (2023) Tourism: responsible industry. https://www.unep.org/explore-topics/resource-efficiency/what-we-do/responsible-industry/tourism. Last accessed 2025/03/20
U.N. Environmental Program (2025) Regenerate. In: Glasgow declaration implementation report. https://www.oneplanetnetwork.org/programmes/sustainable-tourism/glasgow-declaration/implementation-report. Last accessed 2025/03/31
U.N. World Tourism Organization (2011) Tourism towards 2030—global overview. UNWTO, Madrid. https://doi.org/10.18111/9789284414024
U.N. World Tourism Organization (2018) Tourism and culture synergies. UNWTO, Madrid. https://doi.org/10.18111/9789284418978
U.N. World Tourism Organization (2021) Glasgow declaration on climate action in tourism at the UN climate change conference in Glasgow COP26. UNWTO, Madrid. https://www.unwto.org/the-glasgow-declaration-on-climate-action-in-tourism. Last accessed 2025/03/21
U.N. World Tourism Organization (2024) Resource efficiency in tourism. UNWTO, Madrid. https://www.unwto.org/sustainable-development/resource-efficiency-in-tourism. Last accessed 2025/03/22
Vagena A (2021) Overtourism: definition and impact. Acad Lett Article 1207. https://doi.org/10.20935/AL1207
World's 50 Best (2024) Desa Potato Head. https://www.theworlds50best.com/hotels/the-list/desa-potato-head.html. Last accessed 2025/03/24

Pedagogical Value of Reverse Design in the Pursuit of Sustainable Carton Packaging Design

Cristina Salvador and Maria João Delgado

Abstract In this paper we describe an integrated methodology applied to the teaching of materials and technologies in Design degree courses, through the implementation of a practical activity based on the concept of Reverse Design (RD) and the redesign of a cardboard packaging, which prioritizes the critical analysis of the object and its relationship with materials, digital technologies and production processes. A critical analysis of packaging available on the market was carried out in order to get to know, manipulate and analyze all its components, including the visual, formal and functional concept and the materials used. Subsequently, new solutions were proposed for cardboard packaging and the respective prototypes were made using laser cutting and engraving technologies, discussing the feasibility of these proposals in the context of current industrial production systems. From a pedagogical point of view, this activity involves combining teaching methodologies that promote the development of skills in the area of the relationship between design projects and the materials and technologies associated with cardboard packaging production systems. Thus, this article describes the development of the practical exercise based on a pedagogical *continuum* from RD, to digital technologies and industrial production practices in a real context, highlighting both the technical and material complexity in the industrial production of this packaging and the field of sustainability. The results obtained in this study, based on a qualitative and descriptive analysis of a case study, indicate that the students showed significant improvements in their understanding of the functionality and materialization of packaging, and that contact with industrial production and automation contributed to the development of skills based on a material and technological culture of industrial production.

Keywords Reverse design · Technologies teaching · Sustainable carton packaging · Prototyping · Artifacts analysis

C. Salvador (✉) · M. J. Delgado
Lisbon School of Architecture, CIAUD, Research Centre for Architecture, Urbanism and Design, Universidade de Lisboa, Lisbon, Portugal
e-mail: cristinasalvador@fa.ulisboa.pt

M. J. Delgado
e-mail: mjoaodelgado@fa.ulisboa.pt

D. Raposo et al. (eds.), *Perspectives on Design IV*, Springer Series in Design and Innovation 59, https://doi.org/10.1007/978-3-032-00733-9_6

1 Introduction

RD has proven to be a pedagogical practice that enhances creativity in product design and the acquisition of scientific and technical concepts, as well as facilitating the transfer of academic learning to the context of industrial production.

As a practice commonly used in the industrial sector, it has been adopted as a teaching and learning methodology, with the aim of facilitating the understanding of fundamental concepts incorporated into the creation of objects, design principles and technical and technological development and production processes.

This study, framed within a trajectory of innovation in the design of cardboard packaging, aimed to examine the pedagogical potential of RD associated with industrial production practices in a real context, to promote the understanding and acquisition of scientific and technological knowledge about the materials and technologies used in the design and production of packaging. In this sense, it contributed to the investigation of methodologies for teaching materials and technologies in product design, reflecting on the principles and pedagogical feasibility of activities associated with the materialization of a project for cardboard packaging and consequent production in an industrial environment.

This paper presents a case study in the context of the transformation, adaptation and materialization of cardboard packaging, with a theoretical-practical exercise developed in 4 distinct phases. We describe the activities and discuss the results of this practical experience carried out by a group of 47 third-year students on the Design degree course at Lisbon School of Architecture, which aimed to develop skills in critical analysis and experimental exploration of materials, techniques and technologies, enabling them to understand the importance of materiality in design projects.

The exercise, based on learning by doing, used RD as a way of getting to know, manipulate and analyze all the procedures and components involved in the original production of a package, understanding the visual, formal and functional concept and the materials used.

Starting with the analysis of a physical object, referred to by Otto and Kristin (1998) as a "black box", the students were challenged to observe, explore and dismantle different packages, reflecting on the principles underlying the whole process, from the visual concept, the relationship between form and function, to the materials used, components and assemblies. With Reverse Engineering (RE), the collection and systematization of this information, the students, establishing relationships and combining ideas, materials and technologies from various sources, set about developing a new creative language in response to the current requirements for packaging. Friedman (2021) quoting Steve Jobs states: "Creativity is just connecting things" (p. 66). This work proposal included identifying all the steps needed to experiment, test and prototype paper/cardboard packaging solutions, as well as developing morphogenesis using CAD and creating a prototype using laser cutting technology. Computational techniques were used to create, modify and optimize different geometries, encouraging the creation of innovative and optimized shapes that are often difficult to design manually.

Subsequently, the proposals presented by the students were contextualized in industrial production, assessing the constraints imposed by mass production and the final specifications of the project. The assessment was based on the analysis of the proposals and prototypes presented by the students, as well as on self-assessment and the results obtained in the discussion session on the learning achieved with this teaching methodology. With the development of this practical activity, which incorporates RD as a teaching method and combines theoretical and practical knowledge and skills, such as parametric design and digital fabrication for the development of more sustainable packaging, we hoped to contribute to the conceptual, technical and technological growth of these students.

In the following sections, and based on the review of literature, we present: (i) the approach to the principles of RE and the adoption of this concept by design; (ii) the current constraints in the design and production of cardboard packaging; (iii) a case study based on the development of an experimental activity to redesign a cardboard package, based on the RD process in association with digital manufacturing technologies and processes. Emphasis is placed on the methodology adopted for the choice of materials and technologies in the production of cardboard packaging.

This study is set against a backdrop of the ongoing transformation of educational practices, with more dynamic, critical and creative approaches, promoting learning that goes beyond conventional methods.

2 Review of Literature

2.1 Trajectory from Reverse Engineering (RE) to Reverse Design (RD)

The concept of RE, which was born in the field of engineering many decades ago (Raja 2007), has become a recurring practice in various companies around the world, in line with the dynamics of production and consumption. RE, associated with the idea of design deconstruction, has established the principle of discovering the non-obvious aspects that underlie the composition and functioning of an object, which allows us to know the principles that embody the design creation of the object (Conwell et al. 1993).

In 1990, Chikofsky and Cross (1990) described RE as a process for studying the object and the requirements implemented by the system, which does not imply altering or replicating the same object. From this point of view, RE is understood exclusively as a process of analyzing existing products or systems, which starts from a physical object or model and follows a series of measurement and analysis processes to obtain a virtual model. This sequence of tasks makes it possible to understand the conceptual form of the object, how it was built, and to identify the components of the system, their interrelationships and how they work (Zhong et al. 2020).

Understood as a research activity aimed at identifying industrial production methods or, in some cases, at recreating new products adapted to rapidly evolving areas (Iglesias 2014), this method has been widely applied in many segments of industry, in particular when it comes to improving existing physical objects or replicating certain components, when digital or geometric models of these products are not available. It is a method that has been explored for many years by writers, painters, as well as the pharmaceutical or automotive industries (Friedman 2021), with applications in "such as manufacturing, industrial design, and jewelry design and reproduction" (Raja 2007). Iglesias (2014) summarizes the objectives of RE as follows: (i) to cope with the complexity of systems and projects; (ii) to generate new points of view; (iii) to recover lost information; (iv) to detect side effects; (v) to synthesize higher-level abstractions; (vi) and to facilitate reuse.

RE has been used in cutting-edge research in many areas of science and technology, and applied to problem-solving for products and systems in mechanical engineering, electronics, computer software and hardware, or even biology or chemistry. "While you might assume that reverse engineering has limited value outside the world of computing, its applications are surprisingly broad, actionable, and compelling. In fact, reverse engineering is not just a favored tool of business titans—it's one commonly utilized by literary giants, award-winning chefs, comedy legends, Hall of Fame musicians, and championship sports teams. More important, it's one you can apply in your field to learn from your contemporaries, extract valuable ideas, and evolve your work in exciting new directions." (p. 21) (Friedman 2021).

RE is often used by designers who want not only to understand the original design of the product in order to create new products or create replicas of designs, but also to identify errors and flaws, so they have to be deconstructed in order to extract design information from them.

The inventions of the personal computer, the laptop and the smartphone are, according to Friedman (2021), an example of the result of this practice: "Steves Jobs didn't invent the MP3 player or the cell phone. But he led a team that found a way of combining the two, leading to the iPhone" (p. 59). The author argues that many companies consider this practice to be unethical, as it compromises the protection of intellectual property and infringes copyright. However, he maintains that, when used correctly, reverse engineering can bring numerous benefits for innovation, and teaching and learning models based on this method allow for a critical and creative approach in the search for solutions to solve problems.

In fact, the challenges posed to the market to create innovative products that respond to consumer needs, with very limited timeframes, encourage industries to turn to RE as a tool that makes it possible to deal with the growing complexity of systems, comparing products, services and practices with those of their competitors, identifying ways to reduce costs, ensuring compliance with industry standards or legal requirements, improving quality (Dalrymple et al. 2013), while at the same time boosting the innovation process.

However, it has only been in recent decades, with the development of software, that specific and more rigorous methods have been developed to study an object by deconstructing the physical model and the assembly elements, using photogrammetry or scanning and digitizing processes, for example. In this way, it has been possible to adapt the evolution of advanced manufacturing to meet the requirements of society, as well as retrieve information about objects and buildings that are not properly documented (Iglesias 2014). Its importance in retrieving information from physical objects of historical or heritage significance, or in customizing existing products or systems to meet specific needs, stands out.

Therefore, we can understand that the approach to reverse engineering applied to the teaching and learning of materials and technologies leads the student to learn through the original object, how that product works, breaking it down and analyzing each of the parts (Orta 2006). It also makes it possible to acquire and deepen scientific knowledge (Ladachart et al. 2022), developing skills related to problem-solving and creativity (Zhong et al. 2020; Abdüsselam et al. 2022; Grantham et al. 2010), as well as developing complex ideas. This process helps students to progress to the higher levels of Bloom's taxonomy: design, implementation and evaluation (O'Brie et al. 2010), and "to unlock hidden insights acquire new skills and spark their creativity" (p. 21) (Friedman 2021).

Friedman (2021) identifies three main phases in the process: The first phase, which is vital in this process, involves gathering a set of examples, which differ in their particularities, and which are analyzed in detail in order to learn as much as possible about the context of the object. At this stage, all kinds of information are important because, even if they transcend the limits of the study context, they can contribute to the construction of an idea or project. This phase includes various tasks to help evaluate existing objects, even those with more complex shapes, ranging from deconstructing the objects, 3D scanning, assemblies, materials' testing, non-destructive testing, ultrasound thickness measurements, among others, with the aim of gathering as much information as possible and enabling a greater understanding of the objects' structure and function.

The second stage involves an in-depth and critical analysis of the specific characteristics of the different objects, identifying commonalities, differences that make them unique, strengths and weaknesses.

This is followed by a process of systematizing the characteristics, which can then be included in the re-creation of the product in order to improve the performance of the new product. This process involves critical thinking, which is both convergent thinking based on the application of principles and divergent thinking, which involves the development of creative thinking skills to solve problems that lead to the creation of new ideas.

Although associated with the field of engineering, involving knowledge of mathematics, science and computing (Liu et al. 2023), design has often used this methodology to solve problems, in a scientific vision that seeks to understand and reinterpret existing design processes, by re-evaluating, interpreting and redesigning products, systems and services from an inverse analysis.

In this sense, we believe that RD contributes to a deeper understanding of scientific concepts and improves the ability to solve problems involved in the conception or redesign of an object or innovative systems, in response to user needs and the current challenges that sustainability poses to designers (Preece 2014).

In short, pedagogically, this method of deconstructing an object and analyzing it, based on deductive reasoning, facilitates the understanding of a whole by transforming a complex problem into several elements and, as Friedman (2021) states, enhances the discovery of hidden ideas and the development of new skills and creativity.

2.2 *The Approach of Sustainability in Packaging Design*

Throughout history, packaging has played an important role in the "containment, protection, handling, delivery and presentation of goods, from raw materials to processed goods, from the producer to the user or the consumer" (Ragonnaud 2023). Responding to the demands of new products, they have become increasingly important from an economic, social and cultural point of view (DuPuis and Silva 2011). Produced from different materials such as glass, paper and cardboard, metal, plastic, wood, cork, textile fiber or composite materials, they are an integral part of the product, playing a fundamental role in protecting its contents and in the communicational, functional/technical and commercial relationship with its users. "Packaging should protect its contents, provide relevant information for the consumers, differentiate the product from the competition, communicate the brand's values and be cost effective." (Thompson 2012) and become agents of persuasion by enhancing emotional involvement with the product (Preece 2014).

Its design has kept pace with technological developments and has sought innovative ways of addressing environmental sustainability issues (Moura and Banzato 2003), reducing negative impacts and, as the EU explains in its Action Plan for the Circular Economy (EU 2020), aiming to reuse and recycle packaging, reduce the complexity of the materials used and the waste produced. This strategy draws attention to the need to control the excessive use of packaging, simplify its composition and restrict the use of certain materials, particularly plastics, whenever there are viable alternatives.

As Chick and Micklethwaite state (2011) "Design is critical in addressing the various agendas of sustainability, because it can have significant (both positive and negative) economic, environmental and social ripple effects" (p. 102). The Sustainable Packaging Coalition also argues that "Sustainable packaging design considers the full life cycle of the package, recognizes the principle of Shared Product Responsibility, and consequently seeks to minimize the total packaging system cost through efficient and safe package life cycle design." (SPC 2011). To this end, this organization defines criteria for the entire packaging value chain, with a view to a sustainable packaging system, with an integrated and responsible vision for reducing waste and conscious consumption.

Nowadays, with the growing trend towards e-commerce and the direct flow of products to the home, packaging plays an essential role in the packaging, transportation and distribution of the product and is no longer an integral part of the consumer's purchasing decision. This has led the packaging industry to look for new solutions, requiring designers to develop a concept of interaction between the characteristics of the packaging and the product, with a more circular and sustainable approach (Cardenas 2019).

This orientation has revealed itself in many companies that 'are considering how the packaging can be integrated into the manufacturing process and become part of the end product' (Aylesbury Box Co. 2020). In May 2022, Ogawa Makoto, president of the Japan package design association, JPDAD stated: "I think now is an era when the power of design is required more than ever. And package design is at the center of major social keywords such as "innovation", "regional revitalization", "branding", and "environmental problems". Original ideas, ideals and imagination created by package designers take shape, reach the lives of many people, and make a big difference in society. In such an era, we have been and will continue to convey the power of package design to society through the activities of JPDA" (Message from the President JPDA 2025). Carton packaging is an excellent example of production that meets the requirements of the circular economy, based on the principles of sustainability in terms of consumption, production and resources. Carton and paper are used in primary level packaging, i.e. in direct contact with food products in individual units, in grouped packaging (secondary level) for storage and distribution, and in transport packaging for bulk distribution (tertiary level). Their versatility makes it possible to create packaging with a personalized design, containing all the necessary information, as well as offering effective protection for the products, which increases the relationship of trust with the consumer and responds to the current trend in the development of online commerce. They are a promising option for reducing the use of plastics, using raw materials sourced from wood pulp, which is a biodegradable and recyclable material. The increase in more sustainable practices in industrial production is now a reality, and the paper and cardboard packaging industry has challenged traditional production practices in the face of the carbon footprint generated by the paper industry. In this sense, the design of this packaging must contribute to the sustainable safeguarding of natural resources, which are finite, and to reducing the consumption of raw materials, promoting the use of lighter cardboard that meets the necessary parameters of rigidity and mechanical resistance, produced in accordance with environmental standards (Pyr'yev et al. 2024).

3 Case Study

3.1 Research Design and Context

The case study is a method of empirical investigation of a phenomenon, which allows researchers to understand, explore or describe complex events and contexts, developing valid arguments and examining the particular aspects and significant characteristics of the specific research case (Yin 2002; Stake 2012).

Based on this research strategy, and using various sources and evidence (Yin 2002), we sought to deepen our understanding of the development of technical and technological skills in a carton packaging redesign programme for design undergraduates, using an integrated methodology based on a pedagogical *continuum* of RD, digital technologies and industrial production practices in a real context.

Bringing teaching practices closer to the context of industrial production, materialized in study visits and workshops, is nowadays a focus of education, with a view to fostering integrated training in a work context, where students can confront and discuss many of the concepts they have learned throughout their studies. In fact, they come into contact with the industrial and business dimension, not only in terms of production, but also in the area of management and organization, connected to the principles of LEAN Thinking of creating value by eliminating waste, when confronted with the realities of the market and the functioning of the product's commercial circuits, where learning takes on an effective and efficient nature.

The research was carried out with a group of 47 third-year Design undergraduate students at Lisbon School of Architecture, whose interdisciplinary training is based on a project-centered syllabus, which involves a range of knowledge in areas upstream and downstream of the designer's activity, namely materials and production technologies. In the field of technology and materials training for these design students, the didactic-pedagogical contribution has been based on knowledge of the technical and sensory properties and intangible characteristics of the materials best suited to the materialization of the project, highlighting the technical and material complexity in industrial production and in the field of sustainability. 'Creatively combining processes and re-appropriating technologies will lead to new product experiences for the consumer' (p. 9) (Thompson 2012). To this end, a carton packaging redesign activity was promoted, guided by an integrated methodology that includes the RD process, the use of digital technologies, as well as adapting the project to industrial production technologies to solve specific problems.

This activity followed all the necessary steps for conducting an artifact analysis (Martin and Hanington 2012), examining and comparing existing products on the market, as well as material and industrial manufacturing requirements.

In this research we carried out detailed descriptions of the research environment, drew on various data sources, triangulated information resulting from observations, analysis and discussion, and collected the sketches, presentations, reports, prototypes produced by the students, which reinforce the reliability of the study (Yin 2002).

3.2 *Exercise Development—Reverse Design*

The activity began with a theoretical presentation on the complexity of packaging design, particularly in ecological, communicational and functional terms, always bearing in mind the needs of users. With the support of projected images, some examples were highlighted, analyzing functionality, consistent and coherent communication in packaging design, as well as social responsibility and sustainability, as values of the present and the future. Finally, the raw materials and processes used to transform carton were also discussed, taking into account the various factors involved in each project, such as functional, structural, technical, aesthetic, safety, cost, environmental, resistance and performance attributes, among others (Ashby 2005). Chick and Micklethwaite (2011) state that "Designers often begin with materials, and design through materials. Material recycling is therefore a natural instinct for a designer seeking to reduce environmental impacts" (p. 104).

The objectives of the exercise and the respective development phases were defined and are summarized in the table in Fig. 1.

1st Phase—Research

In the first phase, the students, following the RD phases advocated by Friedman (2021), began the RD process, with an initial approach to the design method, artifact analysis (Martin and Hanington 2012) involving themselves in a detailed research into the functional and communicational aspects of two or three small packages (primary or secondary), with different purposes and produced with different materials

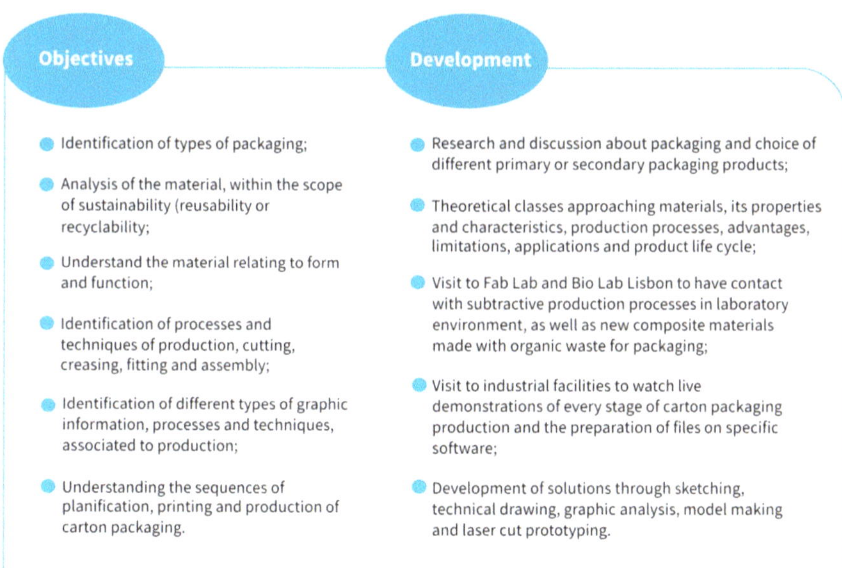

Fig. 1 Objectives and exercise development. *Source* Authors

(carton, plastic, metal, glass, etc.). By associating the types of packaging with the contents and identifying the function, shape, materials, properties, processes and production techniques, it was intended that students would understand the principles of packaging design and the relationships between their functions and materiality. An analysis was also made of packaging versus function versus reuse and recycling. The students not only presented photographic records and hand-made drawings of the packaging studied but also added important notes to characterize them (Fig. 2).

2nd Phase—Production Processes

In addition to this phase, study visits were carried out in order to introduce these students to and integrate them into the manufacturing and digital production processes.

Initially, a visit was made to the FabLab and Bio Lab in Lisbon, with the aim of expanding students' knowledge about subtractive technologies in a laboratory environment, with a focus on laser cutting and engraving. Greater knowledge about the use and regulation of these processes is important for the modeling/prototyping of the final object. In this laboratory space, students also had contact with new ways of producing composite materials, using organic waste and renewable material, well

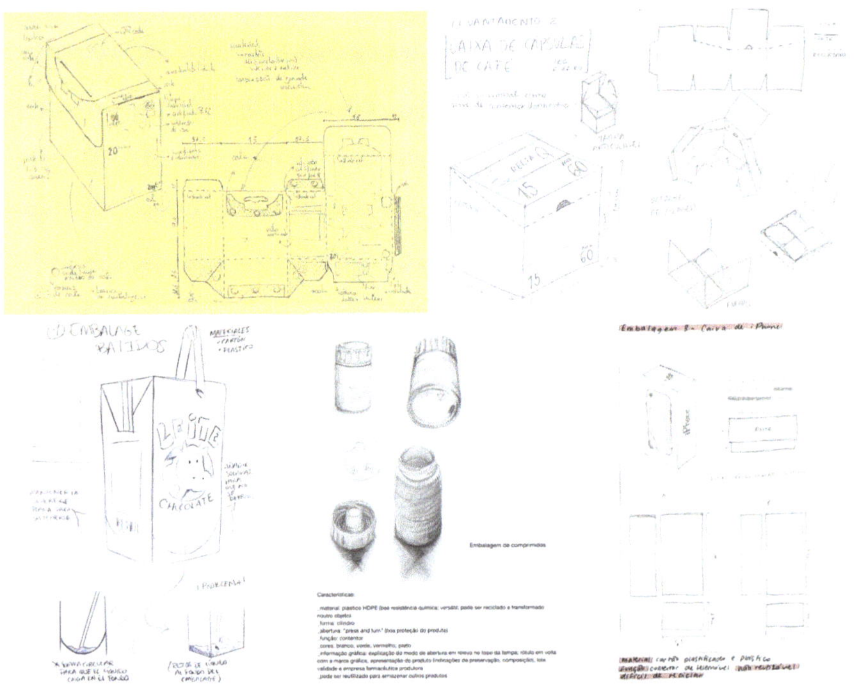

Fig. 2 Work by five students for the 1st phase of the exercise, analyzing tea, coffee, milk, pills and mobile phone packages. *Source* Students

adapted to the world of more sustainable packaging and with a lower environmental impact.

In a second moment of this phase, for better integration into the packaging production process, a study visit was carried out to a printing company that produces carton and cardboard packaging, NOVOTIPO Europa, S.A., whose motto is "helping customers replace current packaging with more sustainable packaging solutions that minimize the impact of the carbon footprint, improve the brand image and add value to their product, creating new consumption habits" (Green Packaging Planet 2021). On this visit to the company, students learned about the various types of carton and boxes, as well as getting in touch with all issues related to pre-pressing, printing, finishing, cutting, folding, creasing, packaging, storage and transport (Green Packaging Planet 2021), in particular about: software; files; number of colors; Pantone colours; screen calibration (RGB vs. CMYK); preparation of tests; control tools; cut; bleed; creases; types of carton; and productive limitations.

Later, in the R&D laboratory, they evaluated a practical case, which consisted of measuring the product, selecting the type of box, building the technical drawing and executing the prototype (plotter cut). This visit made it possible to introduce students to the production process of a carton packaging printing company.

3rd Phase—Reverse Design

In this 3rd phase, the analysis, disassembly and identification of the structure of a previously selected cardboard packaging began. A checklist was created recording specific information: material characteristics, fittings, connections, types of cutting, folding, creasing, packaging frame, finishes (lamination or varnish), colors and graphic information such as logos, barcodes, technical specifications, expiration dates, among others. This phase also included the use of CAD in 2D drawing, to more accurately represent all the details of the packaging (Fig. 3). The relationship between packaging vs function vs reuse and recycling was also evaluated.

4th Phase—Project and Prototyping Experience

Faced with the identification of some problems and the expectation of carrying out a project that would result from the modification/adaptation of the packaging selected for the exercise, this study was conditioned not only by the resolution and improvement of some identified problems, but also by the introduction of two variants: one, associated with the change in scale of the products, which implies formal diversity or variations in the technical characteristics of the material to achieve a new functionality, and the other associated with ecological aspects, either in the elimination of the use of glue points for assembling the components, or in the analysis of wasted material in the cut, according to the planning, as shown in Fig. 3.

This phase dedicated to the project, was based on the principles of know-how, resorted to the development and execution of experimental models and prototypes, understood as a pedagogical strategy of formal and functional exploration. In all the different functions attributed to prototypes, whether in the role of ideation or in the role of transforming the original intentions of their creators, as it is "during the process of materializing the idea in the search for the solution that will finally take

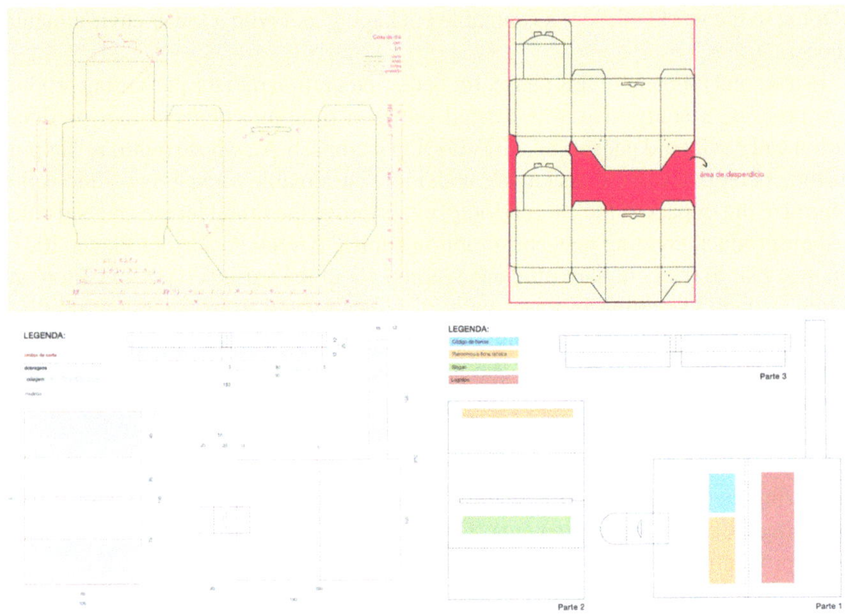

Fig. 3 Work by two students for the 2nd phase of the exercise. *Source* Students

shape in the project" (Lefort and Desvaux 2014) (p. 35), the ideas were deepened, becoming less abstract as they materialized.

After testing several solutions, each student developed a final prototype of the object, as a "tangible representation of artifacts at various levels of resolution" (p. 78) (Martin and Hanington 2012), using CAD with 2D and 3D drawings, together with laser cutting and engraving technology (Fig. 4). All of these tasks were compiled into an individual report of a reflective and critical nature, accompanied by drawings in digital format of the packaging planning, with dimensions, and in accordance with the codes of the operations to be carried out. This phase was completed with the delivery of a prototype of the proposed solution.

3.3 Results and Discussion

In the first phase of this study with the packaging artifact analysis approach, the expected results are quite specific, allowing a deep understanding of the functions of packaging—protecting, preserving, transporting and facilitating the use of the product—and its ergonomics and handling.

The exercises developed by the students were analyzed qualitatively and all interventions and outputs resulting from the study visit were systematically observed. With the aim of evaluating the learning and stimulation that the study visit to the

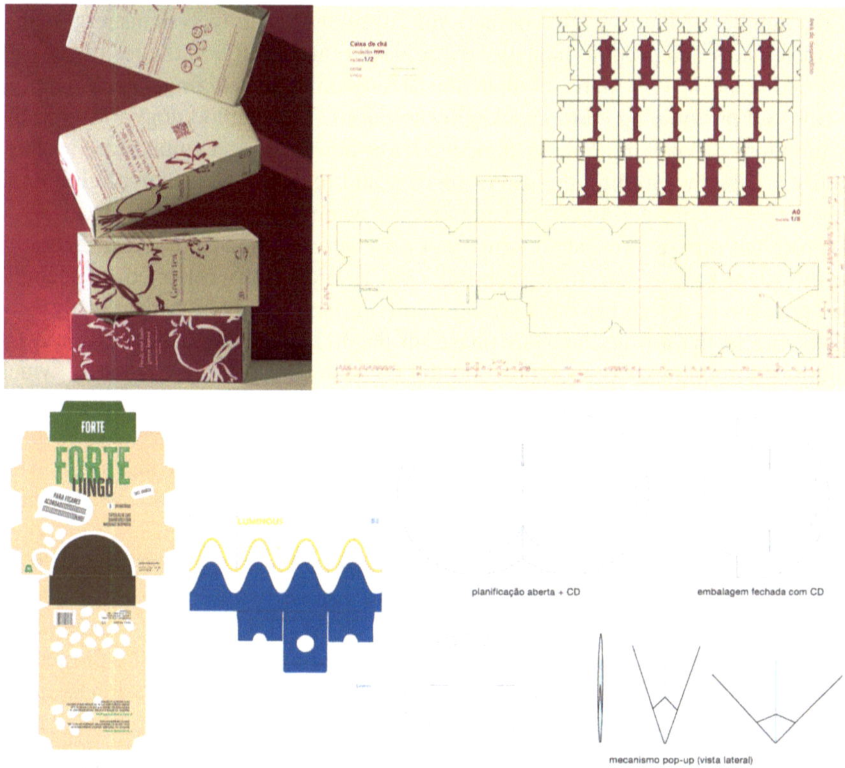

Fig. 4 Work by four students for the 3rd and 4th phase of the exercise. *Source* Students

manufacturing unit provoked in the students, a session was held at the end of the visit with teachers, students, designers and company managers, during which the contributions of this activity to their training were discussed.

The students highlighted the expansion of knowledge needed to carry out the final project, in terms of the technical and aesthetic possibilities of the materials used, the processes, production technologies and the use of resources, in a product life cycle that is less linear and more closed as possible, as suggested by Vieira (2018). They mentioned theoretical and practical knowledge of the functioning of machinery; the systems and quantity of printed colors; the software used; the fold margins; the thicknesses of carton and cardboard; different types of assembly; possible percentage of waste; the diversity of cutting edges; the printing of reliefs in Braille; graphic preparation with perforations, lines and bleed; finishes with lamination and/or varnish; and automatic gluing. They also revealed the importance of understanding product management, waste, the market and customized responses to consumers.

A process of constant formal research and familiarization with packaging and the RD process proved to be very important towards the development of a final prototype. The lessons learned regarding the effective use of the material, reduction of the number of binders (glue points), improvement of fitting systems, ergonomic characteristics and graphic image, were achieved through RD activities mentioned by Friedman (2021), which culminated in disassembling, analyzing, experimenting, testing and prototyping carton packaging solutions. The methodology mirrored the assumptions of knowing how to think and know how to do and emphasized sustainability issues, by avoiding the use of binders and reducing waste, highlighting the importance of considering ecological practices when designing packaging.

From the reports presented, which not only described and presented the entire development path of the final proposals, but also critically reflected on them, it was possible to determine the learning results in the process of critical analysis of packaging in the artifact analysis phase, in the RD process and in the search for solutions through experimentation and the creation of a prototyped object, which "can be used as an effective tool for design teams, clients, and end users" (Martin and Hanington 2012) (p. 78). As reported by the students, this sequence of tasks made the learning process more dynamic and the critical look placed in the first phase in the analysis of objects proved to be fundamental throughout the entire process, allowing the discovery of multiple choices and optimization.

Additionally, the scope of the exercise provided a broad view of the process, from product conception to distribution. Therefore, we can see that the approach to artifact analysis, RD, as well as visits to the company NOVOTIPO and Fab Lab/Bio Lab, opened new perspectives on the potential of packaging design and the production process, giving the possibility of interacting with experienced professionals in the field. The students considered that the specific knowledge learned contributed positively to their training by cultivating a reflective and critical outlook and promoting their autonomy in exploration and decision-making. They also highlighted that in addition to the application of knowledge, this exercise emphasized the importance of in-depth analysis combined with continuous experimentation.

4 Conclusion

This case study emphasizes the pedagogical value of Reverse Design and the importance of structuring design learning in an integrated way with the materials and technologies present in industrial production processes. It highlights the need to implement practical and interactive activities for successful student learning in Design, promoting both technical development and social and environmental adequacy.

The design of any carton packaging is an intentional activity that can only be fully realized when considering which materials and ideas are intrinsically combined in their materialization. In this case study, we were able to determine that RD, complemented with artifact analysis and the creation of prototypes, led to a redesign with a significant impact on the way students learn to think and project in design, as it acted as a tool for reflection and transformation of the original intentions of its creators. Reverse Design, as a design method, helps to decompose reality to better understand it, culminating in a redesign. By proposing integrated solutions, students showed significant improvements in understanding how the project works and in the materialization of more sustainable packaging. We understand that, in order for products to remain competitive in the market, they must consider the evolution of materials, the introduction of new technologies, new production processes and comply with the principles of Eco-design, which dominate many strategies in the field of packaging design. Our study also demonstrates that contact with production in a laboratory and industrial environment, as well as lectures with professionals, strengthen and enhance the development of skills, based on a material and technological culture of industrial production, preparing them to work autonomously or collaboratively in an industrial environment.

Acknowledgements The authors would like to thank the 3rd year students of the Design Degree from Lisbon School of Architecture (year 2023–2024), who participated in this exercise, to Fab Lab and Bio Lab Lisboa and to NOVOTIPO's team, who accompanied us in this study.

References

Abdüsselam M, Turan-Güntepe E, Durukan Ü (2022) Programming education in the frameworks of reverse engineering and theory of didactical situations. Educ Inf Technol 27:6513–6532

Ashby M (2005) Materials selection in mechanical design, 3rd edn. M.F. Butterworth-Heinemann (2005)

Aylesbury Box Co. (2020) When packaging is part of the product. https://abcbox.co.uk/packaging-part-of-the-product/. Last accessed 2025/04/02

Cardenas M (2019) Systems material waste: packaging systems and waste. Harvard University, Cambridge, p 11

Chick A, Micklethwaite P (2011) Design for sustainable change: how design and designers can drive the sustainability agenda. AVA Publishing, Switzerland, p 102

Chikofsky E, Cross J (1990) Reverse engineering and design recovery: a taxonomy. IEEE Softw 7(1):13–17

Conwell J, Catalano G, Beard J (1993) A case study in creative problem solving in engineering design. J Eng Educ 82(1):227–231

Dalrymple O, Sears D, Evangelou D (2013) The relative pedagogical value of disassemble/analyze/assemble (DAA) activities. Adv Eng Educ v3-n4 Sum

DuPuis S, Silva J (2011) Package design workbook: the art and science of successful packaging. Fair Winds Press

EU (2020) Plano de ação para a economia circular. https://ecoeconomy.aeportugal.pt/uploads/publicações/Plano%20de%20Ação%20para%20a%20Economia%20Circular.%20Para%20uma%20Europa%20mais%20limpa%20e%20competitiva.pdf. Last accessed 2025/04/02

Friedman R (2021) Decoding greatness: how the best in the world reverse engineer success. Simon & Schuster, New York

Grantham K et al (2010) A study on situated cognition: product dissection's effect on redesign activities. In: ASME international design engineering technical conferences/computers and information in engineering conference, Montreal, QC, Canada, August, pp 15–18

Green Packaging Planet (2021) https://www.alzamoragroup.com/pt/compromiso-sustentavel/. Last accessed 2025/04/03

Iglesias R (2014) Reverse Design o la deconstrucción proyectual del diseño. In: Proceedings of the XVIII conference of the iberoamerican society of digital graphics: design in freedom. Blucher, São Paulo

Ladachart L et al (2022) Using reverse engineering to enhance ninth-grade students' understanding of thermal expansion. J Sci Educ Technol 31:177–190

Lefort E, Desvaux N (2014) Aprendiendo con maquetas. Pequeñas maquetas para el análisis de arquitectura. EGA Expresión Gráfica Arquitectónica 19(24):62–71

Liu X et al (2023) Effect of reverse engineering pedagogy on primary school students' computational thinking skills in stem learning activities. J Intell 11(2):36

Martin B, Hanington B (2012) Universal methods of design: 100 ways to research complex problems, develop innovative ideas, and design effective solutions. Rockport Publishers, Beverly

Message from the President JPDA (2025) https://www.jpda.or.jp/en/about/message. Last accessed 2025/04/04

Moura R, Banzato J (2003) Embalagem, unitização & conteinerização, 4ª Edição. IMAM, São Paulo

O'Brie S et al (2010) Learning through reverse engineering. American Society for Engineering Education

Orta P et al (2006) Use of reverse engineering as a teaching tools in mechanical engineering education, Section 2546. In: 2006 ASEE annual conference & exposition, June 18–21, Chicago, Illinois

Otto K, Kristin W (1998) Product evolution: a reverse engineering and redesign methodology. Res Eng Des 10:226–243

Preece S (2014) The five things product packaging must do. https://www.forbes.com/sites/onmarketing/2014/07/23/the-five-things-product-packaging-mustdo/#c038e033ac4f

Pyr'yev Y et al (2024) Optimization of the solid cardboard in carton design. Bioresources 19(4):7963–7976

Ragonnaud G (2023) Members' research service PE 745.707. Revision of the packaging and packaging. Research Service (EPRS), p 2

Raja V (2007) Introduction to reverse engineering. In: Raja V, Fernandes J (eds) Reverse engineering: an industrial perspective. Springer Science and Business Media, Heidelberg, p 2

SPC. Sustainable Packaging Coalition® (GreenBlue, Ed.) (2011) Definition of sustainable packaging. https://kidv.nl/media/rapportages/definition_for_sustainable_packaging.pdf?1.2.1. Last accessed 2025/04/03

Stake R (2012) A arte da investigação com estudos de caso, 3ª edn. Fundação Calouste Gulbenkian, Lisboa

Thompson R (2012) Graphics and packaging production. Thames and Hudson Ltd., London, p 9

Vieira P (2018) Estratégia de Design para a Comunicação da Sustentabilidade. Tese de Doutoramento em Design, p 363. https://repositorio-aberto.up.pt/handle/10216/117025. Last accessed 2025/04/03

Yin R (2002) Case study research: design and methods, 3rd edn. Sage Publications, Thousand Oaks, CA

Zhong B, Siya K, Zehui Z (2020) Investigating the effect of reverse engineering pedagogy in K-12 robotics education. Comput Appl Eng Educ 29:1097–1111

Color in Communication Design: How Colorblindness Affects the Perception

M. Madalena G. Ribeiro

Abstract Color is a crucial component of visual design, in different ways, namely, playing a key role in conveying emotions, enhancing usability, and improving user experience. However, individuals with color vision deficiency (CVD) encounter significant challenges in distinguishing and interpreting color-coded information. This limitation can hinder accessibility and reduce the effectiveness of visual communication, making it essential for designers to develop more inclusive solutions. To achieve this, designers must understand how color perception progressively changes as the severity of CVD increases. However, many professionals lack sufficient knowledge about this phenomenon, and its practical implications remain poorly understood. Without this awareness, design choices may unintentionally exclude CVD individuals, leading to usability barriers. This study addresses the issue by using a perceptual model of color representation to simulate how the color domain narrows as CVD severity increases. By demonstrating how individuals with different degrees of color blindness perceive color, the study offers valuable insights about their challenges and provides designers with practical tools to assess and improve accessibility, ensuring that visual messages remain effective for a broader audience. By integrating these findings into the design process, professionals can create more inclusive and user-friendly interfaces, reducing barriers for CVD individuals and promoting more accessible communicational products.

Keywords Accessibility in communication design · Color perception · Usability and user experience · Inclusivity

M. M. G. Ribeiro (✉)
TECHN&ART, Polytechnic University of Tomar, Tomar, Portugal
e-mail: mribeiro@ipcb.pt

Polytechnic University of Castelo Branco, Castelo Branco, Portugal

D. Raposo et al. (eds.), *Perspectives on Design IV*, Springer Series in Design and Innovation 59, https://doi.org/10.1007/978-3-032-00733-9_7

1 Introduction

An important piece in design, color is crucial for visual communication, directly affecting how we interpret messages and shapes our emotional comprehension. On the one hand, a collection of information derived from the contrasts between the shapes in a particular picture can be gleaned from the color composition. On the other hand, colors also have cultural and psychological connotations that influence how the message is understood.

When it comes to communication design, the bold use of color not only adds visual balance and clarity but also reinforces the message, making sure that it is rigorously and effectively felt and understood. Color selection in graphic design can direct the viewer's eye within a composition, establish contrasts to emphasize key information, and strengthen a brand's visual identity. Color has a key role in defining feelings, expressing emotions, and even structuring visual narratives in multimedia design and in audiovisual, all of which contribute to the message's overall meaning.

However, there are factors that might influence how individuals see color, and some people are not able to distinguish colors accurately. In addition to visual illnesses like cataracts, glaucoma, or macular degeneration, age, or interaction with chemicals like pharmaceuticals or other substances, the primary cause of colorblindness is an hereditary trait.

Formally designated as color vision deficiency (CVD, in short), affected by this issue, the colors become less saturated, with less light and/or with different hues, at the eyes of the observer. The greater the degree of severity, the greater the change in the way in which colors are perceived, since the color domain become visualized by a CVD individual decreases gradually, becoming increasingly shrunken.

To compensate for their perceptual limitations and improve their perception of what they see, colorblind people employ a variety of personal strategies (e.g., observing the shadows or changing the observing point). Nonetheless, individuals are repeatedly exposed to digital and printed communication products in their daily lives, whether through computer programs, social media, internet or even in the physical world. In these circumstances, the observer is unable to use those subterfuges.

In order to ensure inclusion for this target audience as well, it is crucial to develop techniques that fall within the purview of Communication Design while also taking into consideration the constraints placed by CVD. The first step in doing this is to gain a better understanding of how the color domain becomes shrunken according to severity degree and yet how the different colors get distorted, the goal here proposed.

2 Color Perception

Rods and cones are cells found in the retina of the human eye. While rods only function in dark situations, mastering human vision in low-light conditions, the color perception is carried out by the cones, which are divided into L-, M- and S-cones, each kind having a different sensitivity. The letters L, M, and S refer to the wavelengths of light that each cone type is more sensitive to. According to the tristimulus theory (Marieb 2001), color perception results from the combination of the signals collected by these three types of cones, and the color can even be represented mathematically in the LMS color space, in which the parameters l, m and s depends on the sensitivity of the L, M and S cone-cells. As each combination of parameters correlates to a different color, this representation is unambiguous.

People that can detect color in three dimensions on a regular basis are known as trichromats, and their condition is known as trichromacy.

2.1 *Color Vision Deficiency*

But not everyone sees color in a regular way. Color vision deficiency or color vision deficit (CVD), commonly known as colorblindness or daltonism, is a visual impairment that causes distortion in the way that the color is perceived.

While other factors such as brain fever (Mollon et al. 1980), cortical trauma (Damasio et al. 1980), disorders, degenerations or dystrophy, diseases of various origins, diabetes mellitus, or yet exposure to toxic agents (Birch et al. 1979) can cause alterations in the regular perception of color, CVD is mainly caused by a genetic inherited condition (Sharpe et al. 1999).

The genetic CVD is designated differently, depending on the severity and on the cone cell type affected (Birch 2001; Case 2003). Anomalous trichromacy is a deficit that occurs when one of the cone-cells type has a partial abnormality; Dichromacy occurs when a type of cone-cells is missing or completely non-functioning; and Monochromacy, when two or even three types of cone cells are missing or completely non-functioning. These people see the world in grayscale and, sometimes, with slight shades of blue (Wu and Fawzi 2012).

Despite the general public's lack of knowledge about CVD, it is fairly widespread, with a prevalence rate of 5% in the general population and 8% in men, having the deuteranomaly subtype a special emphasis of 5% (see Table 1). In anomalous trichromacy, the degree of severity varies from 1 to 99% (continuously), being the loss of acuity performance correlated with the severity. When the degree of severity reaches 100%, the condition is designated it means that the person is a dichromat, the term used for people with only two functional types of cone cells.

2.1.1 Anomalous Trichromacy

Due to a partial loss of sensitivity of cone-cells type (L-, M or S-cone cells), this anomaly is more common in the M-cones (6% of males and 0.41% of females) followed by the one in the L-cones (1% of men and 0.01% of women), tough much less expressive (see Table 1). It leads to a general decreasing in the color's vividness

Table 1 CVD prevalence (Kalloniatis and Luu 2011), in special for anomalous trichromacy

Type of CVD	Male	Female
Overall color vision deficiency (%)	~ 8	~ 0.5
Anomalous trichromacy (%)	6	0.41
Protanomaly (anomaly of the red cone cells) (%)	1	0.01
Deuteranomaly (anomaly of the green cone cells) (%)	5	0.4
Tritanomaly (anomaly of the blue cone cells)	Rare	Rare

and, depending on the type of anomalous trichromacy, some colors are more biased than others.

2.2 *Color Vision Deficiency Simulation*

Color vision deficiency cannot be understood easily by most people with normal vision. For such, there were proposed several proposed methods CVD simulation (Vienot et al. 1999; Kondo 1990; Capilla et al. 2004; Toadere 2009; Lee and Santos 2010; Machado and Oliveira 2011; Flatla and Gutwin 2012; Ohtsuka et al. 2012; Luque et al. 2014), but mostly are not appropriate for our study, because most of them are restricted to dichromacy simulation and some are used only for each specific colorblindness type or subtype.

The method proposed in Machado and Oliveira (2011) for simulating anomalous trichromacy stands out from the others, being widely referred in the CVD scope literature because it allows the simulation of several types of anomalous trichromacy, as well as various levels of severity (10–100%, 10 out of 10). Therefore, this method was our choice be applied in our study and both to generate the images for Fig. 2 (in this section), which depicts how the colors are seen by individuals affected by each of the three types of anomalous trichromacy (with several degrees of severity). As the degree of severity augments, chromatic diversity gets loss progressively. The degradation of hues is analogous for protanomaly and deuteranomaly, reason why the (partial or total) impairment in the M- or L-cone cells, is known as "red-green CVD". In fact, in both anomaly types, the reddish and greenish colors become increasingly similar, leading to confusion (Fig. 1).

There is noticeable the distortion in the hue of the reddish tones, less pronounced in the greens which, in turn, become gradually more grayish, losing vibrancy. As for the reddish, reds and oranges lose saturation and become darker, i.e., fade and become brownish, whereas pinks turn bluish shades. The blues and yellows don't change.

The corruption of colors is less sighted in tritanomaly. The hue of the reddish colors (oranges and pinks) becomes redder, and greens and blues both approach cyan.

Fig. 1 Image with flowers seen by people with regular color vision

The analyses of the images give us a general idea of how the anomalous trichromat people see the color, and how and about how the color degrades as the severity of the disease imposes.

However, to a better understanding about how each color degrades, it is necessary to represent each color in a proper an accurate way. Considering the wide set of alternatives to express the color, the choice must fall onto an intuitive color space, which will be discussed in the next section.

3 Color Representation

Color spaces allow colors to be represented mathematically (usually) using 3 parameters that univocally define each color. By varying these parameters, all represented colors can be achieved, forming a 3D geometric model.

The RGB color space is utilized in display devices, being the colors expressed through the triplet red (R), green (G), and blue (B), which values vary between 0 and 255. Geometrically, this space forms a cube (see Fig. 2b) (Foley et al. 1991; Lopes 2008; Adobe 2000).

However, the RGB color space is not intuitive in the way of specifying and adjusting colors, the reason why alternative color spaces arose since the last seventies, being, at the moment, widely used in the color pickers of graphical applications. Some of these color spaces, designated as perceptual or "phenomenal" (Tkalcic and Tasic 2003), meet Munsell's line of sight (Munsell 1907).

Fig. 2 Image with flowers seen by people with different CVD conditions. Simulation by the method presented in Machado and Oliveira (2011), implemented in JavaScript programming language for HTML5 web compliant browsers. Image in **a** courtesy of Free Wallpapers (http://www.freewalldownload.com/)

3.1 Perceptual Color Spaces

The first mathematical color specification was proposed in the early twentieth century by Munsell with the Munsell Color System (Munsell 1907). This system represents color using three numerical parameters: hue (H), chroma (C) and value (V), which refers to the tint, brightness and vividness of color, respectively. *Hues* are arranged circularly like in the color wheel, *chroma* represents saturation, increasing outward

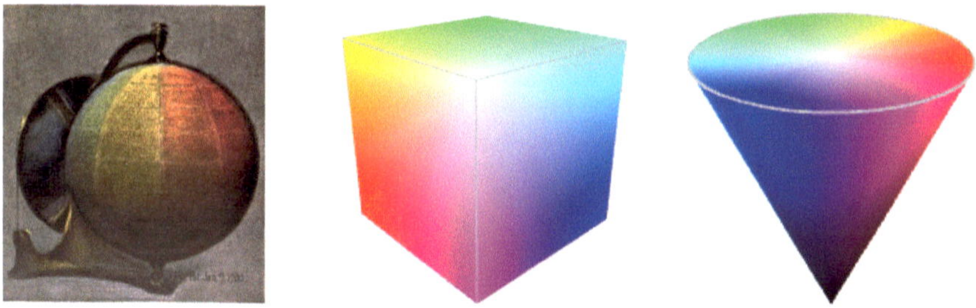

Fig. 3 3D geometric models of the: **a** Munsell color space; **b** RGB color space; **c** HSV color space, being the images in **b** and **c** generated on computer

from the central axis (see Fig. 2) and *value* increases from bottom to top. Geometrically, the Munsell Color System is a globe (Adobe 2000; Sangwine and Horne 1998; Fig. 3).

Inspired on the Munsell Color System, color spaces like HSV (Smith 1978) and HSI emerged as linear transformations of RGB (Tkalcic and Tasic 2003). Their parameter-based approach makes color specification and adjustment intuitive, earning them the label "perceptual." They are also useful for analyzing color disturbances, justifying their adoption.

In order to take advantage of the perceptional nature, we chose the HSV of color space for color representation, in our study.

3.1.1 HSV Color Space

The HSV color space (Smith 1978) is geometrically represented by an inverted cone, as represented in Fig. 2c, being the color specified through the parameters hue (H), saturation (S) and value (V), hence the acronym.

H ϵ [0, 360] denotes the color hue, being distributed around the hue wheel (see Fig. 4). Note that RGB primary (red, green and blue) and complimentary colors (yellow, cyan and magenta) are evenly distributed. S stands for saturation, which is related to the vividness of the color. It increases from the central axis of the cone outwards, meeting on the hull the most saturated colors. The color brightness is represented by V (value), which rises from the apex to the base of the cone, where the brightest colors are. A central shaft connects the gray shades, with the black color at the cone's top and the white color at its base (Foley et al. 1991; Ford and Roberts 1998).

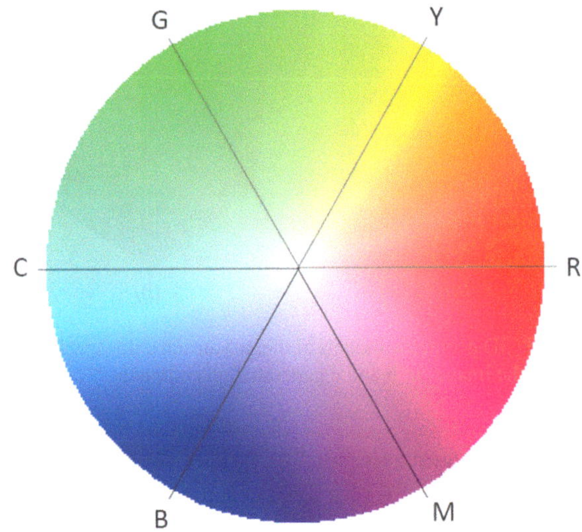

Fig. 4 Representation of the top of the HSV

4 Proposal Approach

To the best of our knowledge, there is no work showing, in a perceptual way, how the colors are gradually adulterated due to the CVD condition, though the shrinking of the color model. Therefore, having in mind providing a better understanding how the color changes as the degrees of severity augment, we drive our efforts in carrying out a study that allows to realize:

- How the color domain reduces, for diverse degrees of severity, evidencing the shrinking of the perceived color space;
- How each color changes, comparing its original position in the color model and the position assumed when affected by the CVD condition. In other words, how each color is altered by a certain degree of severity of the CVD condition.

Since deuteranomaly is the most common subtypes of CVD (see Table 1), with prevalences without parallel compared with the other subtypes of CVD, it will be our focus.

As before expressed, the HSV color model is the alternative chosen for the representation, since it complies with the required intuitive character.

4.1 Methodology

In order to carry out the goals specified above, there were followed a set of purposes.

To achieve the (3D) HSV color model with the colors perceived by individuals with deuteranomaly, we started by considering each color in the RGB color space (by varying each channel R, G and B in their domain), and for each, we:

- simulate how this (R, G, B) color is perceived by CVD individuals;
- convert this adultered color to the HSV color space;
- represent this color graphically (in the HSV color space).

Additionally, it is both a purpose to perceive how each color is adultered. For it, we consider, once again, individually each color in the RGB color space (by varying each channel R, G and B in their domain) and, for each one:

- firstly, convert this color to the HSV color space. Let's call it as "color-point A";
- on the other hand:
  - simulate how the (R, G, B) color is perceived by deuteranomalous individuals;
  - convert the adultered color to the HSV color space, achieving the color-point B;
- represent graphically the line between the color-point A and the color-point B. This line links the original color, seen by individuals without colorblindness and the color seen by a CVD individual.

4.2 Setup

The processes referred in Sect. 4.1 was coded in Processing language, in the IDE environment, due to the amenities provided by the P3D renderer, namely the interaction with the 3D objects, namely, to perform rotation and displacement.

The CVD deuteranomalous simulation is performed by using the method proposed in Machado and Oliveira (2011), being the RGB to HSV conversion carried out through the method presented in Smith (1978).

As result, we achieved the images of Figs. 5, 6, 7 and 8, which allowed us to depict our goals.

5 Results

Figure 5 illustrates how the HSV color space shrinks, as the degree of deuteranomaly increases. As the degree of severity of deuteranomaly augments, the HSV color model becomes more and more shrunken, being the colors close to the 60°–240° plane (i.e., yellow (60°) and blue (240°) half-planes) the most preserved. Thus, we considered it

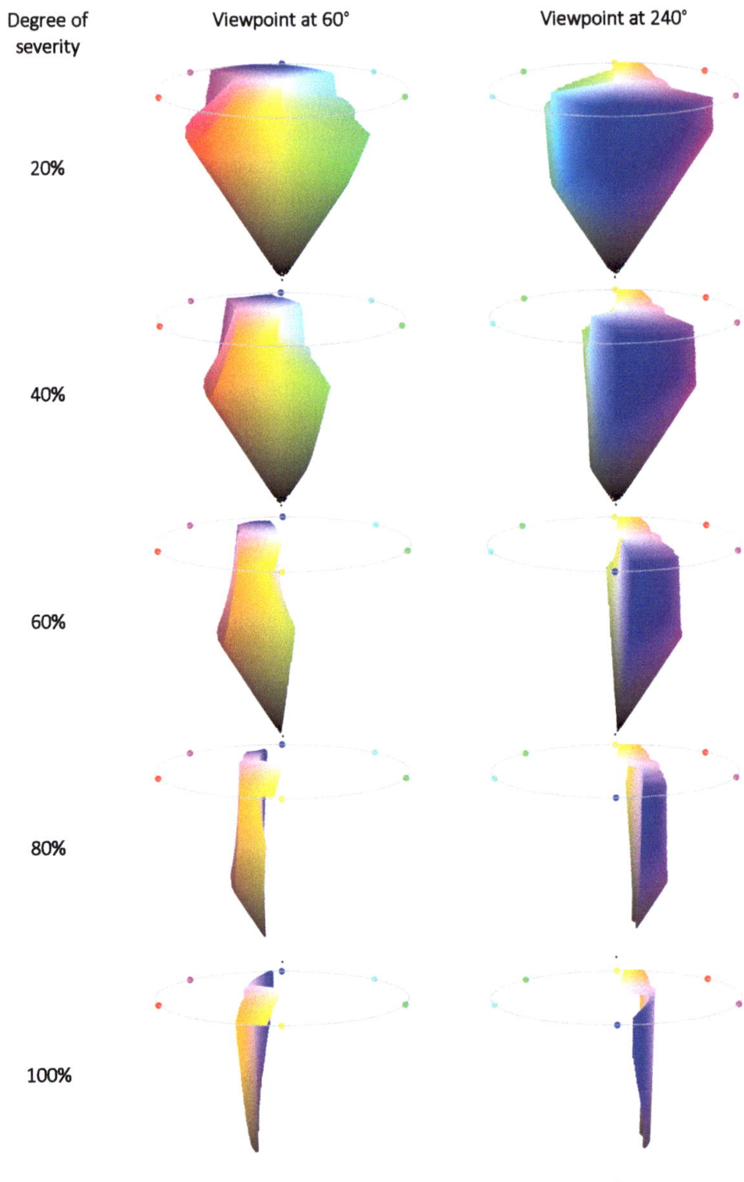

Fig. 5 Simulation of the HSV color model when seen by individuals affected by deuteranomaly, with different degrees of severity (20–100%)

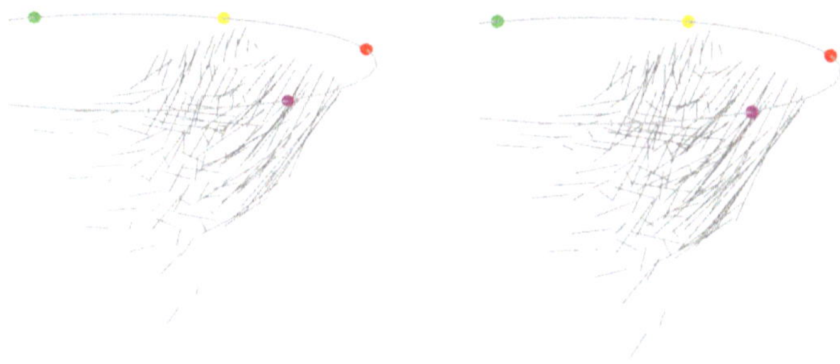

(a) Degree of severity: 40% (b) Degree of severity: 80%

Fig. 6 Lines connecting colors and the same colors when affected by deuteranomaly (in the HSV color model), considering the degrees of severity of **a** 40% and **b** 80%. The colors hue varies between 240° and 360° and yet between 0° and 60°, i.e., the colors between blue and yellow, passing through magentas and reds

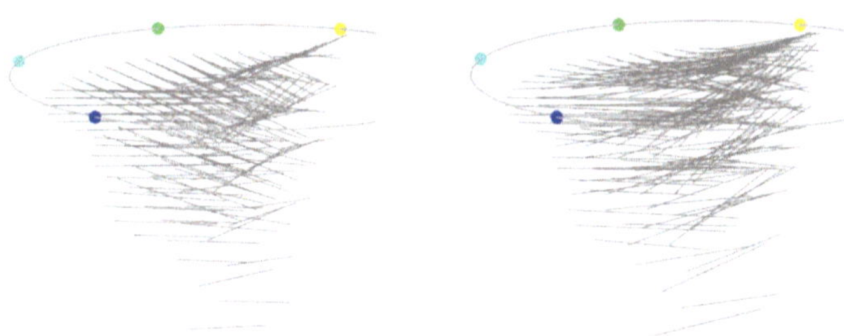

Degree of severity: 40% Degree of severity: 80%

Fig. 7 Lines connecting colors and the same colors when affected by deuteranomaly (in the HSV color model), considering the degrees of severity of **a** 40% and **b** 80%. The colors hue varies between 60° and 240°, i.e., the colors between yellow and blue, passing through greens and cyans

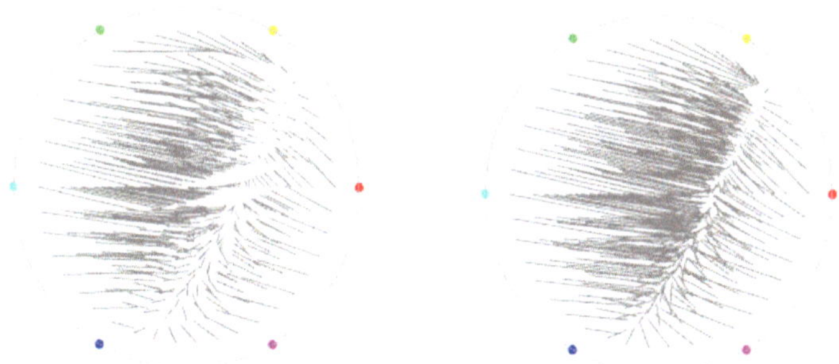

(a) Degree of severity: 40% (b) Degree of severity: 80%

Fig. 8 Top view of the lines connecting the original colors (in the HSV color model) and the same colors when seen by deuteranomalous individuals, with degrees of severity of 40 and 80%

convenient to observe the color model, looking from both sides of these half-planes to a clearer visualization.

Through the images of the first row of Fig. 5, we can see that even in the deuteranomalous case of a severity degree of (only) 20%, the most saturated colors with higher levels of light disappear from the color space (i.e., the colors near the base and the cone hull), except the colors closer to the yellow (60°) or blue (240°). When the CVD severity augments to 40%, the disappearance of saturated colors is more accentuated (see second row of the Fig. 5).

When the severity reaches the 60%, the color space gets reduced significantly. It is evident that all the expressive greens have disappeared completely, as well as vibrant reddishes and even moderate ones. It left only some discolored greens and a few pale purples. The only saturated colors that are still seen are those whose hue is particularly closer to the yellow of 60° or to the blue of 240°.

For 80% of severity, there only outlived the yellows (near 60° of hue) and the blues (near 240° of hue), since the visual system approaches a triangular planar area crossing the yellows (near 60°) and the blues (near 240°).

By observing at Figs. 6 and 7, which illustrates how the colors "move" when affected by deuteranomaly. It is evident that the distortion of colors always follows the same pattern: loss of saturation and light. However, while in the reddish colors (magentas, reds and oranges) it is more notorious the loss in light (see Fig. 6), greens and cyans loss more saturation that light (see Fig. 7), becoming quite desaturated (grayish).

Figure 8 shows that there are not significant changes in the hue of colors. In the cases where it happens, it is slightly noticeable, since the colors become also more desaturated and darker.

For instance, Fig. 8b shows that the hue of the greens close to hue of 120°, change to yellow (of 60°), but they will be seen as desaturated brownish instead of yellow, since their saturation and light fall down in simultaneous.

We would even say that the only cases in which a change in hue is really noticeable is in the case of pinks close to magenta and the cyans, which become bluish, as shown in Fig. 1.

6 Conclusion

Color plays a crucial role in communication design, influencing both the aesthetic and functional aspects of visual messaging. However, the perceptual challenges faced by individuals with color vision deficiency (CVD), particularly deuteranomaly, interferes with the normal way of seeing the color and consequently in the way they perceive digital and printed communication materials with significant implications. The high prevalence of this CVD impairment and the frequency of contact in everyday life with communication content underscore the need for more inclusive design approaches.

Before defining inclusive strategies, it is crucial to have more effective knowledge about how color is adulterated. Given the lack of a proposal that shows, in an intuitive way, how color deteriorates, we set out to do it in this study. Our findings illustrate the shrinking of the color domain, as severity increases, affecting both the saturation and lightness of most colors, being the degradation of reds primarily due to a loss of light, while the greens suffer more from a decrease in saturation.

Our analysis demonstrates that even at a mild severity level of 20%, a significant loss of saturated and bright colors occurs, particularly in the hues close to red and green. As severity progresses to 40% and beyond, the mutation becomes more pronounced, leading to an increasing dominance in desaturated color. When the anomaly reaches 80%, the visual experience of a deuteranomalous individual is largely confined to hues near the yellow (60°) and blue (240°) half planes, with the other vibrant colors fading into a narrow, less distinguishable spectrum.

These findings have significant implications for communication design, ensuring color accessibility is paramount, both for digital and printed materials.

Notwithstanding the adapting proposals existing in the literature, they can be reserved mainly for cases where other strategies are not viable, as in the case of photos. The changes of colors, the use of patterns over image areas and the contours highlight are some of the ways of adaptation alternatives in literature.

As far as design is concerned, designers must prioritize other approaches like contrast based on color brightness differences, adopting color schemes that emphasize hues less affected by CVD, such as yellows and blues. The insights gained from the knowledge presented in this document, understanding how colors are degraded by deuteranomalous condition, can serve as a basis for this endeavor, helping in new color scheme proposals. CVD simulation algorithms are useful tools for assessment of the inclusive color schemas proposals. By integrating these considerations into design practices, we can move towards a more accessible visual communication landscape.

References

Adobe (2000) Color models—technical guides. Technical report. Adobe Systems Incorporated

Birch J (2001) Diagnosis of defective colour vision, 2nd edn. Elsevier Science, Edimburgh

Birch JB, Chisholm IA, Kinnear P, Marré M, Pinckers AJ, Pokorny J, Verriest G (1979) Acquired color vision defects. In: Congetinal and acquired color vision defects. Grune and Stratton

Capilla P, Diez-Ajenjo MA, Luque MJ, Malo J (2004) Corresponding-pair procedure: a new approach to simulation of dichromatic color perception. J Opt Soc Am A-Opt Image Sci Vis 21:176–186

Case B (2003) Color blindness. Technical report

Damasio A, Yamada T, Damasio H, Corbett J, McKee J (1980) Central achromatopsia: behavioral, anatomic, and physiologic aspects. In: Neurology. American Academy of Neurology, pp 1064–1071

Flatla DR, Gutwin C (2012) "So that's what you see": building understanding with personalized simulations of colour vision deficiency. In: Proceedings of the international ACM SIGACCESS

conference on computers and accessibility (ASSETS'12), Boulder, Colorado, USA, Oct 22–24. ACM Press, pp 167–174
Foley J, van Dam A, Feiner S, Hughes J (1991) Computer graphics, principles and practice, 2nd edn.
Ford A, Roberts A (1998) Colour space conversions. Westminster University, London
Kalloniatis M, Luu C (2011) Color perception. In: Kolb H, Fernandez E, Nelson R (eds) WebVision: the organization of the retina and visual system. John Moran Eye Center, University of Utah
Kondo S (1990) A computer simulation of anomalous color vision. In: Proceedings of the symposium of the international research group on color vision deficiencies, Tokyo, Japan, Mar 26–28. Kugler & Ghedini, pp 145–159
Lee J, Santos W (2010) Fuzzy-based simulation of real color blindness. In: Proceedings of the annual international conference on engineering in medicine and biology society (EMBC'2010), Buenos Aires, Argentina, Aug 31–Sept 4, 2010. IEEE Computer Society, pp 6607–6610
Lopes JM (2008) Cor e Luz. IST-UTL, Lisboa
Luque M, Fez D, Acevedo P (2014) Software for simulating dichromatic perception of video streams. Color Res Appl 486–491
Machado GM, Oliveira MM (2011) A model for simulation of color vision deficiency and a color contrast enhancement technique for dichromats. SBC 2011
Marieb E (2001) Anatomy & physiology, 1st edn. Addison Wesley Longman
Mollon JD, Newcombe F, Polden PG, Ratcliff G (1980) On the presence of three cone mechanisms in a case of total achromatopsia. In: Verriest G (ed) Color vision deficiencies V, Chap 3. Adam Hilger, Bristol, pp 130–135
Munsell AH (1907) A color notation—a measured color system, based on the three qualities hue, value and chroma. Geo. H. Ellis Co.
Ohtsuka S, Hira S, Kihara K, Ikegami J (2012) 54.2: novel real-time and bi-directional color simulator for dichromacy and trichromacy on smartphones. SID symposium digest of technical papers, vol 43, pp 729–732
Sangwine S, Horne R (1998) The color image processing handbook. Chapman & Hall, London; New York
Sharpe L, Stockman A, Jagle H, Nathans J (1999) Opsin genes, cone photopigments, color vision and color blindness. In: Color vision: from genes to perception, p 351
Smith AR (1978) Color gamut transform Pairs. Comput Graph 12(3):12–19. https://doi.org/10.1145/965139.807361
Tkalcic M, Tasic JF (2003) Colour spaces: perceptual, historical and applicational background. In: Proceedings of the computer as a tool—region 8 (EUROCON'2003), Ljubljana, Slovenia, Sept 22–24, vol 1. IEEE Computer Society, pp 304–308
Toadere F (2009) Colors vision analysis for a normal viewer and for a daltonist. In: Vlad S, Ciupa R, Nicu A (eds) Proceedings of the international conference on advancements of medicine and health care through technology (IFMBE'2009), Munich, Germany, Mar 26–28, vol 26. Springer, Berlin, pp 243–246
Vienot F, Brettel H, Mollon JD (1999) Digital video colourmaps for checking the legibility of displays by dichromats. Color Res Appl 24:243–252
Wu DM, Fawzi AA (2012) abnormalities of cone and rod function. In: Retina. Elsevier Inc. https://doi.org/10.1016/B978-1-4557-0737-9.00044-8

The Authorial Voice, Witnessing and the Documentary

Kenneth Feinstein

Abstract Central to our understanding of both history in the form of written text to the documentary is the relationship of the authorial voice, the narrative, and our access to the events discussed. In the historical text this has been centred on a linear narrative and the voice of the author. In documentaries our access to events becomes more central. With this our relationship between witnessing the events and the narrative becomes the realm of the authorial voice. This brings forward the questions not just of truth but also of how narrative controls our access to that truth. Central to this question of voice is the relationship between the filmmaker and the witness. In this paper we will look at how new forms of documentary has developed to centre the voice of the witness more into the narrative and the narrative structure. We will see how new forms of presentation are being developed to address these changes. We will also briefly look at the difference between documentary and non-fiction film trying to untangle a confusion between the two that leads to false notions of what a documentary is.

Keywords Testimony · Documentary · New media · Cinéma Vérité · Installation art

1 Introduction

What I would like to look at is a relationship between authorial voice and the documentary tradition. Most discussions on documentary tend to either follow the structural model first laid out be Bill Nichols in *Representing reality* (Nichols 1991) or follow Grierson's dictums on the idea of truth. Grierson, like Dziga Vertov before him where both theorists and practitioners, whilst Nichols like André Bazin are writers looking at the form from the outside. As we look at how documentary as a form has developed, we have to look to both how the structures valorize certain forms of storytelling, what it distinct as a cinematic form unique from the written text, fiction

K. Feinstein (✉)
School of Design, University of Leeds, Leeds, UK
e-mail: k.a.feinstein@leeds.ac.uk

D. Raposo et al. (eds.), *Perspectives on Design IV*, Springer Series in Design and Innovation 59, https://doi.org/10.1007/978-3-032-00733-9_8

and other forms of non-fiction film. Where do we find the core elements that define it as separate from these other forms that it gets grouped with. We need to look at its form in terms of what does it present and how does it do that. If it presents truth, is it a different truth from what other forms present to us or is it its relationship to the denotational idea of truth that is front and center? We tend to center the discussions of documentary around issues if a verifiable idea of truth. In discussions surrounding this type of film we focus on verifiability and authorial voice. What is true and how who tell us that truth. Novels and fiction film can have unreliable narrators who we experience the narrative through, but the documentary is not allowed this luxury. The burden of the documentary as film has been that it represents reality and presents it to us in realistic time, a linear narrative. It is seen as part history, part film part journalism and more recently part entertainment. Within these frames we create rules to judge what makes a documentary. But by have these outside forms define our discussion we have to see what gets missed. Are we missing important movements within the form and how does that change how it develops?

2 How We Read Documentaries Through the History Text

In his definition of the technical image Vilém Flusser states that history is an outgrowth of linear writing (Flusser and Ströhl 2002). What he calls historical consciousness is narrative and linear in form. History takes past events and recasts them so that they can illustrate for us a how and why that explains our situation today. Linearity is central to how history is presented to us. It is past and recounted in a fashion that unfolds before us, but one where we never confuse it for a present. Such a structure can be found in historical fiction, but not history. It is the story of how we got to where we are. Be it the works of William Gibbon or Richard Hofstadter, the point of the historical is the use of the past to reflect on the present. And as we have sub-categorized histories into types of analysis, the historical becomes even more so a methodology to recount events based on a theoretical model.

Because the historical is a narrative, it maintains the authorial voice. The author is either act as if they are witness, but not a participant, who recounts events or they are the one accumulating the details, fact, in order to make sense of events through the filter of their thesis. An example of the former being William L. Shirer's *The Rise and Fall of the Third Reich* and the latter in Hofstadter's *The American Political Tradition and the Men Who Made It*. In both cases the reader is at a remove from the documentation. They are stories told through the author's eyes. They have seen the evidence; they have digested it and turned it onto the tale they tell us. We are at a remove from the evidence. It is only presented in fragments in order to reinforce the authorial voice. The author chooses what we see and most of what we "see" is descriptive. It is connotational even as it attempts to appear objective. The historical text presents itself as an objective unfolding of the events of the past. It follows a logic of cause and effect. It re-presents us with events that are significant. Yet it cannot give us direct access to those events.

What is significant is what history gives us that goes into the documentary tradition and how the two diverge. From the historical text the documentary inherits a linear narrative as a series of causes and effects, the voice of authority (either explicit or implied), and the idea of the presentation of truth as being relevant to the present. These aspects relate to how past events are shaped into an understandable text. The documentary diverges from the historical text because it allows for direct access to events and witnesses. This is the document in the documentary. Linear time can be broken up. Concepts like parallelism are possible in the narrative. We can have people speak in the present whilst images from the past are on the screen. We can repeat scenes, play them back at other speeds, mix past and present and comment on what we see. All the flexibilities of cinema are presented to us. It allows for first and third person narratives to be present shifting our view of events. Third person perspective is central to American Direct Cinema as well as French Cinéma Vérité. Whilst the first person is part of the news interview or film like *The Fog of War* (Morris et al. 2004). By presenting the document, filmed events and witness interviews, to the viewer we start to move away from the neatly packaged narratives and moral messages of the author into a post-historical environment where viewers must determine the meaning of what they see and hear on their own. The authorial voice may still demand its presence, but by having us see and hear the documents of history the grip of the author starts to loosen.

Seeing images brings something extra to the written narrative (Barthes 1981). We are presented with information that we can interpret. Not only can we see events as they unfold, but we also get to hear the voice of witnesses. We get to see how they speak, what feels sincere and what feels ironic or humorous. We can experience the interview as an event instead of just a recounting of events. With a written text there is no tension between what is said how we encounter it. The author is placed between us and what we read. With the documentary, we become present in that interview, we get to see people interact in the world, our presence is central to the entire form. From the Lumiére Brothers' film of their workers leaving the factory, the idea of the camera as denotational device has been central to non-fiction film.

3 Film Theory from Making to Reading: How It Changes How We Understand What We See

According to Vilém Flusser, with the invention of photography we entered into a post-historical age (Barthes 1981). Our relationships to linear narrative and time changed greatly. With the photograph, we became able to experience moments of past time in a direct way. The photograph is not an artifact from the past, it presents am moment form the past to us as a present. I can see the face of my ancestors. It is proof that they existed. It has changed how we experience and understand time. The idea of a past-present comes out of our photographic experience of time. The relationship between an image inscribed a moment in time and our viewing opens up a gap where the past is present before us as if it is present. We get to reexperience what has happened as if it is unfolding before us. We are given access to what in the historical text would only be the eyewitness. It brings us into the process of experiencing events.

4 The Quest for Truth

The documentary tradition places the idea of a truth at the center of the film. It assumes that what is asserted by a director is truth. From there the narrative is formed. The director as researcher and filmmaker posits a theory that is proven by how the images and sounds placed before us reflect their narrative of truth. When watching documentaries such as the *Why We Fight* series we are aware that the work is propaganda, but the narrative voice is still one asserting the truth. Even with American Direct Cinema where the fly on the wall techniques allows us to see events as they appear to unfold, we still have an authorial voice presenting an unquestioning truth. If the camera sees it, then it must be true. The assumption being that what we see cannot be questioned as real, but the documentarian's role it to present the facts to us in a way as to reveal the truth behind what we see. We are not expected to question what we see although we may question what people say. The image presents an unflinching truth even if the voice can lie.

The closest we get to questioning what we see is when the filmmaker presents us with the unreliable witness, someone who tells us a varnished version of events that will be undercut by the filmmaker. Interestingly we tend to find this with the formal interview, where the subject is placed in a position to be questioned regarding their actions. This is what we see in television news shows and also in the work of Errol Morris. In rare cases we may find the scenes where what the camera sees is put up against people's memories such as the editing room scene in *Gimme Shelter* (Flusser 2000). In this scene the filmmakers are in the editing room with Mick Jagger of the Rolling Stones watching the footage of when Meredith Hunter is stabbed by a member of the Hell's Angels. Here Jagger is witnessing something that the camera has caught that he didn't see on stage. Here we are given the camera, representing the view of the director, as the unquestioned view of reality. The mechanical eye because it is presented as denotational cannot be questioned, where the memories of participants can be. Whilst it would seem that we are placing our trust in the mechanical eye over the witness, in reality we are placing the witnessing of the filmmaker over that of anyone else as the directors, David and Albert Maysles are also credited as cinematographers.[1] They are presenting their evidence to us and Jagger. On top of this they are presenting this in an editing room, running the footage forward and backward, slowing it down and stopping the film so as to make their point. This is presented as a dramatic revealing of the truth, not a questioning of it.

[1] There were other camera operators as part of the crew and the Maysles did not shoot that footage themselves, but as cinematographers they take on the responsibility for the existence of the footage.

Truth is extracted from and moulded by the cinematic experience. If what we see on camera is a murder, then the act of filming it becomes evidential. It claims its reality through the fact that we can see it. The context of events stops being the lived experience and get transformed into film, effect rules cause. What is being called truth is the revealing of a process. It is what we find in the end.

When we speak of truth in the documentary we are actually talking about two distinct meanings to the word. We have the truth as the presentation of facts as we find in direct cinema or investigative works like *The Thin Blue Line*, or we have a more abstract or metaphorical idea of truth as the forces that shape events as we might find in *Chronique d'un été* of which we be speaking or of later. The aims and techniques of these different truths are very different. While both do use facts, observation and testimony their objectives are very different. In the first category we can find that the quest for truth is to be actualizable as in The Thin Blue Line which was able to free an innocent man from death row. This comes from the journalistic tradition. The latter idea of truth tends to present itself more either through an essayistic form or by presenting voices in a more open-ended fashion. It looks for causes behind the issues. Jean-Luc Godard's *Ici et Ailuers* is an interesting case in where the work began as an attempt to present the actionable, journalistic truth and transformed itself into the latter metaphorical search for truth (Godard et al. 1976). Its questioning is what is truth is centred on this shift from one sense of concept of truth to the other. Part of this work is to question the authorial voice which is necessary in the journalistic concept of discovering truth. What Godard questions in this work is how the ideas of truth and seeing are defined through the authorial voice of a director.

5 The Voice of Testimony Against the Narrative Voice

Whilst the tendency of documentaries is to use the established models of the fiction film or the essay, this continues to valorise the voice of the author over the voices of witnesses. As we have stated, this tends to place them in a secondary role of support to the author's thesis. This trend has led to a variety of developments within the documentary, most notably the placement of the documentarian as the central character of the work. We have seen this most notably in the work of Micheal Moore and Morgan Spurlock. In these works, the documentarian takes on the role of witness taking on the authority of the voice of the witness while retaining the editorial control of the filmmaker. In the English-speaking world this tendency of the documentary to have a traditional narrative structure and a clear moral ending is tied to the documentarian's demand for the authorial voice. This may be a reflection of the development of the professional documentarian within the broadcast and film industries. Where we find works that move away from this is with documentarians who come to the field from other backgrounds. The most notable of these are Jean Rouch, an anthropologist, his collaborator Egar Morin, a philosopher and sociologist, and Claude Lanzmann, chief editor of *Les Temps Modernes*. If we look at their films *Chronique d'un été* (Rouch/Morin) and *Shoah* (Lanzmann) what we see is that by being open about the process of filmmaking and their role they have opened up the relationship between the witness and the viewer. In both cases the films allow us to see the performativity of an interview or monologue while refusing to impose the authorial voice. They have kept the works opened without a definitive ending giving us a moral message.

5.1 Rouch, Morin's *Chronique d'un été*

Rouch and Morin were founders of what is known as *cinéma vértité* and in its name we see the quest for the presentation of truth (Rouch et al. 2005). Whilst practitioners of *direct cinema* such as Rickie Leacock have tended to refer it as *American CV* they are different in form. Where Leacock and Pennebaker wanted to seem invisible to events, the French directors made the camera and their presence be a central part of the work. They saw their presence as being a provocation towards bringing forward a conversation with the subject prompting them to breakdown their inhibitions and present their testimony. They turned the work form being observational to being collaborative. In this sense there is no authorial voice to *Chronique d'un été*. If we see the authorial voice of the documentary as being declarative, their work is interrogative.

Taking inspiration from Vertov we can see structural parallels to *Man With a Movie Camera* (Vertov et al. 1929). Both start by announcing its intentions and ends with the intended audience viewing the work. In between *Chronique* moves through different classes and workspaces in Paris as well as having round table talks and monologues in between (Godard et al. 1976). It is unique in that it turns the idea of presenting the truth into a question, much as later Godard will do with *Ici and Ailluers* in the 70s. The filmmakers are positioning themselves not as documentarians or but as sociologists/ethnographers. They are looking for the correct form of film to be used for their fields as much as trying to find a new form of documentary. Both interests go hand in glove. Starting with Grierson the documentary as a form has had to answer the question of in who's interest does it exist. Is it a subset of journalism, is it a vehicle for social advocacy, is its function to preserve history or is it an ethnographic recorder? What is unique in this film is that these questions are posed to the subjects of the film. They are shown the finished work and asked their thoughts. They are transformed from mere subjects of observation into active participants in the work. This is a major turn in the documentary. Whilst some film historians will place this film into the historical context of the new technological developments of 60s documentary, the handheld camera, sync sound, etc., allowing them to group this work with American Direct Cinema, this shift of the role of the subject to participant is very significant. It moves the centre of documentaries from the single authorial voice to a collective more opened perspective on events. This places the voice of the witness at the centre of the work. Their experience is best represented by the scene where Marceline walks along the Place de Concorde and talks about her memories about being taken off to Auschwitz-Birkenau and her memories of her father. It is intimate and we can share the emotion. Here the questions of reality and artifice break down. The scene is about Marceline, in essence she controls it through her voice, the camera and the directors allow us to experience the event with them, but they do not control what that event is about. In contrast if we look at the work Errol Morris whilst we find the presentation of the work is based in testimonies, the final voice in the work, the one telling what the truth is still rests with the documentarian. The testimony remains supplemental to the documentarian's quest for the truth. In *Chronique* we find that the very idea of truth is seen as a collective process that Morin described as "an investigation of man himself, without wanting, *a priori,* to favor one or the other of these problems" (Kahana 2016). In the essay 'Chronicle of a Film' (Kahana 2016) Morin refers to Rouch and himself as author-researchers. The very act of having two directors is used to subvert the idea of a singular concept of a reality to depict. Which is best seen at the end of the film when they are discussing if they think this experiment is successful or not. The idea of collaboration is central to the whole project. It centres itself from the idea that the truth is a collective journey and not an historical process. As researcher/authors Rouch and Morin move away from the documentary as journalism or history towards a more nuanced view where we are presented personal experiences and discussions, but no conclusions. That is left for the views to find. As such it turns us into active viewers, participants in the process. The work is not telling us what to understand instead it presents the collaborator/witnesses in the fullness of their voices and lets us interact with them.

Where American Direct Cinema comes from a journalistic impulse, Cinéma Vértité comes out of a more ethnographic and research-oriented perspective. Taking directly from Rouch's background as an anthropologist filmmaker the form is open to bringing the voice of the witness to the fore, making their view of the world the core of the project. This approach is central to Claude Lanzmann's best known work *Shoah.*

5.2 *Lanzmann and Shoah*

Shoah explores history without giving us the historical narrative. In making this film Lanzmann. Chose to use no historical footage or voice over narration. Instead, he tried to find as many voices from as many perspectives as possible at that time. He gives us the voices of victims, perpetrators, participants in the industry of death and those who lived through events, but never thought they were involved. In the latter category we find mainly Polish villagers who lived in Chełmo and people who moved into Jewish homes after the occupant were taken away. Where it is easy to understand the feelings and thoughts of the survivors and the perpetrators, it is the voices of these people that carry the most disturbing resonances for us as viewers. Their presence in the film presents us with an ethical question, how would we act in the same situation? What are our prejudices, and do they blind us to what is happening around us?

By turning away from conventional history Lanzmann has centred the entirely on testimony. He plays the role of interviewer or interlocutor throughout the entire work but never take on the role of main character as more performative documentarians do. Instead, he engages the witnesses in conversation. When he talks to people in Poland, he requires a translator and in historical documentaries the act of translation the conversation would be cut out, but for Lanzmann the back and forth from him to the translator and then to the witness is the act of conversation. It remains in the film, because it is the truth of the conversation. In this work Lanzmann is taking up the path indicated in *Chronique d'un été* and moving it forward. He is centring the work entirely on testimonial. He brings his camera to them in order for them to be able to tell their stories in their own environment. This removes the authority of the interviewer form the situation. It shifts the power relationship away from filmmaker. Instead, we are presented with conversations and the kind of intimate remanences. This doesn't mean that this is an easy film to watch or that we are comfortable with every speaker, rather it means that we are hearing more intimate conversations.

Whilst Lanzmann has structured *Shoah* in rough historical phases two for the two parts of the film, each section contains interviews with survivors, perpetrators, participants and witnesses. They are interspersed so that we see how these events were able to come to pass. He does little in the way of specific commentary, rather allowing the witnesses to present their testimony in their own surroundings. This allows for a more complex view of how the Shoah was able to happen. The stories told within the film present a wide variety of perspectives which he presents with a minimum of editorializing. As such he is letting each person speak for themselves in long uncut takes. One of the reasons for these long uncut takes involving translators is to let us know that these are not manipulated, they are not cut so that what we hear we know has an authenticity. Continuing on from the cinéma vértité approach, Lanzmann eschews traditional narrative structure and the imposition of a constructed dramatic arch. As a form it doesn't see its role to tell us what is right and wrong. It trusts more in the voices it presents than in the views that the film maker brings to the work. As a topic the work is about displacement and then erasure of a people. It is about the site that is not there, all that is left is the testimony of witnesses of this catastrophe. It trusts in the voices of those whose lives have been marked by these events. And this trust in these voices is found in the time that is devoted to their testimony. It creates meaning not just in who and what we see, but also by how these various contrasting parts interact with each other. We are present with each person we feel the weight of their experiences. Their bearing witness to us places us in a position of responsibility to them. We take on an obligation to the other by being witness to them (Lévinas 1998). It is this taking on of the responsibility to the other through hearing their testimony that brings the beyond being a political polemic. It is this being present with the witness, this sharing of the burden of testimony that a documentary contains which the historical text cannot. It is where the extra beyond the image that Barthes sees in photojournalism becomes the post historical that Flusser sees. This is where what we see in *Shoah* that becomes a model for installation-based documentary.

6 What Happens with New Media

As we have seen the main directions of the documentary have been either towards journalism or presenting us in the presence of witnesses in order to experience their testimony. We have seen how as the form of the documentary has moved away from the image illustrating texts the place of testimony takes over as a main part of the form. Film history has tied this to the technological developments that have changed the medium. The power of the voice over lessened once we had portable cameras and sound. This is the era that developed direct cinema. Since then, what has also change how we make documentaries has been the shift of viewing experiences. The first direct cinema pieces were made for television, and this allowed an intimacy in what we say and how we saw it. It also transformed the single documentary into documentary series. The idea of how we experience the work shifted from a single short experience in a darkened room to something that we would see in parts, allowing the work to open up more to multiple perspectives and voices. One off broadcast documentaries continued to maintain a connection to journalism, whilst the series could be historical or more cultural. It could focus on a person or a specific event. As the form started to find different ways of presenting itself the documentary started to develop new forms that related to the formats it was presented on.

As we discuss new media have define what we mean so that we can see how it has changed documentaries. We can roughly break new media down into web-based work, immersive works and physical installation. Each exerted its own demand on the form, but the emphasis on testimonial is a unifying factor.

Social media is based on the idea of the user being the content provider. This tends towards the valorisation of the eyewitness. Cameras on mobile phones have allowed us all to bear witness and present that witnessing as events happen. It is a continuation of the journalistic/direct cinema form of documentary, and it is the timeliness of the images that carries its impact. This footage of witnessing much like other forms of vernacular images it tends to use highly conventionalised composition. We know how to create the images and how to view them. This conventionalisation can mitigate the uniqueness of the voice of the witness. It can be said that this form tends towards viewing events but not bearing witness in that the camera operator is an outside observer of events.

By contrast video sharing sites allows for the presentation of longer work. Original works on video sharing sites can be of any length, but since they do not finance the production of what is seen, the works on these platforms will either be from traditional producers or be self-financed. This comes closer to either traditional documentary, in the first case, or that subset of direct cinema, the autobiographical film, in the latter. Where the uploading of traditional documentaries does allow for greater access, it doesn't change how such works are made or how they are intended to be seen. In the latter case the concept closer to an essay or personal journal. It is drawing from the idea of personal testimony even if the documentarians would not always think of it in this way. This form contains a tension between personal testimony and the need for a narrative structure. While the solutions can be varied, they tend towards conventionalization as much as the social media form does.

Moving away from solely online formats, there is immersive work and installation-based works. The immersive works, whether using VR, 360° video or AR focus on placing the viewer into specific situations. This can be into a unique situation like life amongst peoples from distant locations or by placing oneself into a situation they would never experience otherwise, such as the Guardian's 360° video *6 × 9* about solitary confinement in US prisons (Guardian 2017). As a form this has not been very popular, but as the form develops works tend towards using presence and a form of bearing witness in how they develop their narrative structures.

Finally, installation-based documentary as a form goes back the beginnings of multiscreen filmmaking. We trace it back the Eames' *Think* in the IBM Pavillion of the 1964 World's Fair (Think 2025). These works have tended, much like the VR work that followed, to be travel logs or broad statements on the state of life at a given time. From there multiscreen documentaries have moved into the museum and gallery spaces. In the museum context works may be commissioned for informational institutions such as museums of science or industry. But as the multimedia documentary moves into the non-industrial realm, it takes up the challenges presented to the form by Rouch, Morin, and Lanzmann.

7 Multimedia, Multiple Screens, Multiple Voices and Testimony

Where the technical innovation of direct cinema was 16 mm sync sound recording and for social media, the handphone camera, with multiscreen documentaries the innovations are twofold. There is the programmability of sequencing play and there is the physicality of space. The opening up of the sequencing of scenes is based on databasing and software programming. The databasing of content opens up the narrative structure of works allowing for sequencings that are unique to each performance. This demands that the work be sequenced through programming languages that can generate clip sequencing in real time. While this form is created through a database, because of the randomness of the sequencing it demands images of a different type from cinematically edited documentaries. When creating images in this type of work one has to be aware of their ability to stand alone, have meaning outside of an edited sequence, while knowing that they will be placed into relation to any number of other clips on the database. In works such as Lev Manovich's *Mission to Earth (Soft Cinema Edition)* (2003–2004) (Manovich 2003) or Gary Hustwit's *Eno* (Hustwit 2024) we find that the unique sequencing of the work brings an intimacy to the viewer. It turns the relationship of the viewer to the work into one that is conversational. It carries both the idea of the presence through conversation as espoused by Martin Buber into a tangible form (Buber and Smith 2002) and also is formed by Deleuze's understanding of how meaning is made in cinema (Deleuze et al. 2018). Deleuze states that meaning is found in how tow clips are put together and normally we think of this as how editing creates meaning, hear we find that meaning is still created by the same relationships it is just that we have taken out of the hands of an editor. As we remove the authorial voice it doesn't change the basic function of how meaning is created.

The other defining element is the use of space within the work. Multiple screens mean that we are engaged in a physical space. Our attention can go from screen to screen. In works where that are edited for this multi-screen experience attention is programmed by the filmmaker. This what Isaac Julian has called multiscreen montage (Julien 2014). Multiscreen montage does not need use the database, instead clips just have to follow the sequencing as would any traditional film. This form wants to engage the viewer in a more personal way through the programming of attention but doesn't bring forward a relationship of being-with the other that is found within the documentary form. There is no need engage a presence with the viewer as there is in the documentary. There is no need for testimony In the documentary form as the multiscreen montage is a fictional work and as such it can only mimic the act of testimony, it can never contain it.

By bringing the relationship of the viewer and the clip into such an intimate situation the act of testimony is engaged. This two-way movement found in the act of testifying becomes the centre of what is before us. The act of testifying on screen brings with the act of bearing witnessed to by the viewer. They are drawn together in this act. The nature of this intimacy transforms the documentary from the act of explaining events to a public into the face-to-face of being with the other who is testifying to me. Presenting the voice of the other becomes what the documentary does. When we enter into a space in which multiple voices can be seen or heard in sequence or at once, we are able to move through the space, moving closer of further back form the images, we develop a physicality to the images that enhances the intimacy we have with them. The conversational overtakes the narrative when the work becomes more spatial, and the sequencing is made open ended. This method of presentation also encourages the documentarian to concentrate on bringing the voice of testimony to the fore. We want to hear the voice of the other in direct address to us. Where the handheld camera tended to want us to observe what is happening before us, so too multiple screens and sounds tends to want us to see the face of the other in a more intimate form. As such the multiscreen documentary can only make sense as a form of testimony. As we open up how we can experience the work we also allow our own presence to define this experience. We are allowed to determine and define how we see what we see and then make meaning form this. It is in this act that we are engaged in conversation with the work and presence as being-with the other is enacted. As this is enacted the dual face of bearing witness comes into its fullness and the idea of the documentary as the act of testimony becomes real.

References

Barthes R (1981) Camera lucida: reflections on photography. Hill and Wang, New York
Buber M, Smith RG (2002) Between man and man. Routledge, London; New York
Deleuze G, Tomlinson H, Habberjam B (2018) Cinema I: the movement-image, Paperback edn. Bloomsbury Academic, an imprint of Bloomsbury Publishing Plc, London; New York
Flusser V (2000) Towards a philosophy of photography. Reaktion, London
Flusser V, Ströhl A (2002) Writings. University of Minnesota Press, Minneapolis
Godard J-L, Gorin J-P, Mieville A-M (1976) Ici et Ailleurs. Gaumont Film Company, p 50 minutes
Guardian T (2017) 6×9: a virtual experience of solitary confinement. The GuardianVR, Guardian News & Media Limited. https://www.youtube.com/watch?v=odcsxUbVyZA
Hustwit G (2024) Eno. Film First Co, Tigerlily Productions, p 100 minutes
Julien I (2014) Theatrical fields: special brunch and screening with Isaac Julien and Mark Nash. NTU CCA Singapore Digital Archive
Kahana J (2016) The documentary film reader: history, theory, criticism. Oxford University Press, Oxford; New York
Lévinas E (1998) Entre nous: on thinking-of-the-other. Columbia University Press, New York
Manovich L (2003) Making art of databases. V2 Pub./NAi Pub., Rotterdam
Morris E, Williams M, Ahlberg JB, McNamara RS, Glass P, @Radical.media (Firm), Globe Department Store (Firm), SenArt Films (Firm), Columbia TriStar Home Entertainment (Firm), Sony Pictures Classics (Firm) (2004) The fog of war eleven lessons from the life of Robert S. McNamara. Columbia TriStar Home Entertainment
Nichols B (1991) Representing reality: issues and concepts in documentary. Indiana University Press, Bloomington
Rolling Stones, Coca-Cola Company Collection (Library of Congress), LC Purchase Collection (Library of Congress), Miscellaneous Collection (Library of Congress) (1970) Gimme shelter. Cinema V, Columbia Pictures Home Entertainment, RCA SelectaVision, RCA/Columbia Pictures Home Video
Rouch J, Morin E, Dauman A, Lifchitz P, Loridan-Ivens M, Parolini ML, Arte Video (Firm), Argos Films (2005) Chronique d'un été Paris (1960). EDV 236 Arte Video, Arte Video
"Think," 30 Mar 2025. https://www.eamesoffice.com/the-work/think-film/
Vertov D, Kaufman MA, Vertova-Svilova EI, Vertov D (1929) Man with a movie camera. British Film Institute, Kanopy Streaming, p 1 streaming video file (69 min)

Design Education and Global Strategies for Social Impact

Carla Cadete

Abstract This article presents a pedagogical practice building on the positive outcomes of the Erasmus+ Teaching Mobility Program at Higher Education Institutions (HEIs) in February 2023. Inspired by the initiative at Porto Lusófona University in 2021, a similar briefing was conducted with second-year Graphic Design students at Tomas Bata University in Zlín, Czech Republic. The project aimed to implement the most effective graphic solution at the host university while also generating funds for the same proposed NGO, The *Big Hand*. The article is organize in four parts: Introduction, conceptual framework, methodology and conclusion. The development process was structured through the *Design Thinking* methodology. A practice that emphasize the importance of social design innovation as a key application area in the field of design, encouraging a socially conscious approach that prioritizes human needs. This practice offers valuable insights and innovative strategies for enhancing teaching and learning in design education while effectively motivating and engaging students. The practice also explores opportunities to implement these methods on a larger scale, involving international students to tackle global challenges through design. Furthermore, it connects academia with international community, showing how universities can promote social change while enhancing students' educational experiences. Although all students showed dedication and interest in real-world initiatives—especially those with a civic focus—significantly increased their involvement and motivation.

Keywords Socially responsible design · Pedagogical practices · A pencil for a school · Erasmus+ teaching mobility

C. Cadete (✉)
Lusófona University, Porto, Portugal
e-mail: carla.cadete@ulusofona.pt

D. Raposo et al. (eds.), *Perspectives on Design IV*, Springer Series in Design and Innovation 59, https://doi.org/10.1007/978-3-032-00733-9_9

1 Introduction

The article highlights the significance of collaboration between two academic institutions and a stakeholder, such as a non-governmental organizations (NGO). In 2022, second-year Communication Design students created an advertising campaign to raise funds for a school in Matsinho, a small village in Mozambique's Manica province. This campaign was relaunched in 2023 as part of the Erasmus+ Teaching Mobility Program, in partnership with students from Tomas Bata University in the Czech Republic. The solidarity campaign, titled "A Pencil for a School," aims to finance the construction of a school in Matsinho. The *Big Hand* is a non-governmental organization (NGO) dedicated to improving the wellbeing of children in challenging situations, particularly orphaned girls. The organization is responsible for organizing a campaign focused on several key goals: ensuring access to clean water, sanitation, nutrition, healthcare, and education. The *Big Hand* adopts a child-centered approach, believing that children who grow up in a healthy environment have the potential to change the world. The organization collaborates with communities, local organizations, and governments to create "friendly schools" inspired by UNICEF's best practices. This project resulted from a collaboration between the *Big Hand* and Lusófona University of Porto, which was formally established through a collaborative agreement. These techniques increase self-assurance and drive, which promotes increased cooperation in learning settings. With partners like *Big Hand* and other NGOs like *Associação Ajudar Moçambique*, the collective *Now You See Me Moria*, and Pedro Hispano Hospital, pedagogical practices are frequently implemented in a curriculum-based setting. To improve student achievement and expand their knowledge and abilities, we include guest lectures, seminars, studio tours, and exhibitions as integral parts of our Communication Design Bachelor curriculum. In order to prepare students to work with both vulnerable populations and for-profit organizations, this course involves collaboration among students, teachers, and other stakeholders on a variety of topics. Students who complete these programs will have the abilities needed to thrive in both academic and professional settings. This project shows our dedication to their growth by providing challenges that go beyond typical coursework and fostering a sense of urgency and purpose that goes beyond merely getting good grades.

2 Conceptual Framework

The design field has significantly transformed, shifting from a product-oriented discipline to a people-centered practice addressing societal challenges (Howell et al. 2016; Ramirez 2011). This evolution aligns with the values of the millennial generation and emphasizes the importance of context-oriented problem-solving in design education (Howell et al. 2016). Universities adapt their curricula to include social responsibility and community engagement, preparing students to co-design solutions with disadvantaged communities (Ramirez 2011). The transition from "greed to need" in design highlights the necessity of understanding affiliation in interdisciplinary processes to confront global societal challenges (Harland and Santos 2009). A social justice-oriented approach to design has emerged, focusing on six key dimensions: transformation, recognition, reciprocity, enablement, distribution, and accountability. This approach requires commitment to conflict, reflexivity, and personal ethics, acknowledging the political responsibility of addressing complex social issues through design (Dombrowski et al. 2016).

Recent research highlights the importance of preparing design students for addressing complex social challenges through human-centered approaches. Napier (2015) describes a pedagogical approach that teaches students to collaborate with stakeholders, deploy participatory design methods, and develop solutions for social issues.

Design educators play a crucial role in shaping responsible and sustainability-oriented designers who can contribute positively to society. By integrating the concept of design responsibility into education, students can be encouraged to adopt sustainable design thinking and innovative solutions (Ashour 2020; Leerberg et al. 2010). This approach involves employing both practice-based and theory-based learning methods to promote critical and reflective design practices (Leerberg et al. 2010). Educators must also address professional immunity claims that designers use to minimize their moral culpability, helping students develop ethical awareness (Owens 2006). Furthermore, design education should go beyond teaching technical skills and aesthetics, focusing on helping students understand themselves, their world, and empowering them to tackle social problems innovatively (Goodman 2022). By instilling a holistic approach that combines technical proficiency with ethical responsibility and social awareness, design educators can prepare students to become agents of change.

3 Projects and Methodologies

3.1 A Pencil for a School, *a Design Project Held on Lusófona University (Porto—Portugal)*

In the Winter semester of 2022, Communication Design students, from the 2nd year of the Bachelor were tasked to creating an advertising campaign to raise funds for a school in the village of Matsinho, in Manica, Mozambique. The campaign targeted all residents of Portugal, with the goal of generating money for the construction of a school. The client for the project was the *Big Hand,* an NGO based in Lisbon. The briefing consist to create three A2 posters, a roll-up, a T-shirt, a donation can design, apply the design on a pencil, and an Instagram post for the campaign. The project was completed over four weeks, with two three-hour classes per week. However, the time frame proved insufficient, so five teams of two students each were formed, and a one-day workshop was added to supplement the design sessions. "Teamwork and the workshop were additional benefits – a creative learning space where students with their peers and the tutor work together to find design solutions. A practice that generated students' engagement with the project." (Cadete 2022a, p. 57).

Students were engaged in brainstorming sessions using tools such as mind maps, mood boards, and sketchbooks. Divergent thinking was used to investigate a variety of design approaches, which stimulated a vast array of creative possibilities. Portuguese students at Porto Lusófona University.

The ideas were then refined by assessing the advantages and disadvantages of each approach. Design elements such as color, contrast, balance, typography, legibility, composition, and scale were carefully considered to ensure the campaign's message was impactful and accessible. One challenge arose with the roll-up's narrow width and tall height, which made visual composition difficult. To address this, an intermediate presentation and evaluation were conducted, encouraging group discussion and feedback. A collaborative process was established in the classroom, allowing students to engage in constructive discussions, share insights, and benefit from collective problem solving. This approach helped refine the campaign's design while ensuring that ideas were fully explored and, when necessary, revisited. The active participation of the teacher was also crucial in guiding the students, providing feedback, and highlighting aspects they might have missed. This project ultimately demonstrated the importance of collaboration, ideation, and iteration in the creative process.

A workshop held on December 17th, 2021, at Lusófona University, was a one-day event aimed at enhancing the design students' project experience. It took place during the ideation phase and provided students with the opportunity to engage with the President of the NGO by a *Zoom* video conference. The students could ask questions, share their ideas, and hear feedback from the client, fostering a closer connection to the real-world project. The workshop allowed students to explore various techniques, both analogic and digital, under the guidance of their form teacher. This experience was valuable in promoting teamwork and collaborative learning, as students worked together to find design solutions. The engagement with a real-life client, particularly one with a humanitarian purpose, further motivated students to dedicate themselves to producing quality work. By encouraging a collaborative learning environment, the workshop mirrored the dynamics of a design studio, where both students and lecturers actively participate in solving real-world problems. The hands-on approach fostered creativity, increased the number of potential solutions, and ultimately helped students find the most effective design. Though the workshop was optional, it received enthusiastic participation from all students, demonstrating the positive impact of such proactive initiatives on student commitment and responsibility toward the projects.

3.2 A Pencil for a School, *a Design Project Held on Tomas Bata University (Zlín, Czech Republic)*

Building on the positive outcomes of the Erasmus+ Teaching Mobility Program at HEIs and inspired by the initiative at Porto Lusófona University in 1st semester of 2022 (Cadete 2022b), in the 2nd semester of 2023, a similar briefing was conducted with second-year Graphic Design students at Tomas Bata University in Zlín, Czech Republic. The project aimed to implement the most effective graphic solution at the host university while also generating funds for the same proposed NGO. The workshop's opening featured a remote appearance by David Fernandes, the NGO's President, who joined from the village of Montechimoio.[1] During his presentation, he provided an local overview of the village and the Campaign's objectives. After that, the NGO's anthem[2] was shared with the class to deepen their engagement with the topic in an immersive process, emphasizing the urgency of initiatives like this and demonstrating how design can act as a powerful agent of change within an academic setting.

The methodology followed was consistent with that used for Portuguese students, employing the *Design Thinking* methodology (Ambrose Harris 2010; Lupton 2011), consisting of *Problem Definition, Ideation, Prototyping, and Implementation.* The first two stages of the methodology—*Problem Definition* and *Ideation*—were conducted in person at Tomas Bata University.

Problem Definition: The initial step involved analyzing the briefing, provided by The *Big Hand,* which included a variety of materials: photos from the Matsinho community and children, a dossier about the NGO's history, and logos of its partners. The key components of the briefing were reviewed, including the target audience (children aged 6 to 18), the timeline and the goal of raising funds for the school's construction (Fig. 1a–c). This approach was chosen due to its proven success in the initial case. During the *Ideation* phase, students engaged in brainstorming sessions using tools such as mind maps, mood boards, and sketchbooks. Various design concepts were explored through divergent thinking, which encouraged a wide range of creative possibilities (Fig. 2).

However, with only five face-to-face days available, the project was adapted to a blended mobility model, incorporating hybrid work and evaluation methods. The *Prototype* phase was guided by the Czech Design Professor during classes held at the university from February 17th to March 14th. The methodology employed was similar to that used in Portugal, with the key difference being that there was no need to done the workshop, as the classes were sufficient for completing the work. During the Erasmus mobility period, the entire process was explained to the Design Professor at Tomas Bata University. The ideas were then refined by assessing the advantages and disadvantages of each approach. Design elements such as color, contrast, balance, typography, legibility, composition, and scale were carefully considered to ensure

---

[1] https://www.youtube.com/watch?v=kXhDe8Un7Zk.

[2] https://www.youtube.com/watch?v=QWw0nzGI3Zo.

(a) (b) (c)

Fig. 1 **a** left, **b**, and **c** right**.** The workshop's opening featured a remote appearance by David Fernandes, the NGO's President (left); The creative process during the workshop in Tomas Bata University (2nd, 3rd and 4th). *Source* Cadete (2023)

Fig. 2 During the *Ideation* phase, students engaged in brainstorming sessions using the sketchbook and tools such as mind maps, and mood boards. *Source* Cadete (2023)

the campaign's message was impactful and accessible. One challenge arose from the roll-up's narrow width and tall height, which made visual composition difficult. To address this, an intermediate presentation and evaluation were conducted, fostering group discussion and feedback. A collaborative process was established in the classroom, allowing students to engage in constructive discussions, share insights, and benefit from collective problem-solving. This approach helped refine the campaign's design while ensuring that ideas were fully explored and, when necessary, revisited. The active participation of the teacher was also crucial in guiding the students, providing feedback, and highlighting aspects they might have missed. This project demonstrated the importance of collaboration, ideation, and interaction in the creative process.

On March 19th, an online *Zoom* meeting was held for all student groups to present and analyze their proposals. The session, attended by the Design Professor from Tomas Bata University, provided an opportunity for both professors to offer feedback on each presentation, highlighting strengths, identifying areas for improvement, and suggesting adjustments. Most changes were minor, as students had already received guidance during Design classes with the Czech professor (Fig. 3). The final delivery and assessment took place during another *Zoom* session on March 31st, attended by the NGO President, the Design course Dean, and the two professors overseeing

Fig. 3 Design solutions from three different groups. The jury selected the left solution. *Source* Tomas Bata students, 2023

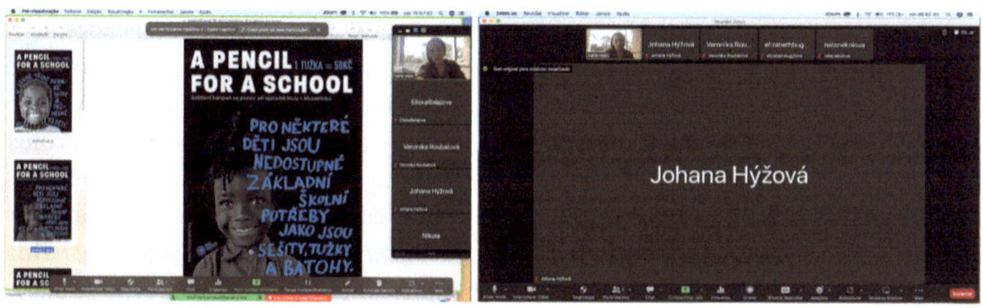

Fig. 4 The final delivery and assessment took place during a *Zoom* session on March 31st. *Source* Cadete (2023)

the project from Tomas Bata University and Lusófona University (Fig. 4). After the presentations, the panel evaluated and selected the best proposal for implementation. In May, the winning project was printed and implemented as a campaign on the premises of Tomas Bata University. By July, the funds raised from the campaigns conducted at both universities were transferred to the NGO.

4 Discussion of Results

The value of collaboration in these pedagogical practices lies in their ability to unite academic institutions, students, and stakeholders, fostering a multidisciplinary approach to design that addresses real-world social challenges. By engaging with organizations like Big Hand, students participate in co-creative processes that emphasize critical and innovative thinking while working towards tangible outcomes.

The benefits of collaboration include: (i) Direct Contribution: Students contribute to initiatives like *A Pencil for a School*, understanding how their design skills can drive social change and promote global sustainability; (ii) Strengthening Community Connections: Partnerships between NGOs and the connections between academia and the community were vital for fostering a shared purpose; (iii) Enhanced Communication Skills: Working with stakeholders improves students' abilities to communicate, co-create, and preparing them for diverse professional environments; (iv) Global

Responsibility: By addressing international challenges, such as access to education in Mozambique, students develop a sense of responsibility toward global issues, aligning design education with the broader goal of social innovation.

The international collaboration between Porto Lusófona University and Tomas Bata University, facilitated by the Erasmus+ Teaching Mobility Program, offers several advantages: (i) Cultural Exposure: Students experience diverse cultural perspectives, which broadens their understanding of global contexts and challenges; (ii) Collaborative Learning: Working on the same project across different countries provides students with insights into various approaches and methodologies; (iii) Real-World Impact: The collaboration directly benefits a real-world NGO, highlighting the practical application of design for social change; (iv) Networking Opportunities: Students and professors form connections that may lead to future academic or professional opportunities. Since then, every year, students from the Design course at Tomas Bata University have gone on Erasmus Mobility at Lusófona University. Also, a joint project between both universities is currently being considered by the professors; (v) Flexible Learning Model: The methodology combines teaching methods, enhancing learning and project value. The hybrid educational model combines the benefits of in-person and online learning, ensuring interactive engagement while providing flexibility; (vi) Enhanced Project Quality: Input from professors and stakeholders across multiple institutions improves the quality of projects and promotes cross-disciplinary expertise; (vii) Increased Student Engagement: Collaborating on global projects with tangible outcomes, such as funds raised and campaigns implemented, boosts student involvement and satisfaction; (viii) Proven Methodology: The successful use of *Design Thinking* in both educational contexts showcases its adaptability and effectiveness; (xi) Meaningful Change: Students gain experience in driving meaningful change, which fosters a sense of purpose and responsibility in their professional practice.

This collaboration not only enhances education but also addresses global challenges through socially responsible design practices.

5 Conclusion

This article demonstrates that design education could extend beyond aesthetics and functionality to include ethical, cultural, and societal impacts. This article illustrates that design education can encompass more than aesthetics and functionality, including ethical, cultural, and societal impacts. Developed at two European universities in different countries, has successfully raised money to help build a school on another continent. It demonstrates that design educators can empower students to become agents of change by adopting a holistic approach that integrates technical skills with ethical responsibility and social awareness. A pedagogical practice that aims to foster the development of skills across various domains, including design methodology, critical and creative thinking, teamwork in multidisciplinary settings, and the generation of innovative ideas. Additionally, these initiatives provide design students with immersive experiences that focus on specific social issues, allowing them to apply the tools and methods learned throughout their curriculum to create a positive impact. Students are encouraged to view design as a tool for addressing social challenges rather than merely pursuing profit. This initiative provides students with the opportunity to engage in innovative practices, addressing real-world problems and preparing them for future challenges while allowing them to make meaningful contributions to society. It emphasizes the importance of going beyond traditional university methods to develop essential workplace skills and practical benefits. The project promotes human-centered design and social innovation, inspiring other institutions to adopt similar strategies that enhance student engagement and encourage societal contributions. The methodology used incorporates a variety of complementary teaching methods that enrich learning and add value to the final project.

For Portuguese students, this approach allows them to interact directly with clients, emphasizing teamwork while focusing on the needs of others. It also includes a workshop as an additional component. International students similarly have the opportunity to connect with clients, collaborate in teams, and participate in both face-to-face and online classes. They can work on real projects under the guidance of instructors from various educational institutions across different countries. All students showed dedication and interest in real-world initiatives—especially those with a civic focus—significantly increased their involvement and motivation.

Moving forward, we will document and disseminate these projects through detailed case studies and presentations. These will be shared at academic conferences, and workshops, and published in journals that emphasize design education and social innovation; cross-university collaboration through new partnerships with institutions across Europe to replicate and adapt these practices in diverse academic settings; curriculum integration by encouraging other universities to include similar projects into their curricula, particularly in programs emphasizing creativity, problem-solving, and social responsibility; scaling up by explore opportunities to apply these methods on a larger scale, involving international stakeholders to address global challenges through design. By sharing and expanding these pedagogical projects, academia can better harness the potential of design education to inspire students, foster innovation, and contribute meaningfully to society.

Acknowledgements This work was funded by Fundação para a Ciência e Tecnologia (FCT) under HEI-Lab R&D Unit (UIDB/05380/2020).

Conflict of Interest The author declares that there is no conflict of interest.

References

Ambrose G, Harris P (2010) Design thinking. Ava Publishing

Ashour AF (2020) Design responsibility and sustainability in education. Int J Des Nat Ecodynamics 15:129–133

Cadete C (2022a) The role of design: a humanitarian approach and an opportunity to prepare students for the real working world. In: The European conference on arts, design and education (ECADE), IAFOR, Official conference proceedings, pp 53–61. ISSN: 2758-0989. https://doi.org/10.22492/issn.2758-0989.2022.4
Cadete C (2022b) Social design in a curricular context: a collaborative practice with a humanitarian purpose. Paper presented at 4th interdisciplinary and virtual conference on arts in education (CIVAE), Spain, Conference proceedings CIVAE 2022. Spain Adaya Press, Madrid. https://doi.org/10.58909/adc22705971
Cadete C (2023) A design project aimed to promote social change: from the classroom to the community. In: The Paris conference on education 2023: official conference proceedings, pp 33–41. ISSN: 2758-0962. https://doi.org/10.22492/issn.2758-0962.2023.5
Dombrowski LS, Harmon E, Fox SE (2016) Social justice-oriented interaction design: outlining key design strategies and commitments. In: Proceedings of the 2016 ACM conference on designing interactive systems
Goodman AG (2022) Who they become is as important as what they know. We must teach students to become agents for change so they can bring design thinking to causes they care about. ZOO! Investigación en Diseño y Comunicación visual
Harland RG, Santos MC (2009) From greed to need: notes on human-centred design. https://hdl.handle.net/2134/5378
Howell BF, Stark CG, Christiansen TJ, Hofstrand RH, Pettit JL, Slooten SN, Willett KJ (2016) Introducing new design disciplines into a traditional industrial design program. DS 85-2: Proceedings of NordDesign 2016, Volume 2, Trondheim, Norway, 10th - 12th August 2016
Leerberg MB, Riisberg V, Boutrup J (2010) Design responsibility and sustainable design as reflective practice: an educational challenge. Sustain Dev 18:306–317
Lupton E (2011) Graphic design thinking: beyond brainstorming. Princeton Architectural Press
Napier P (2015) Design for social change: a pedagogical approach to prepare students for human-centered design practice
Owens KM (2006) Creating responsible designers: recognizing and responding to professional immunity claims. Vis Commun Q 13:152–165
Ramirez M (2011) Designing with a social conscience: an emerging area in industrial design education and practice. Conference: ICED11 18th International Conference on Engineering DesignAt: Copenhagen

Reimagining Wicked Problems Through Speculative Design: Collaborative Classroom Experiments

António Gorgel Pinto, Paula Reaes Pinto, Filipe Figueiredo, and Joana Ramalho

Abstract Design can be explored by various agents—professional and non-professional designers—in different areas, through processes, which represent the creative and social territories inhabited by those agents. Design thinking principles have been increasingly adopted by public and private organisations, from various sectors, to tackle complex problems and envision solutions. Speculative Design can play a fundamental role in experimenting with unconventional approaches. The DXT ideation tool, hereby presented, is a methodological tool, which engages students and creative practitioners in a participatory design process that challenges conventional human-centred perspectives in Human-Computer Interaction. Rooted in the principles of Augusto Boal's Theatre of the Oppressed, it aims to respond to unsustainability, through societal innovative, ecological and inclusive solutions. As an exploratory pedagogical process, it explores dystopian future scenarios, with the help of generative AI. In doing so, it confronts the students with inexhaustible possibilities and the amplification of image generation, with few constraints. However, it also confronts them with the unpredictability of the outcomes of AI tools. For instance, while AI models can help diminish biases and prejudices, they can also inversely enhance them. Simultaneously, while speculative scenarios stimulate critical thinking around dominant narratives, revealing the possibilities of current trends, they challenge the conventional linearity between past, present and future.

Keywords Speculative design · Generative AI · Theatre of the oppressed

A. Gorgel Pinto (✉) · F. Figueiredo · J. Ramalho
UNIDCOM/IADE, Universidade Europeia, Lisbon, Portugal
e-mail: antonio.gorgel@universidadeeuropeia.pt

A. Gorgel Pinto · P. Reaes Pinto
CHAIA, Universidade de Évora, Évora, Portugal

F. Figueiredo
CET, Universidade de Lisboa, Lisbon, Portugal

D. Raposo et al. (eds.), *Perspectives on Design IV*, Springer Series in Design and Innovation 59, https://doi.org/10.1007/978-3-032-00733-9_10

1 From Conceptual Repositioning to Problem Framing

In *Wicked Problems in Design Thinking*, Buchanan (2016) points to the existence of multiple possibilities for conceptual repositioning in design. These dynamics occur systemically and in different situations, which can be recognised through signs, objects, actions, and thoughts. The author identifies four major areas in which design is explored by both professional and non-professional designers: symbolic and visual communication, material objects, activities and organised services, and complex systems or environments. Although these four areas of design thinking are structured around their respective spheres of knowledge, interests, and dominant trends, they are not confined to categories of visual communication, objects, services, systems, or environments, resulting from specific design processes; rather, they also represent the creative territories inhabited by all designers without exception, where the "dimensions of design thinking by a reconsideration of problems and solutions" are explored.

However, the idea that design is an activity shared by various social agents, whose initiatives are based on the same principles of creativity and reinvention of ways of interacting with the environment, is widely accepted nowadays. In this context, some designers with scientific knowledge, particularly those more attuned to this issue, have sought new ways of engaging with *empirical designers*, aiming for both to participate in the creation of objects and functions that contribute more effectively to society. Equally relevant is the increasing adoption of design thinking principles by both public and private organisations to tackle complex problems, acknowledging the effectiveness of different design methodologies and their capacity to generate solutions across various scenarios—an approach that is now recognised not only by businesses but also by a wide array of stakeholders (Manzini 2015).

When interconnected, the four areas of design identified by Buchanan (2016) do not prioritise any particular one, reflecting a dynamic interaction where signs, things, actions, and thoughts merge, driving innovation forward. Designers navigate these areas by reconsidering problems and solutions, often shifting their focus to uncover new dimensions. Since the late twentieth century, designers have been expanding their scope by incorporating into their equations less conventional variables within their field of action, fostering the emergence of new questions regarding the form and function of products, user experience, and holistic systems. This conceptual repositioning is evident across all design practices, driving innovation and addressing contemporary concerns such as sustainability, technology, and ethics.

In the process of reflecting on problems and solutions, speculative thinking plays a fundamental role in design, as it allows for the questioning of assumptions, the imagining of alternatives, and the exploration of possibilities that go beyond immediate and conventional solutions. Through experimental and unconventional approaches and testing predefined limits, this expanded logic enhances the creation of innovative scenarios for social, environmental, and technological challenges, providing a wider range of possible responses. Thus, speculative thinking in relation to *wicked problems* not only drives innovation but also cultivates a critical discourse on the impact of design decisions on society, encouraging a more reflective and transformation-oriented practice.

While the different areas of design have fixed meanings that are recognised through theoretical foundations, serving as a basis for analysing existing knowledge, repositioning seeks an expanded field without insurmountable limits. Recognising the distinction between a design area and a positioning is essential, not only to ensure that design thinking is regarded as both a field of knowledge and a set of practices but also to highlight its capacity for uncovering new perceptions and possibilities to be tested. In this sense, a consistent approach to new design positionings, as well as their replication in new situations, constitutes a fruitful means of understanding the logic employed by various designers focused on the invention of new possibilities (Buchanan 2016).

The concept of *wicked problems* emerged as an alternative to linear design approaches characterised by the definition of a problem followed by its phased resolution—a reductionist process that fails to reflect the true relationship between thinking and execution, given that this dynamic is complex and varies according to each situation and its idiosyncrasies. Beyond being complex, design problems are indeterminate and may pertain to any aspect of human or more-than-human life. A significant characteristic of the design approach lies in the fact that designers begin this process by identifying or creating a "quasi-subject matter, tenuously existing within the problems and issues of specific circumstances." This is followed by the stages of ideation, prototyping, and testing, which ultimately lead to the design outcome. "A quasi-subject matter is not an undetermined subject waiting to be made determinate. It is an indeterminate subject waiting to be made specific and concrete" (Buchanan 2016).

Design practice is not centred on generating singular solutions for a given problem, although there are areas of design with a strong technical rigour where highly optimised and objective solutions are applied, approaching a more exact form of reasoning. Particularly in its more creative and interdisciplinary dimensions, design rarely focuses on a single solution to a problem. In order to generate more meaningful projects, designers have developed working methods to manage the indeterminate nature that characterises the search for a solution, applying reasoning patterns such as *deduction*, *induction*, (normal) *abduction*, and *design abduction*. *Deduction* and *induction* are essential for predicting and explaining real-world phenomena, yet they are insufficient for developing new ideas. In design and other productive professions, *abduction* is the predominant reasoning pattern, with *design abduction* being even more specific (Dorst 2015).

On the one hand, *deduction*, regarded as the most solid form of reasoning, follows a logic of cause and effect, where the elements of a given situation and their interactions are understood, allowing for the prediction of outcomes. While the calculations may be complex, deductive reasoning is reliable because it employs known elements and relationships to forecast results, which can be verified through observation. *Induction*, on the other hand, entails discovering patterns by observing elements and their interactions without prior knowledge of the underlying laws. Hypotheses are proposed to explain phenomena, and subsequently tested through predictions and observations, thereby combining *induction* (the creation of hypotheses) and *deduction* (the testing of predictions). Although *induction* is often presented as a purely deductive form of reasoning, it is, in fact, a creative act essential for scientific progress.

Abduction is the process of introducing a new element into a problematic situation to achieve a desired outcome. It begins with a basic understanding of the intended result and aims to generate value through the new outcome. It is especially well-suited for problem-solving that is based on experience, in which the desired outcome and the method for accomplishing it are clear, but the specific solution must still be developed. This method operates within established relational patterns and does not query the method itself. Although it is effective in numerous instances, it is predicated on a conventional reasoning framework that does not inherently encourage innovation (Dorst 2015).

Design abduction seeks new solutions by questioning existing methods. In this approach, designers simultaneously develop new elements and the relational pattern required to achieve a given objective, testing and refining both aspects in parallel. Rather than employing conventional processes, the methods used in project development must be created and evaluated together, iteratively leveraging alternative solutions. The use of an experimental and speculative methodology, the expansion of a concept to frame the problem, and the establishment of social interactions are distinctive principles when compared to other forms of reasoning. The expansion of a concept to frame the problem is a moment of ideation that considers multiple possibilities before arriving at something more concrete, which is then evaluated through various design processes. The use of an experimental and speculative methodology aims to simulate different possibilities for addressing the problem, relying on the development of research mechanisms that facilitate reflection on the issue and the testing of solutions. Lastly, the creation of a dynamic of social interactions is an integral part of the design process itself, through which the stakeholders involved share different understandings of the problem-solution (Dorst 2015).

A central characteristic of *abductive design thinking*, particularly in questioning the recurrent interactions within a problematic situation, is the creation of a *frame* that enables an alternative perspective on the problem and enhances the potential for innovative solutions (Dorst 2015; Schön 1983). In order to construct an innovative and appropriate *frame*, the outlining of a problem establishes the conditions necessary to generate the desired solution, taking into consideration the significance of the material to be developed and the existing information about the problematic situation. This process is comparable to *induction*, in which a relational pattern is proposed and experimented with, followed by the introduction of a new element, resulting in the development of a feasible and realistic solution (Dorst 2015).

Among the most relevant aspects influencing the efficiency of a *frame* are the resultant image of the problem and its social dimension. The development of an appropriate *frame* for a given problem produces an image that incorporates the diversity of issues under analysis, potentially even raising new questions that may have emerged in the process. The understanding generated through this image functions as a common denominator for discussing the problem and possible solutions. Whether for a team of designers or for project stakeholders, it is crucial that a *framing* be "thought-provoking and lively, engaging people's imagination so their thoughts readily move along in the proposed direction." The shared understanding within a coalition of designers working towards problem resolution must be collectively recognised, with its effectiveness depending on the alignment of logical approaches. This dynamic involves the generation of metaphors and patterns of interaction that come to be designated as a *frame* once they begin to make sense as a shared comprehension of a given nexus between problem and solution (Dorst 2015).

The search for a conceptual repositioning in *framing* a problem, emerging from the expansion of a particular point of view, stands out as a key element in design thinking when dealing with complex problems, enabling the designer to speculate, *frame*, and *reframe* the issues under analysis. This is a process through which the designer envisions new actions, functionalities, and relational patterns to achieve a previously identified objective, following an exploratory and iterative trajectory. The creation of a dynamic of social interactions for exploring and applying concepts, analysing different perspectives, and generating new ideas is a highly significant aspect of the design process, through which all participants share different understandings of the problem-solution while simultaneously fostering innovation. In parallel, the development of metaphors that represent a speculative approach to the problem and its social dimension functions as a form of *punctum* (Barthes 2020 [1979])—the unexpected detail resulting from a divergent perspective, whose effect is to trigger an emotional response capable of driving disruptive solutions.

2 A Speculation-Based Pedagogical Framework Beyond the User/Designer Dichotomy

A methodological framework for speculatively exploring *wicked problems*, particularly in the field of UX/UI design for sustainable Human-Computer Interaction (HCI), is the Digital Experience Theatre (DXT) ideation tool,[1] which was developed as part of a research initiative at IADE, Universidade Europeia, with the active participation of students who helped test and improve its first versions. Through this iterative process, DXT evolved as a pedagogical and research-driven tool that aims to empower designers to rethink the role of technology in creating more inclusive, sustainable, and community-centred digital experiences.

Rooted in the principles of the Theatre of the Oppressed (TO), the tool engages students and creative practitioners in critical reflection and participatory design processes, challenging conventional human-centred perspectives in HCI. By integrating elements of societal innovation, the DXT encourages the conceptualisation of digital solutions that respond to ecological and sociocultural dynamics, fostering more inclusive, community-driven, and humanity-centred outcomes. Through speculative scenarios generated with different text-to-image AI models, the tool not only

---

[1] *DXT—The Digital Experience Theatre*, https://dxt.cargo.site/

questions prevailing assumptions about digital interactions but also empowers participants to envision alternative futures that prioritise holistic sustainability and collective well-being.

The TO methods developed by Augusto Boal, including the *newspaper theatre*, *invisible theatre*, and *image theatre*, were the main inspirations for the DXT research, through which designers and creatives of digital interfaces are considered user/*spect-actors* rather than merely user/spectators. Boal's dichotomy between *spect-actor* and spectator, to which we associated the user figure, facilitates the transformation of the individual, who, as a witness, begins to assume a more pivotal role as a protagonist.

The spectator traditionally transfers responsibility and agency to the actor-protagonist, who subsequently thinks and acts on their behalf. This process of transforming the spectator into *spect-actor* represents a form of liberation for the former, who, no longer confined to passive reception, is able to establish a relationship of critical freedom with the spectacle and assume a transformative role within social reality (Peixoto 1980).

The DXT ideation tool works as an exploratory pedagogical process guiding design students to reflect on their roles as users/spectators, which is an artificial division that reduces the complexity of design interactions. Based on the TO, which emphasises participation, reflection, and transformation, the DXT involves students in speculative interactions with generative AI models to demonstrate the fallacy of this dichotomy. The idea of a clear separation between users and designers is based on the idea that the latter are the only ones with agency for creative projects and the former only interact or passively consume their products, services, or systems. This viewpoint, however, misses the role of users as empirical designers (Manzini 2015) and mainly the role of professional designers as both creatives and users. The intrinsic nature of the designer role is revealed by this entanglement of agencies, in which meaning, functionality, and creativity are generated as a result of social interaction between the diverse actors and contextual reframing efforts.

This is a praxis that aims to be both a reflection of the human and more-than-human conditions to be transformed and a form of reflection in action that mobilises designers and creatives of digital interfaces to develop disruptive ideas. In working with generative AI, students speculate collaboratively between themselves and with the AI, reinforcing the idea that design is always shaped by distributed agencies—including technological, social, and ecological forces. This methodology for digital interface design through a specific ideation tool like the DXT, combining critical reflection to reframe *wicked problems*, is a model for provoking the *oppressed consciousness* and thus generating the desired societal transformation toward a sustainable future for HCI. This pushes students to move beyond linear problem-solving and embrace design as an evolving dialogue, where authorship is fluid, roles are interchangeable, and the act of creation is not about control but about negotiation, responsiveness, and adaptation within complex systems. Through this pedagogical process, students learn to see design not as a unilateral act of producing for users but as an emergent practice of co-construction that extends beyond human-centred paradigms, fostering more inclusive, participatory, speculative, and humanity-centred approaches to technological development.

After conducting user research to gather meaningful insights into a *wicked problem*, students reframe the design question using the DXT ideation tool, employing dystopian speculation to discover fresh viewpoints and question established beliefs. This aligns with the concept of *agonism*, a framework that deepens the understanding of dissent by cultivating diverse perspectives that highlight both the urgency and fragility of social contexts while also proposing potential alternatives. Unlike antagonism, which frames conflicts as circumstances that cannot be reconciled and with no common ground, *agonism* cultivates a space where opposing viewpoints can engage in dialogue, allowing for the development of a shared understanding (Mouffe 2009; Disalvo 2012). Manipulating these tensions creatively to boost innovative solutions, particularly through dystopian speculation, DXT's agonist approach emphasises that disagreement and the existence of diverse perspectives are fundamental to understanding humanity. This process seeks to promote dialogue, negotiation, and collaboration among students with diverse perspectives, ultimately resulting in more comprehensive and inclusive design outcomes during the subsequent design solution phase.

The speculative co-creation of imagery utilising text-to-image AI models leads to a systematic process where student groups are categorised into *conductor-designers* and *agonist-designers*, which are essential roles for the effective functioning of the ideation tool. The *conductor-designers* initiate the process of creating the first image representing a *wicked problem* worst-case scenario, which triggers the image-based dialogue with the *agonist-designers*. The *agonist-designers* engage with the *conductor-designer's* dystopian speculation, reinterpreting the same image or concept, conducting both design players in the sequential co-design of a dystopian visual representation and respective problem reframing. After exchanging ideas or opinions in order to reach an understanding about the selected problem, in a process that takes approximately an hour and a half, the group is supposed to gain awareness of a specific issue and be better prepared to concentrate on designing solutions addressing the *wicked problem* that initiated the discussion.

3 Creating Dystopian Future Scenarios With Text-to-Image AI Generation

In the DXT co-design process, students engage in the co-creation of speculative dystopian scenarios through text-to-image AI generation. This co-design of imagery is combined with *The Image of the Rainbow of Desire* method (Boal 1994). In this Boal's method, after the selection of the problem to be tackled by the group, the protagonist creates the images and presents them to be subsequently reproduced by the participating characters who are playing the role of agonists. The latter can then produce new images from the protagonist's point of view. The adaptation of Boal's method results in the division of students participating in the DXT co-design process into two roles: *conductor-designer(s)* and *antagonist-designers*.

A dialogue develops between them based on dystopian ideas embodied in images. The *conductor-designer(s)* lead the process and present an AI generated dystopian image to the *antagonist-designers*. The latter then manipulate the presented image, ideally deepening its dystopian nature. In this step, each participant can create their own image, trying to do so through what they imagine to be the perspective of the *conductor-designer(s)*. As in Boal's original method, the images are presented together – in this case in a grid – and the *conductor-designer(s)* intervene in them with keywords that aim at identifying critical problems and possible solutions, replacing the original "confidential monologues" envisioned by the dramaturgist, which brought to light the protagonist's desires. The antagonist-designers then incorporate the new ideas by creating new images.

AI tools applied to the domain of images have sparked highly controversial responses, ranging between distrust and fascination. These two reactions stem from the same condition: the unpredictability of responses alongside a relationship of similarity/dissimilarity. In this regard, it would be appropriate to recall Walter Benjamin's position in his *Doctrine of the Similar* (Benjamin and Tarnowski 1979 [1933]), where he advocates for greater attention to the reproduction of processes that generate similarities rather than to the similarities themselves. That is, from Benjamin's reading, we can recover an interest in the relationships that stimulate new formalisations, making them a principle of validation.

The way AI tools enable the conception of future scenarios—which in the DXT method are embodied in dystopias—is based on straight connections within their processes, producing results of unpredictable dissimilarity, while still ensuring internal coherence. It is within this interval—between the expected and its realisation in an unforeseen form—that lies the potential for discovery in the ideation process. Moreover, the challenge arises from a textual formulation aimed at visual identification. In this process of converting text into image, Dürer's famous engraving of a Rhinoceros (1515) is often recalled, as it was created solely based on a textual description from an Indian painter, without Dürer ever having observed a specimen of the animal. The imaginary world he generated, while not completely severing its connection to a real-world reference, is shaped by a set of elements that could be considered speculative. Yet, the image was accepted as a faithful representation for centuries. W. J. T. Mitchell, who dedicated significant attention to the transformation of textual descriptions into images in his books (Mitchell 1986, 1994), including Dürer's engravings, emphasised the power of images to construct new territories in the realm of possibility. The collage practices advanced by Dadaist, Surrealist, and Constructivist artists further expanded the imagination by integrating into their discourse the *uncanny*, the notion of cut, rupture, and fragmentation, but also montage as an exercise of re-signification and the conquest of a new order.

By making use of AI technologies, this exercise is amplified, broadening the frontiers of possible image generation. As Manovich (2023) argues, rather than being based on a *skilled artistic act*, AI tools derive their strength from an inexhaustible and powerful network of images and data, making them the ideal domain to create without any kind of constraints. Students who engaged in the experimentation of the DXT ideation tool are confronted with outputs that are not so free from any kind of constraints. While experimenting with different text-to-image AI models, they were faced with the problem that AI-generated content tends to amplify societal stereotypes. Even though there are variations between various models and tools, there is a tendency for inequality in terms of gender, age, and race representations that reinforce social biases. This is also seen when specifically addressing personality types, professions or occupations, and geographical locations. Nevertheless, findings from these models also indicate that making use of prompt expansion schemes can effectively increase diversity (Naik 2023). Prompt expansion techniques, complemented with the use of appropriate semantics, allow broadening or refining the scope of a prompt to generate more relevant—in this case, more diverse—outputs from the AI model. However, expanded prompts, which take into account biases such as those related to gender, age or race, might not always work in resolving prejudices and stereotypes and might inversely introduce new biases, deliver wrong and weird responses, or produce outputs with lower image quality and inconsistencies (Naik 2023). Despite the many possibilities outlined, it is crucial to consider these responses not as final results—where they might be deemed limited by a lack of detail or rational grounding—but as provocative instruments within ideation processes.

4 Speculation and Temporality Within the Digital Experience Theatre

The integration of speculative design and temporality within the Digital Experience Theatre (DXT) challenges the conventional linear trajectory of technological progress. Traditional digital transformation presupposes a future that extends from the present in a predictable way, as time is understood as a chronological succession of events, a structured progression from the past to the future. However, the non-linear approach based on speculative pragmatism (Savransky 2017) and situated speculation (Dunne and Raby 2013) are strategies for realising transformative digital experiences through interaction between participants, uncertainty, and multiplicity. Likewise, other theoretical perspectives challenge this assumption, proposing a more complex, non-linear temporality, which is contingent, folded, and shaped by unpredictable events (Bergson 2002 [1896]; Foucault 1972 [1969]; Rovelli 2016; Serres and Latour 1995). Incorporating the approach of speculative pragmatism, situated speculation, and a practice with a performative dimension, DXT acts through a dynamic, complex non-continuous temporality in the development of digital experiences (Bleecker 2009; Candy and Dunagan 2017). It uses experimental images to drive innovation. The generation of dystopian scenarios through AI promotes critical reflection and creative conception based on problematic design situations. The ideation of innovative micro-utopian futures, through speculative experimentation, is a lever for ideation and transformative solutions that can contribute to a more sustainable future.

In relation to the idea of time in the context of the DXT project, Henri Bergson's concept of *la durée* (duration) questions mechanical and linear time by arguing that time is not an objective and measurable entity but rather a subjective and fluid experience. It establishes the difference between "mechanical time", which is divisible and quantifiable, and "lived time", which is elastic and shaped by perception (Bergson 1926). At DXT, this distinction becomes particularly relevant as the participatory method of UX/UI, metaphorically described as a rubber-band problem-focused ideation tool that shifts from dystopian to micro-utopian thinking, driving innovation in socio-technological design across multiple temporalities. In this regard, it is also important to highlight the fact that digital environments sometimes dissolve the limits between past, present, and future, allowing the coexistence of different temporalities in a single experience. Digital products developed as part of DXT's participatory workshops mirror this strategy, as user interactions shape the evolving narrative of digital platforms.

Foucault too, in his work *The Archaeology of Knowledge* (Foucault 1972 [1969]), considers that time is discontinuous and that it is through its ruptures, transformations, and contingencies that knowledge is constructed. His perspective stems from the defence of a discursive and fragmented temporality to the detriment of historical continuity. Similarly, through a speculative design approach and the creation of AI images, DXT unfolds in a temporality that is shaped by imminent experiences, interruptions, and interactivity rather than a linear sequence of events. This approach is crucial for designing digital products that promote an integrated model of economic, social, cultural, and environmental sustainability, thus responding to the needs of the community. The model of non-linear time advocated by Serres and Latour (1995) resembles a scarf that can be stretched or folded so that it can deal with moments that are far apart. In DXT, this non-linear temporality manifests itself through the creation of dystopian scenarios that allow for layered and intersecting temporalities (Manzini 2015), allowing users to experience past and future scenarios as co-present, moulded through emergent interaction.

Physicist Rovelli (2016) questions the traditional notion of time as a crucial structure of the universe. In his work *The Order of Time*, he argues that time is not a fixed entity but an emergent phenomenon and bases his conviction on the fact that he considers that the perception of time depends on relational interactions in quantum terms. Similarly, in DXT, time is perceived through the unpredictability of events resulting from interaction with participants, technological contingencies, and algorithmic processes. This approach utilises user-centred design methodologies that promote flexibility and contributions to the community.

Speculation, understood as a strategy and cognitive process, presupposes confrontation with the unknown, a concept explored by Gilles Deleuze (Halewood 2017) in his work *Difference and Repetition* (Deleuze 1968). The author values the power of difference, uncertainty, and becoming to the detriment of knowing reality only through linear temporality and representation. For him, the unknown is a space that generates ideas and futures, i.e. knowledge. In this sense, his speculative thinking promotes creative and innovative approaches to uncertainty. This vision is closely related to speculative design, speculative pragmatism, and, naturally, with DXT, since they all act with uncertainty, experimentation, and dynamic processes of possibilities, enhancing experience, new ways of thinking, social organisation, and the imagination of alternative futures. In all instances, the unknown is seen as a creative stimulus for transformation and innovation.

The UX design process within DXT approaches situated speculation to develop ideation tools that reflect the contingencies and complexities of real-world digital interactions. The speculative narratives developed in DXT are shaped by users' interactions with sociocultural contexts and the limitations of digital tools. This approach is related to the concept of situated speculation, defended by Halewood (2017), who considers that speculative thinking always depends on context—institutional, technological, material, and social. By engaging with situated speculation, DXT focuses on imagining futures resulting from interaction, uncertainty, negotiation, and improvisation. This perspective is further developed by Wilkie et al. (2017) by framing speculative research as a pragmatic inquiry based on the possibilities and uncertainties that emerge from specific socio-technical arrangements. Moreover, DXT employs speculative design principles, which engage with possible futures rather than adhering strictly to a linear past-present-future structure.

As well as making it possible to explore how temporality shapes digital products, this approach also enables the investigation into how temporality acts on the expanded socio-cultural impact of creating digital future scenarios. Manzini (2015) emphasises the importance of social innovation in design, advocating participatory and sustainable solutions that address contemporary challenges.

Speculative design, as defined by Dunne and Raby (2013), is a forward-thinking approach that challenges normative assumptions by envisioning alternative futures. Unlike traditional design, which focuses on immediate problem-solving, speculative design fosters critical discussions about potential technological trajectories by encouraging people to question existing socio-technical paradigms. This method invites exploration of "what could be", enabling users, designers, and theorists to examine how emerging technologies might reshape cultural, ethical, and temporal dimensions of life. Opposing the goals for efficiency of traditional design, speculative design aims to provoke debate and reflection.

Dunne and Raby (2013) stress that speculative artefacts stimulate critical reflection to rethink dominant narratives, offering future scenarios that reveal the potential consequences of current technological trends. This engagement with alternative possibilities and uncertainty counteracts the deterministic view. This conceptualisation aligns with the broader theoretical framework of speculative pragmatism (Wilkie et al. 2017), which also challenges linear assumptions about progress and explores futures as emergent and contingent, grounding speculation in the material conditions of an unfinished present. This aligns with DXT, which doesn't just present a digital future but immerses participants in an evolving experience in which the past, present, and future interact dynamically (Auger 2013). Instead of having a solely receptive, passive attitude, the DXT user co-creates and negotiates possible futures within the performance of the experience itself. Speculative pragmatism recognises that digital futures are co-emergent with the present (Tonkinwise 2015), thus opposing the idea that digital transformation is a teleological process. Together, speculative design and speculative pragmatism create a fertile space for questioning not only what is likely to happen but also what kinds of futures are desirable, ethical, or possible, based on current conditions.

5 Conclusion

The Digital Experience Theatre (DXT) ideation tool stands as an approach to tackle wicked problems through speculative design and collaborative classroom experiments. The proposed methodology engages students in participatory design processes inspired by Augusto Boal's Theatre of the Oppressed, particularly *the Image of the Rainbow of Desire* method, calling into question conventional human-centred perspectives in HCI, as well as leveraging critical reflection on the vocation of technology in creating inclusive and sustainable digital experiences. In this context, the concept of wicked problems is central, emphasising within the design thinking process the expansion of the designer's field of action to explore other territories and reconsider problems and solutions creatively.

Through a specific practice based on speculative co-creation with generative AI models, DXT allows students to conceptualise alternative futures that prioritise holistic sustainability and a sense of humanity. This collaborative process transcends the traditional user/designer dichotomy, highlighting the importance of distributed agencies in shaping design outcomes, where meaning, functionality, and creativity are generated by designers and users in an intersubjective effort. Thus, promoting dialogue, negotiation, and the exploration of diverse perspectives and sensibilities, as well as moving beyond linear problem-solving to embrace a more inclusive and humanity-centred practice, DXT facilitates the development of innovative solutions that address complex societal challenges.

The TO works as a mechanism to engage students in speculative interactions to boost participation, reflection, and the emergence of disruptive ideas, which is a dynamic that is facilitated with generative AI. The integration of dystopian scenarios and agonist approaches using text-to-image AI models enables the DXT to guide students to question established solutions to wicked problems and discover uncommon viewpoints. This use of AI tools amplifies the ideation process, despite challenges related to societal biases in AI-generated content, which are mitigated by using prompt expansion techniques to increase diversity and equity.

Ultimately, within collaborative classroom experiments, the DXT represents the transformative potential of speculative design in addressing wicked problems, in which students learn to understand design as an evolving dialogue where authorship is diluted in interchangeable roles, and the act of creation is about reflection, negotiation, and adaptation to complex systems.

References

Auger J (2013) Speculative design: crafting the speculation. Digit Creat 24(1):11–35
Barthes, R.: Camera Lucida: reflections on photography. Vintage, London (2020 [1979])
Benjamin, W., & Tarnowski, K.: Doctrine of the similar New Ger Crit, 17, 65 (1979[1933])
Bergson H (1926) Durée et Simultanéitée: À propos de la théorie d'Einstein. Libraire Félix Alcan, Paris
Bergson, H.: Matter and memory. Zone Books, Cambridge, Massachusetts (2002 [1896])
Bleecker J (2009) Design fiction: a short essay on design, science, fact, and fiction. Near Future Laboratory
Boal A (1994) The rainbow of desire: the Boal method of theatre and therapy. Routledge, London & New York
Buchanan R (2016) Wicked problems in design thinking. Des Issues 8(2):5–21
Candy S, Dunagan J (2017) Designing an experiential scenario: the people who vanished. Futures 86:136–153
Deleuze G (1968) Difference and repetition. Presses Universitaires de France, Paris
Disalvo C (2012) Adversarial design. The MIT Press, Massachusetts & London
Dorst K (2015) Frame innovation: create new thinking by design. MIT Press, Cambridge, Massachusetts
Dunne A, Raby F (2013) Speculative everything: design, fiction, and social dreaming. MIT Press, Cambridge, Massachusetts
Foucault, M.: The archaeology of knowledge. Pantheon Books, New York (1972 [1969])
Halewood M (2017) Situated speculation as a constraint on thought. In: Wilkie A, Savransky M, Rosengarten M (eds) The lure of possible futures. Routledge, London, pp 52–64
Manovich L (2023) Seven arguments about AI images and generative media. In: Arelli E, Manovich L (eds) Artificial aesthetics. http://manovich.net/index.php/projects/artificial-aesthetics
Manzini E (2015) Design, when everybody designs: an introduction to design for social innovation. MIT Press, Cambridge, Massachusetts
Mitchell W (1986) Iconology image text ideology. University of Chicago Press, Chicago
Mitchell W (1994) Picture theory: Essays on verbal and visual representation. University of Chicago Press, Chicago
Mouffe C (2009) The democratic paradox. Verso Books, London & New York
Naik R, Nushi B (2023) Social biases through the text-to-image generation lens. arXiv, Cornell University
Peixoto F (1980) O Teatro do Oprimido invade a Europa. Encontros com a Civilização Brasileira, vol III (19). Rio de Janeiro, pp 165-185
Rovelli C (2016) The order of time. Penguin Books, London
Savransky M (2017) The Wager of an unfinished present: notes on speculative pragmatism. In: Wilkie A, Savransky M, Rosengarten M (eds) The lure of possible futures. Routledge, London, pp 25–38
Schön DA (1983) The reflective practitioner: how professionals think in action. Temple Smith, London
Serres M, Latour B (1995) Conversations on science, culture, and time. University of Michigan Press, Ann Arbor, Michigan
Tonkinwise C (2015) Design fictions about critical design. Modes of Criticism 1
Wilkie A, Savransky M, Rosengarten M (eds) (2017) The lure of possible futures: speculative research and pragmatic inquiry. Routledge, London & New York

From Non-places to Places of Belonging: Design, Identity and Community in the Contemporary Urban Development

Cátia Rijo and Helena Grácio

Abstract This article explores the concept of "plural forms of design" as a strategic approach to addressing contemporary challenges of sustainability and social inclusion. It highlights the importance of co-creation in design, involving communities, experts and diverse agents in the construction of local identities and meeting spaces. It analyses initiatives such as the New European Bauhaus and UNESCO's Creative Cities Network, underlining the role of design as a catalyst for social change and sustainable urban development. Grounded in authors such as Augé, Bianchini, Florida, Hall and Lynch, the research examines how contemporary urban spaces affect the construction of identities and feelings of belonging. The study concludes that social responsibility in design education is fundamental for preparing professionals capable of acting as cultural agents and transformers in a globalised society that simultaneously values local identities.

Keywords Design · Plurality · Co-creation · Sustainability · Local identity · Social responsibility · Urban space

1 Introduction

Contemporary design faces a moment of redefinition in the face of complex global challenges, from environmental crises to growing social inequalities. In this context, there emerges the need to understand design not merely as an aesthetic or functional tool, but as a strategic discipline capable of articulating plural forms of imagining, learning and constructing more equitable and sustainable futures (Veneris 1990).

The vision of "plural forms of design" recognises that design practice should not be shaped by a single or linear vision, but rather by a collaborative approach that integrates diverse perspectives, cultures and knowledge systems. As Santos (2000)

C. Rijo (✉) · H. Grácio
iA* Arts Research Unit, University of Beira Interior, Covilhã, Portugal
e-mail: crijo@eselx.ipl.pt

Lisbon School of Education, Polytechnic University of Lisbon, Lisbon, Portugal

D. Raposo et al. (eds.), *Perspectives on Design IV*, Springer Series in Design and Innovation 59, https://doi.org/10.1007/978-3-032-00733-9_11

suggests, current globalisation should not be viewed as a homogeneous phenomenon, but as a process that can be reconfigured through a universal consciousness that values diversity and plurality.

This perspective aligns with Hall's ([1992] 2004) thinking, which argues that identities are not fixed, but rather constructed over time, based on historical, social and cultural factors. In his work "Cultural Identity in Post-Modernity", Hall explores how identities are deeply linked to territory and to cultural and social changes, especially in a globalised context. This understanding is crucial for contemporary design, which must recognise the fluidity and complexity of local identities in its creative processes and interventions.

In the urban context, design assumes a particularly relevant role. As Lynch ([1960] 1998) observes, the way in which city inhabitants perceive and interpret urban spaces, creating a "mental image" of the city, directly influences their feeling of belonging and identity. Lynch highlights five fundamental elements that structure these perceptions: paths, edges, districts, nodes and landmarks. The clarity and legibility of urban space, according to the author, are essential for the development of an emotional and symbolic connection between citizens and their territory.

This article proposes to analyse how design can function as a tool for co-creation, contributing to the development of local identities and meeting spaces, whilst simultaneously responding to contemporary challenges. To this end, we shall examine initiatives such as the New European Bauhaus and UNESCO's Creative Cities Network, and discuss the importance of social responsibility in design education, incorporating perspectives from various authors who study the relationship between space, identity and belonging.

2 Design as a Process of Co-creation

The concept of co-creation in design emphasises the importance of involving diverse actors—communities, experts from different disciplines and other agents—in the creative process. Rather than offering definitive answers, this approach aims to open spaces for dialogue and experimentation, where solutions emerge from the intersection of diverse perspectives (Manzini 2015; Thackara 2005).

This collaborative and dialogical approach to design is particularly relevant in the context of post-modernity, as Hall ([1992] 2004) suggests, in which identities have become more fragmented and fluid. According to the author, cultural identities in contemporaneity are continuously formed and transformed in relation to the ways in which we are represented or addressed in the cultural systems that surround us. In a globalised world, this fragmentation of cultural identities can be seen both as a challenge and as an opportunity for design, which must work with this complexity to create meaningful and inclusive solutions.

Brito (2005) complements this vision by analysing the relationship between the city and social identity, with emphasis on urban mobility and the transformations of contemporary urban spaces. The author examines how changes in territory, resulting from urbanisation and city growth, affect the construction of local identities. These constant transformations of urban spaces and social dynamics require design to be flexible and adaptable, capable of responding to the emerging needs of communities.

An exemplary initiative of this approach is the New European Bauhaus, which draws inspiration from the original Bauhaus school to address contemporary challenges such as sustainability and social inclusion. According to the President of the European Commission, Ursula von der Leyen, this project represents "a hope designed to explore how we live better together after the pandemic", combining "sustainability with style, bringing the European Green Deal closer to people's minds and homes".

The New European Bauhaus establishes a "bridge between the world of art and culture, on one hand, and the world of science and technology, on the other", ensuring "the participation of society as a whole: artists, students, architects, engineers, academics and innovators", as stated by Mariya Gabriel, Commissioner for Innovation. This interdisciplinary and participatory approach aims to trigger systemic change, creating spaces for dialogue to imagine new ways of living that are sustainable, inclusive and aesthetically enriching.

Florian (2002) highlights how urban transformations influence local identity and how changes in land use, infrastructure and urban policies can alter the relationship of inhabitants with their territory. In this sense, initiatives such as the New European Bauhaus represent an attempt to preserve historical and cultural aspects of urban spaces, even in transformation processes, allowing the population to maintain an affective and symbolic bond with the place. This preservation, as Florian argues, strengthens belonging and local identity, fundamental elements for the social and cultural sustainability of cities.

3 Design and Sustainable Urban Development

Urban areas today constitute the main focal points for developing strategies, policies and initiatives that make culture and creativity engines of sustainable development and urban regeneration. As UNESCO recognises, these approaches stimulate growth and innovation, promoting social cohesion, citizens' well-being and intercultural dialogue (van Leeuwen et al. 2004).

The "place"—whether city, territory or urban space—is defined by the fusion of its diverse characteristics: heritage, culture, religion, urban planning and history. As Augé ([1992] 2005) observes, the anthropological place is "simultaneously a principle of meaning for those who inhabit it and a principle of intelligibility for those who observe it", constituting an identity-based, relational and historical space. However, Augé also develops the concept of "non-places", referring to contemporary urban spaces (such as airports, shopping centres and underground stations) which, despite being extremely frequent in daily life, do not possess a distinctive cultural or historical identity. These spaces, according to the author, are characterised by the absence of a shared history and a common identity that inhabitants can retrieve, contributing to people's disconnection from these places.

This tension between "places" and "non-places" represents a significant challenge for contemporary design, which must seek strategies to humanise and attribute meaning to urban spaces, even those that tend towards homogenisation and anonymity. Augé also explores how globalisation and consumer culture transform cities and territories, leading to a uniformisation of urban spaces and the emptying of local meanings. This phenomenon can be countered by design that values the cultural and historical specificities of each locality, contributing to the construction and preservation of authentic and meaningful local identities.

Bianchini (1999), in his studies on cities and cultural identity, analyses how modern cities are spaces of strong symbolic and cultural construction. The author highlights the importance of local history and memory in the construction of urban identity, arguing that urban regeneration projects can be more effective when they incorporate and value local cultural identities, creating a sense of belonging and social cohesion amongst inhabitants. Bianchini also explores how cities should be understood as dynamic processes, in which social and cultural relations interweave with urban transformations. This perspective reinforces the idea that design for urban spaces must be adaptive and sensitive to local sociocultural dynamics, promoting interventions that respect and potentialise existing identities.

In the context of globalisation, local diversity acquires special meaning. One observes a diffusion of local culture at a global level, which implies that local specificities—unique and distinctive—are transmitted to third parties, promoting collective recognition of them. The preservation of places and their material and immaterial culture thus becomes a crucial aspect for the formation of feelings of local belonging.

We witness the proliferation of places with transnational identity, adjusting to the description proposed by Bianchini (1999), whilst various authors highlight the need to construct innovative local identities, considering essential the attraction of the creative class (Florida 2002). Florida argues that the success of modern cities depends on their capacity to attract highly qualified individuals, and this attraction is intimately linked to the quality of urban spaces and these cities' capacity to promote a vibrant and inclusive sense of identity. The author highlights that fostering creativity and cultural diversity can strengthen local identity and belonging, in addition to making the city a more dynamic and innovative place. This perspective is fundamental for contemporary urban design, which must consider not only the physical and functional aspects of spaces, but also their capacity to attract and retain talent, promoting creative and culturally stimulating environments.

Creativity, when combined with territorial development, manifests itself not only in individuals' intellectual aptitudes, but also in social relations and the specific territorial context (Robison et al. 2002). As Brito (2005) suggests, changes in territory, due to urbanisation and city growth, profoundly affect the construction of local identities. The author also highlights the impact of migrations and social mobility on how inhabitants identify with urban spaces, exploring how modernity and globalisation redefine the perception of urban space, creating both new opportunities for belonging and challenges for local identities. This duality requires design to adopt a sensitive and integrative approach, capable of reconciling tensions between the global and the local, the traditional and the innovative, the permanent and the transitory.

4 Design as Creator of Meeting Spaces and Local Identities

Design plays a fundamental role in creating meeting spaces that strengthen local identities and promote feelings of belonging. These spaces function as points of convergence for different cultures, traditions and histories, allowing communities to celebrate their singularities whilst finding common solutions to urban challenges.

Lynch ([1960] 1998), one of the most influential scholars of the relationship between the city and human perception, offers a valuable perspective on how urban design can influence the way inhabitants experience and relate to city spaces. By analysing how people perceive and interpret urban spaces, creating a "mental image" of the city, Lynch identified five fundamental elements: paths (routes along which people move), edges (limits and barriers), districts (areas with identifiable characteristics), nodes (focal points or junctions) and landmarks (external references). These elements, when well articulated by design, contribute to the legibility of urban space, facilitating orientation and, consequently, inhabitants' feelings of security and belonging.

The revitalisation of urban spaces through design transforms them into places that narrate local stories and reflect communities' aspirations. Architecture, urban planning and public art are powerful tools for reinforcing the connection between inhabitants and their territories, as well-conceived spaces foster dialogue, cultural exchange and collaborative creativity.

Florian (2002) examines how urban transformations influence local identity, focusing on how changes in land use, infrastructure and urban policies can alter inhabitants' relationship with their territory and, consequently, their identity. The author argues that, although cities evolve, it is possible to preserve historical and cultural aspects that allow the population to maintain an affective and symbolic bond with the place, which strengthens belonging and local identity. This perspective is crucial for contemporary design, which must balance innovation and transformation with preservation and valorisation of historical and cultural heritage.

As Costa (2003) states, "identity can be the same physical presence, or the evocation in another's mind, of something that exists outside them". Thus, the identity of place is what is shown, a representation that makes the subjective and complex visible, adapting to the social, cultural, economic and technological transformations experienced by citizens.

Hall ([1992] 2004) complements this vision by discussing how identities are constructed in relation to an "other", through a process of differentiation that establishes symbolic boundaries. In the context of urban spaces, this means that the identity of a place is defined not only by its intrinsic characteristics, but also by its relationship with other places and by the way it is perceived and experienced by its inhabitants and visitors. Design, in this sense, must consider both physical and material elements and the symbolic and relational aspects of spaces, promoting interventions that respect and potentialise these complex identity dynamics.

Brito (2005) explores the impact of urban mobility and migrations on the formation of local identities, highlighting how constant transformations of urban spaces and population flows create a scenario of fluid and multiple identities. This perspective challenges design to create flexible and adaptable spaces, capable of accommodating diverse cultural expressions and fostering intercultural dialogue, thus promoting an inclusive and comprehensive sense of belonging.

The designer, as an active cultural agent in the construction of meanings, has the responsibility to transmit local identity to the community. As Victor Margolin clarifies, "the designer's role is value creation, including social, cultural, environmental, political and symbolic values". This responsibility implies constant evaluation of their intervention as an agent of change.

5 Social Responsibility in Design Education

Design education plays a crucial role in preparing professionals committed to social responsibility. Faced with challenges such as climate change, social inequalities and accelerated urbanisation, educational programmes in design must transcend aesthetics and functionality, incorporating principles of sustainability, inclusion and ethics.

Dreeben (1976) highlights the implicit values in social and educational processes, a perspective that inspires us to view design as an instrument of cultural and social transformation. Florida (2002) argues that creativity is an essential resource for socioeconomic development, and design education has the potential to channel this creativity to promote innovations that benefit local and global communities. Florida's "creative class" theory offers valuable insights for design education, suggesting that creative professionals play a fundamental role in creating innovation and in cities' economic development. This perspective encourages educational programmes to cultivate not only students' technical and aesthetic skills, but also their capacity to contribute to the economic and cultural vitality of the communities in which they are embedded.

Action research, advocated by Serrano (2004) and Tripp (2005), is fundamental in design educational processes, encouraging students to work collaboratively and engage with real contexts. This approach strengthens the connection between theory and practice, promoting deeper and more transformative learning (Yee and White 2016).

Lawson (2011) demonstrates that designers' thought processes are unique and interdisciplinary, and that education can foster systemic thinking, enabling students to manage the intersection of social, cultural, environmental and economic aspects in their projects. To this end, institutions must promote connections between academia and local communities, fostering co-creation work.

This approach aligns with Hall's ([1992] 2004) vision of the fluidity and complexity of cultural identities in post-modernity. Design education must prepare students to understand and work with these multiple and constantly transforming identities, developing solutions that respect and value cultural diversity. As Hall suggests, identities are constructed through discursive and representational practices, and designers, as creators of visual and material representations, have significant responsibility in how identities are articulated and expressed.

Lynch ([1960] 1998) offers a complementary perspective, highlighting the importance of legibility and coherence in urban spaces for inhabitants' well-being and identification. Educational programmes in design must incorporate these principles, preparing students to create spaces that are not only aesthetically pleasing and functional, but also meaningful and orienting, contributing to the development of a sense of place and belonging.

Bianchini (1999), by analysing how urban regeneration projects can incorporate and value local cultural identities, offers valuable perspectives for design education. Students should be encouraged to adopt a context-sensitive approach that considers local history, memory and traditions in their projects, promoting interventions that strengthen, rather than dilute, existing cultural identities.

Brito's (2005) work on the relationship between city, identity and mobility suggests that future designers must be prepared to deal with the complexity of contemporary urban dynamics, characterised by constant flows of people, information and goods. Design education must, therefore, develop students' capacity to create flexible and adaptable solutions that respond to the needs of diverse and constantly changing populations.

Florian (2002) reinforces the importance of preserving historical and cultural aspects of urban spaces, even in transformation processes, to maintain the population's affective and symbolic bond with the place. This perspective must be integrated into design educational programmes, encouraging students to balance innovation with preservation, change with continuity, creating interventions that respect the past whilst projecting the future.

Finally, but no less importantly, Augé's ([1992] 2005) perspective on the "non-places" of supermodernity challenges design education to prepare professionals capable of humanising and attributing meaning to contemporary urban spaces, even those that tend towards homogenisation and anonymity. Students should be encouraged to develop strategies that counter this tendency, creating spaces that promote identity, relationship and history, elements that, according to Augé, define an anthropological place.

In summary, social responsibility in design education implies preparing context-sensitive professionals, committed to ethics and sustainability, and capable of acting as cultural agents and transformers. This commitment must be reflected in curricula, teaching methodologies and research practices, training designers who recognise their role in constructing a more equitable, sustainable and culturally diverse future.

6 Conclusion: A Collective Path to the Future

Thinking, learning and practising design in contemporaneity implies reimagining practices and paradigms. The New European Bauhaus and UNESCO's Creative Cities Network demonstrate that it is possible to unite art, science, technology and inclusion to transform challenges into opportunities. It falls to designers and educators to collaborate plurally and openly to create a future in which all can prosper.

This research has made it possible to highlight the importance of design as a strategic tool for confronting complex contemporary global challenges, from environmental crises to social inequalities. The approach of "plural forms of design" proposes a collaborative and interdisciplinary vision that recognises the importance of diversity of perspectives, cultures and knowledge systems in the creative process.

The theoretical contributions of authors such as Augé ([1992] 2005), Bianchini (1999), Brito (2005), Florida (2002), Hall ([1992] 2004), Lynch ([1960] 1998) and Florian (2002) have provided a rich conceptual framework for understanding the relationships between design, urban space, identity and belonging. The tension between "places" and "non-places", the importance of urban space legibility, the impact of territorial transformations on local identities and the role of the creative class in urban development are fundamental concepts that should guide practice and education in design.

Initiatives such as the New European Bauhaus illustrate how design can act at the intersection between art, culture, science and technology, promoting solutions that are sustainable, inclusive and aesthetically enriching. Similarly, UNESCO's Creative Cities Network demonstrates the importance of creativity and cultural diversity as engines of sustainable urban development.

The designer's role as cultural agent and transformer has been extensively discussed, highlighting their responsibility in transmitting local identities and creating meeting spaces that promote dialogue, cultural exchange and collaborative creativity. As Margolin states, value creation—social, cultural, environmental, political and symbolic—is central to the designer's role, implying constant evaluation of their intervention as an agent of change.

Design education assumes, in this context, crucial importance, preparing context-sensitive professionals committed to ethics and sustainability, and capable of acting as cultural agents and transformers. Educational programmes must transcend aesthetics and functionality, incorporating principles of social responsibility and promoting connections between academia and local communities.

Through the creation of meeting spaces that project local identities, design can serve as a connecting thread between past, present and future. As Godin (2021) states, "attitudes are competencies, which is good news for all of us, because it means that, if we care enough, we can learn".

This is, ultimately, a call to action for designers, artists, scientists and communities to articulate themselves in a collective effort, recognising that the desirable future depends on our capacity to imagine and build together. Plurality in design is not merely a theoretical concept, but an invitation to co-creation and a collective future, more sustainable, inclusive and culturally diverse.

References

Augé M ([1992] 2005) Não-lugares: Introdução a uma Antropologia da Sobremodernidade, 2nd edn. Letra Livre, Lisboa

Bianchini F (1999) Cultural planning for urban sustainability. In: Nystrom L (ed) City and culture: cultural processes and urban sustainability. The Swedish Urban Environment Council, Karlskrona

Brito JP (2005) Cidade, Identidade e Mobilidade. Edições Universitárias, Lisboa

Costa J (2003) Imagen Corporativa en el Siglo XXI. La Crujía, Buenos Aires

Dreeben R (1976) The unwritten curriculum and its relation to values. J Curric Stud 8(2):111–124

Florian B (2002) Cidades e Transformações: Dinâmicas Urbanas e Identidade Local. Edições Afrontamento, Porto

Florida R (2002) A Ascensão da Classe Criativa: E como ela está a transformar o Trabalho, o Lazer, a Comunidade e o Quotidiano. Editora Elsevier, Rio de Janeiro

Godin S (2021) The practice: shipping creative work. Portfolio

Hall S ([1992] 2004) A Identidade cultural na Pós-Modernidade. DP&A, Rio de Janeiro

Lawson B (2011) How designers think: the design process demystified. Architectural Press, Oxford

Lynch K ([1960] 1998) A Imagem da Cidade. Martins Fontes, São Paulo

Manzini E (2015) Design, when everybody designs: an introduction to design for social innovation. MIT Press, Cambridge

Robison L, Lindon S et al (2002) Is social capital really capital? Rev Soc Econ 6(1):1–24

Santos BS (2000) A Crítica da Razão Indolente: Contra o Desperdício da Experiência. Edições Afrontamento, Porto

Serrano GP (2004) Investigación Cualitativa. Retos e Interrogantes. La Muralla, Madrid

Thackara J (2005) In the bubble: designing in a complex world. MIT Press

Tripp D (2005) Action research: a methodological introduction. Educ Pesqui 31(3):443–466

van Leeuwen J, van Gassel F, den Otter A (2004) Teaching collaborative design. In: Proceedings of the international workshop on construction information technology in education, Istanbul, Turkey, pp 1–13

Veneris Y (1990) Modeling the transition from the industrial to the informational revolution. Environ Plan 22(3):399–416

Yee J, White H (2016) The Goldilocks conundrum: the 'just right' conditions for design to achieve impact in public and third sector projects. Int J Des 10(1):7–19

Museum of Brands: An Exploratory Reading of the Collection Through Design

Marlene Ribeiro and Paula Azevedo Rodrigues

Abstract This chapter presents an exploratory analysis of the Museum of Brands Collection (Caramulo, Portugal), focusing on the intersection between design and memory. The twentieth century Portuguese consumer objects featured in the collection are examined not only as historical records of social life but also as communicative artefacts of notable graphic and cultural significance. Drawing on an object-based critical approach, the chapter investigates the formal and symbolic dimensions of a selected sample of graphic communication artefacts from the Sacor, Nestlé, and Regina brands. The analysis is grounded in theoretical frameworks from Material Culture and Visual Culture studies, offering insights into the Design Culture that the Museum of Brands embodies and can further promote. This contribution highlights the importance of this unique project within both national and international museographic contexts. It also reinforces the institutional collaboration between the Museum of Brands and the Viseu School of Education, while identifying opportunities for future research.

Keywords Museum of Brands · Brands of memory · Design culture

1 Introduction

Our relationship with brands goes far beyond the economic imperative of consumption; it is emotional, social and identitarian. In a world full of graphic signs, where brands appear and disappear in large numbers and at a rapid pace, those that become iconic, that enter our imagination and persist, help us to build a common memory that

M. Ribeiro (✉)
ID+ Research Institute for Design, Media and Culture, University of Aveiro, Aveiro, Portugal
e-mail: marlenefribeiro@ua.pt

M. Ribeiro · P. A. Rodrigues
School of Education, Polytechnic Institute of Viseu, Viseu, Portugal

P. A. Rodrigues
CI&DEI Center for Studies in Education and Innovation, Viseu, Portugal

D. Raposo et al. (eds.), *Perspectives on Design IV*, Springer Series in Design and Innovation 59, https://doi.org/10.1007/978-3-032-00733-9_12

we cling to as a guarantee of belonging (Lopes 2024). Considering that design allows ideas to be communicationally visible, brand design presents itself in this context as an important cultural mediator (Ribeiro and Providência 2022; Ribeiro 2021). In this way, brands present themselves as mechanisms for re-signifying everyday life and allow us to establish new meanings and individual and collective experiences of desire, credibility and well-being, whether through nostalgia or shared values (Herrera and Iñurritegui 2018).

The collection at the Museu das Marcas, Artes Gráficas e Publicidade/Museu das Marcas (Museum of Brands, Graphic Arts and Advertising/Museum of Brands) brings together an important and distinctive collection highlighting the brands that marked the twentieth century in Portugal through consumer objects and graphic artifacts that rarely become collectible museum pieces.

The Museu das Marcas, which is only digital for now, features a rich collection for design research, appeals to community participation, offers varied experiences for education, enjoyment, reflection and knowledge sharing, in line with the contemporary understanding of museums (ICOM International Council of Museums 2022).

In this chapter we propose an approach centered on the relationship between design and memory, starting by contextualizing the Museu das Marcas and its associated collections. We present communication strategies, critically discussing, through objects from three brands in the collection (Sacor, Nestlé and Regina), formal aspects of design and the interest that these artifacts currently raise, namely the overlapping of symbolic value with functional value. We also mention the example of the Museum of Brands (London, UK), which we briefly discuss.

Considering the recontextualization of objects from the collection through the Museu das Marcas, we explored the concept of design culture against the backdrop of the assumptions of the expanded museum (Bautista and Balsamo 2016).

Through an exploratory research approach motivated by the desire to contribute critically to an institution under construction and recognizing the graphic interest of the Museu das Marcas collection, we propose in this contribution, an inaugural reading through Design with the aim of (1) disseminating the collection and mapping the ecosystem of graphic production and brand communication through consumer products in the twentieth century in Portugal; (2) explore the interaction between design and collective memory; (3) analyze the contemporary significance of brand artefacts, questioning the convergence of symbolic and functional values of the objects in the collection; (4) critically examine the role of the Museu das Marcas as a fundamental contribution to the field of material and visual culture.

In line with the objectives of the Museu das Marcas, the aim is also to promote the museum and its collection within the framework of the institutional cooperation protocol established between the Museu das Marcas and the Escola Superior de Educação de Viseu (Viseu School of Education).

2 The Museum of Brands, Caramulo, Portugal

2.1 Context

Located in Caramulo (Viseu, Portugal), a picturesque mountain village renowned for its natural beauty and historical heritage, the Museu das Marcas, Artes Gráficas e Publicidade positions itself as an institution dedicated to preserving and promoting the material memory of Portuguese or international consumer brands in Portugal, with a special focus on the twentieth century.

Although the Museu das Marcas has a founding link with the Museu do Caramulo (Museum of Caramulo)—a national cultural institution respected and recognised for its eclectic art collection (ancient, modern and contemporary) and its rare collection of classic cars—it operates as an independent museum with its own mission and identity.

Presenting itself as a place of memory, the Museu das Marcas was born in the digital environment with the launch of its official website: https://museudasmarcas.pt, which offers public access to a significant part of the collection and curatorial initiatives. The museum also has an active presence on social media platforms such as Instagram, Facebook and YouTube, interacting with a wide and diverse audience and promoting critical reflection on the cultural, economic and social dimensions of branding.

In the near future, the collection will be housed in its own physical space, further consolidating its role as a museological and educational institution. It aims to establish itself as a reference point for the study, conservation and dissemination of the history of brands in Portugal, making a significant contribution to contemporary museology and the safeguarding of intangible cultural heritage.

2.2 Associated Collections

The Museu das Marcas has its origins in the collection of Salvador Patrício Gouveia, its main driving force. Raised in a family of passionate collectors (of art, cars, toys, motorbikes and bicycles), he said in a conversation to prepare this article, "I've always been a collector". This spirit, together with a particular interest in graphic communication, design and advertising, created the perfect ecosystem for Salvador Patrício Gouveia to recognise at an early age the graphic interest of ordinary objects as a means of brand communication.

Created as a space for sharing, the Museu das Marcas searched for similar collections in Portugal to map the brands and consumer objects of the twentieth century that are an important part of the national collective memory. As a result of the contacts and research carried out, the A Primorosa Collection, the Vida Portuguesa Collection, the Frederico Sanches da Cunha Collection, the Manuel Paula Collection, the Museu das Marcas Collection, the Museu do Caramulo Collection and the Salvador

Patrício Gouveia Collection are currently the associated collections of this project. The Museu das Marcas collection has been enriched by donations from individuals and companies from different sectors, which have been identified and catalogued.

2.3 *Featured in the Collections*

The Museu das Marcas Collection is an open project that invites the public (individuals, companies, private or public organisations) to donate packaging, products or advertising material from Portuguese or international consumer brands in Portugal, with a particular focus on the twentieth century. The ambition to open its doors in a physical space has aroused the interest of the public, who can currently follow the collection via digital media (Museu das Marcas 2025a, b, c, d), where the project has been very well received.

For now, around a thousand pieces have been catalogued, photographed and digitised, which is still a modest percentage of what the collection must show. As a recent project (since 2022), it has only reached a limited number of people, so it is expected to grow. The ability to mobilise the public, starting with the donation of artefacts, is essential to establish the collection base for the future curatorial treatment of the digital and physical museum project.

The donation of the Litografia de Portugal (Lithography of Portugal) to the Museu das Marcas is one of these important moments, confirming and reinforcing the graphic relevance (in terms of design, authors and printers) of the collection gathered there, which stands out on the national and international scene in the field of consumer brands.

As part of the dissemination strategy, and taking advantage of the link with the Museu do Caramulo, the partnership established with Google Arts & Culture (Museu do Caramulo 2025) has allowed the Museu das Marcas to take high quality photographs of artefacts from its collection, and to be digitally present and shared in a space that brings together more than 2000 cultural institutions from 80 countries, including the most prestigious art museums in the world.

Given the initial processing of the Museum of Brands collection, the selection and brief reading of the design elements, representational arguments and nostalgic activation of memory through brands that we bring to this document serves the exploratory purpose of publicising the project through the example of the diversity of artefacts available (packaging, labels, claims, posters, stickers, gifts). At this stage, Sacor, Nestlé and Regina have been identified by director Salvador Patrício Gouveia as the most relevant brands (among those already catalogued).

Sacor

Sacor was founded in 1938 and operated in the oil sector in Portugal, benefiting from the protectionist policies of the "Estado Novo" (Portuguese dictatorship), which guaranteed it 50% of the national market share, and by the end of the 1960s it had a network of around 800 service stations. The political and economic changes that followed 25 April 1974 (the date of the revolution that ended the dictatorship in Portugal) led to the nationalisation of the company, which in 1976 became Petrogal—Petróleos de Portugal EP (Museu das Marcas 2025e).

The strategy of promoting the brand through advertising campaigns, lighting, packaging and gifts made it part of the national collective imagination. From the collection of the Museu das Marcas, Sacor provides an obvious link to the important car collection of the Museu do Caramulo.

The poster promoting the Sacor brand (Fig. 1) uses the geometric representation of shapes with volumetry through chromatic variation within the same tone and explores the idea of a journey across Portugal (through the built and natural heritage that stretches from the sea to the mountains), including a diverse fleet on a well-defined and safe path formed by the initial S. The poster promoting an event (Fig. 2) is based on symbolic elements of the city of Porto, in a symbiotic evocation of tradition and innovation (the Arrábida Bridge, inaugurated in the year of the event, 1963, with the largest arch in reinforced concrete, represents construction innovation; and the Clérigos Tower, an inescapable landmark in the city's history and the centrepiece of the composition, which allows a double reading of Oil Platform/Clérigos Tower due to the formal similarity and gestural graphic treatment). The advertisement associated with the strategic brand TAP Transportes Aéreos Portugueses (Fig. 3) uses a watercolour drawing to illustrate the campaign's headline and reinforces Sacor's proximity communication strategy by using the boarding pass provided to each passenger on TAP's Caravela VI-R jet as a support. Finally, the traditional Sacor engine oil can (Fig. 4) has its various sides filled with graphic elements derived from the visual identity, but with typographical variations. This set is varied in terms of supports and graphics, but the colours green and red (the predominant colours of the national flag) are constant elements of communication.

Nestlé

Over 150 years old and present in 188 countries, the Swiss global brand Nestlé (registered in 1868) has been a part of our lives for as long as we can remember. Today it has a very diverse range of products and associated brands, but it was the invention of dairy milk by the pharmacist Henri Nestlé that began the story of the brand that changed infant nutrition around the world (Nestlé 2025). The innovation of this food product and the strategy of product diversification (initially the production of chocolate and condensed milk) led to the rapid conquest of international markets. It was in the 1920s that the brand was introduced in Portugal and began production through the Sociedade de Produtos Lácteos in Avanca, Aveiro (Museu das Marcas 2025f).

Among the brand's advertising campaigns, the Nestlé Baby Contest (since 1935) had a major impact in Portugal, as in other countries, in terms of brand recognition and public loyalty (Nestlé Portugal 2023). Between the 1950s and 1960s, Nestlé launched iconic brands such as Nestum, Cerelac, Nescafé and Nesquik on the Portuguese market; in the following decades, Mokambo, Pensal, Estrelitas, Chocapic and, more recently, Nespresso, Nescafé Dolce Gusto, in a much wider portfolio that continues to grow and diversify, particularly through animal feed and specialised nutrition.

Fig. 1 Poster Sacor, 1950, original painting donated by Carlos Besse Peres (2011), Museu do Caramulo collection

Fig. 2 Poster II Congresso Sacor, 1963, Museu do Caramulo collection

Fig. 3 TAP ticket with SACOR advertising, 1963, Salvador Patrício Gouveia collection/Museu das Marcas

Fig. 4 Sacor engine oil can, Salvador Patrício Gouveia collection/Museu das Marcas

From the Museu das Marcas collection, we selected the Nestrovit Nestlé information booklet (Fig. 5), whose cover shows an illustration with an elegant layout, in soft tones and depicting a scene of maternal connection in a natural floral setting. The Milky Flour packaging (Fig. 6), printed in two colours, features the Nestlé graphic brand in the centre and, in addition to the product description, communicates the vitamin composition and instructions on how to use it in a very direct way, graphically reinforcing the message of instant preparation through the lines of movement typical of cartoons applied to the typography. The use of classic cartoon characters, or characters of their own, gave the brand a strong emotional connection with the children's audience. In the school calendar (Fig. 7), the brand uses the Cangurik scout to communicate an empathetic message of environmental awareness. Aimed at the young-adult audience, the Nescafé Nestlé label (Fig. 8) has the usual structure with text describing the product, how to use it and where it was produced, with the brand highlighted in yellow, red and black colours with gold accents as a way of positioning Nescafé Nestlé in the high quality/luxury product segment, even though it corresponds to a mass consumer product.

Fig. 5 Nestlé Nestrovit (food supplement) booklet, 1954, illustration by Pizzi/Goni, printed by Litografia de Portugal, Salvador Patrício Gouveia collection/Museu das Marcas

Fig. 6 Nestle milk flour packet, prepared at the Sociedade de Produtos Lácteos, Avanca, Portugal, 1980s, Salvador Patrício Gouveia collection/Museu das Marcas

Regina

The almost century-old Regina, founded in 1927, is one of the Portuguese's favourite chocolate brands. The brand's communication strategy is based on simplicity and proximity, emphasising its national origin through headlines such as 'a pleasure proudly ours'. Some of its products, such as milk, hazelnut or fruit-flavoured chocolate bars, chocolate umbrellas, Easter almonds or the hole machine, have become iconic and part of the Portuguese collective memory. Regina, which even ceased production, was acquired and relaunched at the turn of the millennium by Imperial

Fig. 7 Nestlé Nesquik school timetable with Cangurik mascot, 1986 European Year of the Environment, Salvador Patrício Gouveia collection/Museu das Marcas

Fig. 8 Nescafé decaffeinated instant coffee packet, 1960s, Salvador Patrício Gouveia collection/Museu das Marcas

(owner of other emblematic national brands in the same sector such as Pintarolas, Jubileu and Pantagruel), which was recently acquired by the Spanish group Valor (Imperial 2025).

The Regina chocolate box (Fig. 9) explores the randomness of gambling as a way of interacting with the public for a moment, often involving different generations of consumers. This machine was a common sight in shops and fairs. Each hole dropped a coloured ball that identified the prize (a Regina product). It was a skilful way of presenting a constantly updated product range, while at the same time acting as an instrument to encourage repeat purchases. A simplified version of this chocolate box can still be found in shops such as A Vida Portuguesa. In 1963, for example, one of the products that could be obtained from this game was the milk chocolate bar (Fig. 10), the wrapper of which identified the white colour of the corresponding ball. In the spirit of making shopping a fun experience, the Regina brand created numerous numbered collections for the chocolate bars (with typical Portuguese regional figures—Maria Helena mascot, Portuguese tourist destinations, models of cars, models of aeroplanes, Walt Disney cartoons, etc.), which stimulated the public's collecting impulse. The tins of cayenne, chocolates and drops are also an important means of promoting the brand, with graphics that vary between visual narratives using a clean design (Fig. 11) or a geometric typographic arrangement, in contrast to the cursive typography that characterises the visual identity (Fig. 12) of the Regina brand.

Fig. 9 Regina punch box, chocolates, dropes, bonbons, 1940s, Salvador Patrício Gouveia collection/Museu das Marcas

Fig. 10 Regina milk chocolate wrapper from the Walt Disney collection no. 184/63 (Pluto), white ball from the Regina punch box, 1963, printed by Litografia Vitória, Salvador Patrício Gouveia collection/Museu das Marcas

Fig. 11 Regina cayenne chocolates can, 1030s, printed by Litografia Vª Ferrão, Salvador Patrício Gouveia collection/ Museu das Marcas

Fig. 12 Regina fruit sweets packet, 1030s, Salvador Patrício Gouveia collection/Museu das Marcas

2.4 *The Example of the Museum of Brands (London, UK)*

Consumer historian Robert Opie established the Museum of Brands in 1984, following a successful exhibition in 1975 at the Victoria and Albert Museum, called The Pack Age (Museum of Brands 2025a). The museum offers a nostalgic journey into British history through the consumer brands featured in the permanent exhibition 'Time Tunnel'. This exhibition features more than 12000 original objects, including packaging, toys, posters and household products, which offers a fascinating insight into life in the United Kingdom since Victorian times. In addition to the permanent exhibition, the museum hosts temporary exhibitions that look at the public's relationship with brands today, such as the annual exhibition of the most popular brands and products in British supermarkets. This is also a space open to the debate of ideas and research with the aim of educating the public about design particularly the themes of advertising and packaging. In the field of research, we highlight the 'Living Brands' project, a multi-sensory outreach programmed for people living with dementia and their careers designed to encourage reminiscing with the museum collection artefacts.

As a way of promoting its activity and anticipating the experience in the physical space, the museum is present on social media and offers interaction with the public on its website through activities such as the Brand Challenge Quiz. The immersive virtual tour and access to the collection is also available on the Google Arts & Culture platform (Museum of Brands 2025b).

The London Museum of Brands and the Caramulo Museu das Marcas share the aim of preserving, reflecting and celebrating the history of brands and consumption, using everyday objects to tell broader stories about social, cultural and technological change.

3 Design Culture

The activity of design and its research must be understood beyond the technical and aesthetic domain, as it has a significant cultural dimension that this discussion seeks to highlight. Fallan (2010) proposes that the cultural history of design should integrate interdisciplinary perspectives that include studies in the field of material culture, bringing Design closer to Anthropology and Sociology and, considering the analysis and interpretation of artifacts, also bringing it closer to the History of Art. In this sense, Material Culture Studies highlight the complex interactions between artefacts and human beings, emphasizing "the use and meanings of objects" (Woodward 2007, p. 4). It also helps to overcome dichotomies such as art/design, erudite/popular, allowing for an inclusive analysis that considers both established design icons and anonymous everyday objects (Attfield 2000).

Reflecting on the relationship between Design and Culture, Margolin (1989) proposes that Design Studies should include its observation as a cultural expression—design in culture—and as a cultural agent—design as culture. The former (design in culture) assumes that design objects and systems reflect values, traditions, social needs and historical contexts, meaning that design arises from and is shaped by the culture in which it is produced. The second (design as culture) considers that design products can become cultural symbols, influence lifestyles or even create new social practices, thus shaping our behaviors, habits and perceptions, meaning that in addition to reflecting culture, design presents itself as culture and contributes to its construction.

It is also important to decode how we perceive these objects (three-dimensional, two-dimensional, physical and digital) and design systems. Berger (2008) reminds us that looking is an act of choice and, in this sense, when we see, we are not just looking, we are reading the language of images (and objects). But in a world saturated with images, Mirzoeff (2015, 2023) invites us to think about how we see, what we see, and how the act of seeing is linked to power, politics and identity, arguing that all visuality is constructed, meaning that the way we see the world is influenced by cultural values, ideologies and power structures. It is therefore important to be alert because Visual Culture can also be a tool of resistance and design can be critical, active and socially responsible, remaining a political act, never neutral.

Regarding the layers of reading, metaphors, cultural values and relationships we establish with objects, Appadurai (2006) argues that an object is never just a form, but a symbolic interface with multiple meanings that depend on the cultural context and the way they are seen. The author explores that materiality is produced culturally, so objects don't have a fixed meaning, they become meaningful through the social practices and narratives that surround them, so it's not the thing itself that matters, but what it becomes in interaction with humans.

From the contributions of Margolin, Mirzoeff and Appadurai we can conclude that design actively participates in the construction of culture, influencing the way we see, feel and understand the world, and therefore objects, namely those found in the collection of the Museum of Brands, cannot be studied only in terms of visual characteristics and qualities, but interdisciplinary perspectives from Cultural Studies, Design Studies, Image Studies, Anthropology and Visual Culture must be considered for a richer and more complete observation.

Regarding the collection under study, nostalgia is a good term for its presentation. We're talking about a "search for spontaneous and unpretentious voices - voices that belong to the idealized aura of a romanticized past (the crazy 1920s, the exuberant 1950s). Nostalgia, a key ingredient in increasing the market value of a vernacular style, is not a return to history, but a repackaging of history. It treats the past not as the roots of the present, but as a distant other" (Lupton and Miller 1999, p. 157) and, at the same time, as a memory of something we miss and attracts us as a testimony to a perfect time.

Regarding memory brands, considering the Portuguese context, we can refer to the doctoral research by Pedro Carvalho de Almeida about Sanjo (since 1933) or by Nuno Coelho about Saboaria e Perfumaria Confiança (since 1894).

Memory and nostalgia motivate a transition from valuing the functional domain of objects to their symbolic valuation, or Brand valuation. The idea of the Brand as a territory of belonging, based on the human need to establish connections, is highlighted by Millman 2013) who draws on various authors: Ollins states that a tribe is a brand, a religion is a brand; Postrel uses terms such as promise, consistency, continuity, associating the value of the Brand more with its meaning (the symbolic value) than with the attributes of a product or product line; Seth Godin speaks of brands as a possibility of materializing expectations; Swanson highlights the emotional dimension that underpins the Brand, stating that brands are totems, they provide us with narratives about our place in a given socio-cultural context and it is also this narrative dimension that Tom Peters underlines, saying that, as humans, we communicate through stories.

The interest that the Museu das Marcas has aroused among the public is related to the revaluation of the stories activated through the images and everyday objects of past generations—memory for some, fantasy for others—but which today help to build an idea of national identity. A revaluation of consumer objects that is not new and that, for example, in the 1960s and 1970s, in the fervor of the Pop movement through designers such as Milton Glaser of Push Pin Studio, combined elements of consumer culture, imposing itself as avant-garde and breaking with the rigidity of modernist grammar.

The Museu das Marcas collection can also be a starting point for a design project through the resignification and transcreation of contemporary designers and artists.

4 Conclusion

The exploratory reading of the Museu das Marcas collection, based on the communication artifacts that we brought to this document, confirms the relevance of design research focused on consumer products in Portugal in the twentieth century. Contact with the collection allows us to validate the importance of its dissemination for the flourishing of a design culture capable of engaging new audiences. We are convinced that the symbolic graphic relevance of this collection justifies its inclusion in the national and international museographic panorama, thus allowing the preservation of the collective memory of consumption in Portugal and, from future readings, the possibility of resignification and transcreation from design, thus allowing innovation.

The brands that we are calling for in this contribution are the beginning of a sensitive observation (Martins 2021) of possible readings of design which, as an example, can be based on specific criteria of the syntax of visual grammar (color, typography, grid, scale, movement, etc.), the identification and semantic interpretation of symbolic elements of representation (natural, cultural and patrimonial heritage, personalities, etc.) or the pragmatics of communication (effectiveness of the message, ideology, intersectionality, etc.). The potential of the Museu das Marcas collection as an object of study for research in design and other areas of knowledge such as anthropology, sociology and image theory is obvious. This repository of brand artifacts thus offers an opportunity to understand the different layers of meaning and design options that converge in these memory-activating objects and make us travel nostalgically back in time.

We also identified as a research challenge and opportunity the museographic design of the Museu das Marcas, which should be a living, interactive and immersive ecosystem in a digital and physical environment.

Considering this fertile field, we intend to continue studying the formal, symbolic and communicational aspects of this collection and the museography of the Museu das Marcas, based on design. The motivations for continuing this research are our personal research interests in the field of brands, the alignment with the museum's objectives—to study social history through the consumption of brands that have shaped our collective experience, contributing to the preservation of objects and generating specialized knowledge for future generations—and the intention to strengthen the institutional partnership between the Museu das Marcas do Caramulo and the Escola Superior de Educação de Viseu.

Acknowledgements The authors would like to express their gratitude to the Museu das Marcas, Caramulo for their availability and prompt response to our requests, as well as for sending us the images and respective information that we present in this document.

References

Appadurai A (2006) The thing itself. Publ Cult 18(1):15–22. https://doi.org/10.1215/08992363-18-1-15

Attfield J (2000) Wild things: the material culture of everyday life (issue 2). Berg Publishers. https://academic.oup.com/jdh/article-abstract/14/3/248/578728

Bautista SS, Balsamo A (2016) Understanding the distributed museum: mapping the spaces of museology in contemporary culture. In: Boddington A, Boys J, Speight C (eds) Museums and higher education working together: challenges and opportunities. Routledge. https://www.book2look.com/embed/9781317092919

Berger J (2008) Ways of seeing. Penguin Books
Fallan K (2010) Design history: understanding theory and method. Berg Publishers. https://www.bloomsbury.com/uk/design-history-9781847885371/
Herrera E, Iñurritegui LF (2018) Historias que marcan: origen y significado de 50 marcas gráficas. Editorial Gustava Gili, SL
ICOM International Council of Museums (2022) Museum definition. https://icom.museum/en/resources/standards-guidelines/museum-definition/
Imperial (2025) Imperial – A nossa historia. https://www.imperial.pt/historia/
Lopes MV (2024) Marcas Que Fazem Portugal. Fundação Francisco Manuel dos Santos. https://ffms.pt/pt-pt/books/marcas-que-fazem-portugal
Lupton E, Miller A (1999) Design writing research: writing on graphic design. Phaidon. https://ellenlupton.com/Design-Writing-Research
Margolin V (1989) Design discourse: history, theory, criticism. The University of Chicago Press. https://press.uchicago.edu/ucp/books/book/chicago/D/bo3774738.html
Martins MOC (2021) Arquivo Poético Portuense: da conceção do modelo a instrumento de transcriação. Universidade de Aveiro. https://ria.ua.pt/handle/10773/33330
Millman D, Walker R (2013) Brand thinking and other noble pursuits. Allworth Press
Mirzoeff N (2015) How to see the world. Penguin Randon House, UK
Mirzoeff N (2023) An introduction to visual culture. Routledge. https://doi.org/10.4324/9780429280238
Museu das Marcas (2025a) Museu das Marcas—YouTube. https://www.youtube.com/@museudasmarcas
Museu das Marcas (2025b) Museu das Marcas—Instagram @museudasmarcas. https://www.instagram.com/museudasmarcas/
Museu das Marcas (2025c) Museu das Marcas | Caramulo—Facebook. https://www.facebook.com/museudasmarcas
Museu das Marcas (2025d) Museu das Marcas—Website. https://museudasmarcas.pt/
Museu das Marcas (2025e) SACOR - Museu das Marcas. https://museudasmarcas.pt/marcas/sacor/
Museu das Marcas (2025f) Nestlé - Museu das Marcas. https://museudasmarcas.pt/marcas/nestle/
Museu do Caramulo (2025) Museu do Caramulo, Caramulo, Portugal—Google Arts & Culture. https://artsandculture.google.com/partner/museu-do-caramulo
Museum of Brands (2025a) Museum of Brands—what's on? https://museumofbrands.com/
Museum of Brands (2025b) Museum of Brands, London, United Kingdom—Google Arts & Culture. https://artsandculture.google.com/partner/museum-of-brands
Nestlé (2025) Discover our rich history | Nestlé Global. https://www.nestle.com/about/history
Nestlé Portugal (2023) Concurso Bebé Nestlé regressa com Edição Especial | Nestlé. https://empresa.nestle.pt/media/comunicados/2023-concurso-bebe-nestle
Ribeiro M (2021) Design da marca como mediação territorial dos agentes científico, económico e político. Ph.D. thesis, Universidade de Aveiro. http://hdl.handle.net/10773/33348
Ribeiro M, Providência F (2022) The brand as mediation through design. In: Raposo D, Neves J, Silva J (eds) Perspectives on design II: research, education and practice. Springer, pp 245–256. https://doi.org/10.1007/978-3-030-79879-6
Woodward I (2007) Understanding material culture. Sage. https://uk.sagepub.com/en-gb/eur/understanding-material-culture/book225857

For a Humanized Smart City: Interrogating Technological Narratives Through Regenerative, Proximity and Care

Gabriel Patrocínio, José Mauro Nunes, and Américo Mateus

Abstract This article explores the concept of a humanized smart city, emphasizing the integration of technology with social values to enhance urban life. It discusses the balance between digital transformation and human-centric urban planning, highlighting the importance of inclusivity, sustainability, and ethical governance in smart city development. The study addresses key challenges such as data privacy, equitable access to technology, and environmental sustainability. By examining literature and some documented real-world case studies, the article presents strategies for creating intelligent urban spaces that prioritize citizen well-being. The research underscores the need for interdisciplinary collaboration among policymakers, technologists, and urban planners to ensure that smart cities remain adaptable and responsive to human needs. The paper concludes with recommendations for fostering resilient, people-focused smart cities that leverage innovation without compromising ethical and social principles.

Keywords Humanized smart cities · Urban sustainability · Ethical technology · Digital transformation · Inclusive urban planning

G. Patrocínio (✉) · A. Mateus
ISMAT/CIAUD, Portimão, Portugal
e-mail: gabrielpatrocinio@gmail.com

G. Patrocínio
IFHT/UERJ, Rio de Janeiro, Brazil

J. M. Nunes
LABCONSUST/IFHT/UERJ, Rio de Janeiro, Brazil

D. Raposo et al. (eds.), *Perspectives on Design IV*, Springer Series in Design and Innovation 59, https://doi.org/10.1007/978-3-032-00733-9_13

1 Introduction

In recent years, the concept of "smart cities" has emerged as a futuristic vision for urban development, driven by the integration of digital technologies into urban infrastructure and services. However, this approach often neglects fundamental considerations about the human and societal impact of technology on urban life. This study proposes a critical reflection on the intersections between the ideals of smart cities and the philosophy of technology, while exploring the concepts of the city of proximity and the city of caring, outlined by Manzini (2019, 2022). Those concepts—city of proximity and city of caring—offer an alternative framework for the conception and governance from the perspective of its citizens and communities. The proximity city emphasizes the importance of interpersonal relationships and the local community, promoting social cohesion and urban resilience. Meanwhile, the city of care highlights the shared responsibility for environmental sustainability and human well-being, encouraging conscious consumption practices and circular economy. Eugene Morozov's (Morozov and Bria 2018) critique of the narrative of smart cities highlights that technology can be used as a tool for control and surveillance, restricting individual freedom and exacerbating social inequalities. Instead of focusing exclusively on efficiency and digital connectivity, a more holistic approach is proposed, focusing on citizens' needs and experiences. Groys's (2022) philosophy of care expands upon these notions, emphasizing the importance of nurturing relationships and fostering empathy within urban environments. He argues that a truly caring city is one where individuals feel valued and supported, leading to a more compassionate and resilient community. By integrating these concepts into the discussion of smart cities, a more humanized and participatory vision of urban development emerges, implying a collaborative approach between governments, companies, local communities, and technology experts, with an emphasis on inclusion, diversity, and equity, as well as data privacy and security. This calls for a paradigm shift regarding the role of technology in urban life, prioritizing human well-being and sustainability over mere technical efficiency. By adopting this approach, cities can become not only "smarter" but also more livable, inclusive, and resilient for all their inhabitants.

2 Smart Cities

2.1 Historical Origins and Definitions: A 400 Years Old Smart City

It seems fair to relate the optimistic concept of a smart city to a long historical tradition of philosophical, social, or political utopias that goes back as far as one likes it to go. From classical Thomas More's Utopia, published in 1516, and Francis Bacon's New Atlantis, from 1627 (Fig. 1), to the science fiction literature of authors such as The Dispossessed, a 1974 anarchist utopian science fiction novel by the American author Ursula K. Le Guin—the idea of a new city for a new society is brought up. Or perhaps in the urbanistic thoughts of the great architects of the early twentieth century, such as Le Corbusier, Frank Lloyd Wright, Jane Jacobs, Lewis Mumford. Cugurullo (2019) offers an ample historic panorama of smart urbanism, and a broad review of the literature concerning the subject.

Fig. 1 New Atlantis map of Francis Bacon's 'smat city' from the seventeenth century (1627)

Isn't design itself concerned with this utopian thought of changing the environment to adapt to our needs? Herbert Simon, in his book The Science of the Artificial, alludes that design creates "courses of action aimed at changing existing situations into preferred ones", or "devising artifacts to attaining goals" (Simon 1996, pp. 111, 114). The convergence of the computer's ability to process data and to control systems, the continuous demand for new iterations in the urban environments, and also the need for new innovative tools for public administration to deal with the dynamics of economics, health, safety, mobility, education, employment—it all started to make sense in the early 1970s. In the Chilean Cybersyn project of 1971, Stafford Beer devised a futuristic control room—with the design concept team led by Gui Bonsiepe—for the central government of Salvador Allende, bringing the latest computer technology to public administration (Espejo 2014). Around the same period, the city of Los Angeles created the Community Analysis Bureau binding urban planning and computational analysis—considered by some as an early attempt to a smart city, where the big data meets urban planning and administration (Cugurullo 2019, p. 118).

The Digital City (De Digitale Stad, also known by the acronym DDS), launched in Amsterdam in January of 1994, is considered to de be the first project of a smart city, based on the concept of a digital, virtual city, where citizens would be '*netizens*' (UNESCO 2021). Although the original DDS project was phased out in 2001, Amsterdam has today its own smart city project ongoing (see it online here: amsterdamsmartcity.com). Many others followed in several continents in the following decades, and it is fair to say that the twentieth first century became the era of the smart cities. Nam and Pardo (2011) show a list of 84 smart cities located in Asia, Africa, Europe, Americas (North, Central and South) and Oceania, with the title awarded by the Intelligent Community Forum (ICF). The same forum lists 212 '*intelligent communities*' around the world today, but it is important to note that these are only listed according to the ICF criteria for affiliation—many more would certainly claim the title of smart cities today (ICF 2025).

2.2 A Panacea to Solve Urban Problems

Smart cities, digital cities, intelligent communities—are some of the names by which it is referred. All sought as a panacea to solve a growing list of urban problems. Covering the field, there is a plethora of books and articles, from where can be found diverse definitions of the smart cities, such as:

> The effective integration of physical, digital and human systems in the built environment to deliver a sustainable, prosperous and inclusive future for its citizens. (British Standards Institute 2014)

> An urban area (encompassing possibly different areas and scales of the city–street, plaza, neighborhood, or, ultimately, an entire city) that uses electronic data collection sensors located in infrastructures, buildings, vehicles, institutions, and devices (IoT, Internet of

> Things) to supply real-time information of the main cities' operating systems. (Moura and Abreu e Silva 2023)

> The term "smart" refers to any advanced technology implemented in cities with the goal of optimizing resource use, creating new wealth, changing user behavior, or promising new types of benefits related to, for example, flexibility, safety, and sustainability. These benefits essentially arise from the feedback loop inherent in the implementation and use of smart devices equipped with connectivity, sensors, and/or screens. (Morozov and Bria 2020, pp. 20–21)

However, Cugurullo points out that "A single definition, image, and vision of smart urbanism does not exist" (Cugurullo 2019, p. 113). Dameri (2017) explains that since most smart city-related projects have risen from the bottom-up experiences related to specific problems, it becomes difficult to have a generalizable definition.

Reichental (2020) proposes that "a *smart city* is an approach to urbanization that uses innovative technologies to enhance community services and economic opportunities, improves city infrastructure, reduces costs and resource consumption, and increases civic engagement" (p. 44).

Despite the absence of a standard complete definition, there is a common thread among the different characterizations of a smart city, which is the use of technologies to solve quality of life-related problems (Galati 2018) and the centrality of information and communication technologies (ICT) resulting from its evolution from the concept of information city (Lee et al. 2013).

Kozłowski and Suwar (2021) compile six main dimensions of a smart city:

- **Smart economy**: efficiency in energy use; innovation, economic impacts, and return on investment; circular economy
- **Smart mobility**: intelligent transportation and transport systems, intelligent parking solutions, traffic management, mobility-as-a-service
- **Smart environment**: water and waste management, monitoring of environmental indicators, sustainable processes and urbanization, hybrid approaches to manufacturing
- **Smart people**: connected citizens, workers, and visitors; e-health; e-learning
- **Smart living**: smart spaces, advanced materials, public security and urban resilience
- **Smart governance**: digital automation of processes, open data, citizen participation.

Other authors detach 'Smart Energy' and 'Smart Safety' from the economy and living dimensions, as appointed by Nam and Pardo (2011).

In the 2010s both IBM and Cisco launched public-private partnerships programmes to support the establishment of smart cities. They were soon followed by other big information technology companies such as Microsoft, Intel, and Siemens (Halegoua 2020). The absence of a clear definition for smart cities leads to a dispute of which company can offer the smartest solution, turning it into a commercial dispute above all. Comparing two versions of IBM smart city models, from 2013 and 2014 (shown on Fig. 2), we may notice some thought-provoking changes. The models have nine (2013) or ten topics (2014) around three dimensions: Planning and

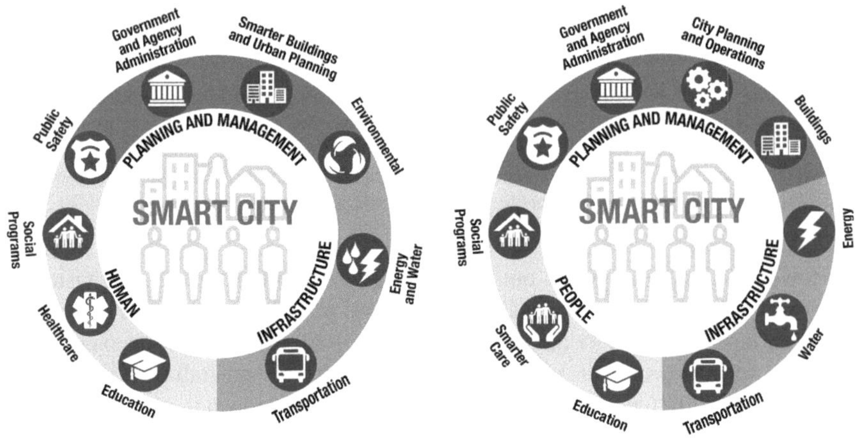

Fig. 2 Two IBM's smart city models (based on IBM 2013, 2014)

management; Infrastructure; and Human (2013) or People (2014). The 2013 model holds an environmental topic, setting the interface between planning and management and infrastructure—which has been cut off of the 2014 model. Public safety, interfacing the human and planning and management dimensions on the first model, becomes a management topic on the second model. The topic "smarter buildings and urban planning" in the first model of 2013 is rendered as just "buildings" in 2014. If we believe that a picture says more than a thousand words, there seems to be a change from an human-centred and environmentally concerned model towards a more bureaucratic and technological one.

In spite of that, Halegoua cautions that big IT companies frequently, following their own primary interest, "position technologies as more important than the outcomes or impacts of using these tools" (Halegoua 2020, p.8). Above all, the main concerns of any city administration should concentrate around their citizens and the environment, not on technology. Technology is a tool—an important tool for a smart city—but should not be an end in itself.

Reviewing the factors that contribute to the city's smartness, Dameri (2017) states that it lies on its core components: **territory**, **infrastructure**, **people**, and **government**. The author conditions the level of smartness implemented in each of these components as a product of three characteristics:

- **Effectiveness**, since a smart city has to create value for its citizens;
- **Environment** concernment, preventing at least further environmental degradation;
- **Innovation**, using technology to reduce environmental impacts and deliver better services.

2.3 *Data and the City: Digital and Smart Cities*

A digital city strategy goes beyond the goals of e-government, and the communicational aspect is at its core. In this sense, the implementation of a digital city relies on ICT, and most specially on **data availability**, **information diffusion** and **e-services**. The attributes of a digital city are referred to by Yovanof and Hazapis (2009) as a combination of: "broadband communications infrastructure; a flexible, service-oriented computing infrastructure based on open industry standards; and, innovative services to meet the needs of governments and their employees, citizens and businesses."

The infrastructure of a digital city is seen by Dameri (2017) as a **competitive-driven** and **technology-oriented strategy**. The author also highlights the importance of "the educational level of citizens in using smart devices or ICT" as one of the main drivers to achieve an efficacious operation.

Academic and empirical views of a smart city differ in fundamental aspects, as both empirical and business-oriented views of smart cities are firmly based on the role of technology, making it closer to the concept of the digital city (Dameri 2017).

A digital city is based on **soft technologies**, and a smart city is based on **hard technologies**. While a digital city has its focus on ICT, a smart city integrates various innovative technologies, including ICT, smart grids, renewable energy sources, integrated transport modes with sustainable fuels, advanced building materials, and others.

Moreover, smart cities ought to be creative. Creativity is a key driver to smart cities, and Yovanof and Hazapis (2009) emphasize that therefore "people, education, learning and knowledge" have paramount importance. As for environmental concerns, Dameri (2017) highlights the aim of smart cities "to reduce CO_2 emissions and to govern energy consumption, waste treatment and building efficiency."

3 Cities of Proximity and Care: Scaled to Its Citizens and for Community Living

Social scientist Carlos Moreno is credited with the concept of the '15-Minute City'—an urban environment planning model conceived to be people centered. "The concept advocates for human-centered urbanism, where aspects such as socialization, self-actualisation, cultural demand and health among others, are accessible in a short commute." (Allam et al. 2022).

Ezio Manzini, in his book Liveable Proximity: Ideas for the City that Cares, defines a city of proximity as

> a city on a human scale, that is dense and diversified in its functions, characterized by public spaces and a mix of residential and productive activities; a city in which the value of proximity is evident in both its functional and relational dimensions; a city that is livable because (…) the proximity found there is amply diversified, since everything you can want, and want to do, is close by. (Manzini 2022, p. 27)

Groys (2022) discusses Foucault's **biopolitical state**, whose "main function is to take care of the physical well-being of their population." "**Medicine has taken the place of religion**, … the body rather than the soul is the privileged object of institutionalized care: **health replaced salvation**."

Groys also invokes Heiddeger's concept of **the human being as Dasein**, or 'being-there', which makes it impossible to oppose the Dasein as a 'subject' to the world as an 'object'. "The existence of Dasein is a project directed towards the future—we permanently plan something for the future. This planning presupposes that we further exist—and have to **care** about our existence" (Groys 2022).

The subject of "**Care**" is often associated with **public health issues** only, and not usually related to a more encompassing aspect of care, considering the daily needs of the citizens, individually and family-wise. It follows, somehow, the paradox of the biopolitical state as discussed by Groys, that has the objective of keeping society healthy and productive, but privileges a system of hierarchies that may not be fair to the common citizen, and those in most need to be cared—sheltered, fed, educated, entertained, and also treated.

Manzini offers a broader vision of care, and of a **city of care**. For the author, the concept of a city of care permeates the idea of proximity—to care and be cared for, one must be close. His **city of proximity** (or 15-minute city) embraces **care** and **digitization**, and **innovative services** built with a bottom-up reverse logic, developed in and by the community. According to him, proximity can essentially be **Functional** (physical) or **Relational** (human relations). He identifies **three levels of care** towards someone or something: dedicating **attention**, taking **responsibility**, and **therapy**, or treating and healing. Cities of care should be community-driven: "given the multiple crises we are facing it is necessary to **re-establish the social fabric and (re)construct communities**" (Manzini 2022).

Transforming the careless **city of services** (where citizens are reduced to users/ consumers) into a **city of proximity and care** is a constant concern for Manzini, and he proposes five points to facilitate this shift:

- **Localization**: "bringing services and activities close to citizens";
- **Socialization**: "favouring the construction of communities";
- **Inclusion**: "extending the network of actors involved";
- **Diversification**: "involving actors initially not considered";
- **Coordination**: "horizontally connecting different areas of intervention" (Manzini 2022).

3.1 Nature-Integrated Care

Victor Papanek, an author and polemist with some of the most significant books on design, ecology and consumerism, is reputed as "one of the twentieth century most influential pioneers of design's politics" (Lamperti 2022). In his last work The Green Imperative, Chap. 5, The Biotechnology of Communities, delves into how design can foster sustainable and interconnected communities (Papanek 1995). Papanek stresses the importance of understanding and respecting the interconnectedness of living systems within communities, and advocates for design that works in harmony with natural processes. He argues that communities should be empowered to address their specific needs and challenges, and develop their own sustainable solutions. These solutions should mimic and integrate with natural processes. According to him, sustainable communities must address the needs of all residents, particularly those who are marginalized. In essence, he calls for a shift towards design that fosters ecologically sound and socially equitable communities, drawing on both scientific understanding and traditional wisdom. His view can be recognized as a pioneering proposition for a nature-integrated community of care.

4 Critical Narratives of the Smart City

The model of smart cities has been subject to a lot of criticism, with concerns about their implementation and underlying principles. It is argued that while smart cities offer potential benefits, they also carry risks related to social equity, governance, and individual freedoms. Substantially supported by technology, smart cities are criticised by the ensuing dependence, recognizing that technologies are not neutral: they can be designed to favour relations or eliminate them, to centralize power or distribute it, to recognize individual contributions or extract value from them.

The effectiveness of smart cities is also frequently measured by the innovation originating from the territory, accessed by the number of patents registered, as appointed by Caragliu and Del Bo (2018), usually overlooking social and environmental costs.

A key critique revolves around the alignment of smart city development with **neoliberal governance** and **privatized Keynesianism**. Authors like Crouch (2009), Krippner (2011), and Birch (2020) argue that smart city initiatives often prioritize private investment and digital consumption over public welfare, potentially leading to increased inequality and financial vulnerability. This approach can shift the burden of economic stimulation from the state to private entities and individuals, mirroring the dynamics of privatized Keynesianism.

Another critical perspective addresses **surveillance capitalism**. Authors like Zuboff (2019) highlight how smart cities, by collecting vast amounts of citizen data, can become arenas for data extraction and behavioral manipulation by private companies. This raises concerns about privacy, commodification of daily life, and the potential for increased social inequalities.

Furthermore, the techno-solutionist discourse of smart cities can obscure structural problems such as income inequality, housing access, and digital exclusion. Critics like Morozov and Bria (2020) caution against prioritizing technology over citizen well-being and democratic control, advocating for alternative models that emphasize digital justice and public ownership of urban data.

4.1 Privatized Keynesianism

The concept of *privatized Keynesianism* refers to a shift in the mechanisms used to stimulate aggregate demand in advanced capitalist economies, particularly from the late twentieth century onward. Unlike classical Keynesianism, which relied on direct public investment and fiscal expansion, privatized Keynesianism involves stimulating consumption through the expansion of private credit. This transformation is rooted in broader neoliberal restructuring, where the state's role in welfare provision is diminished, and the burden of economic stabilization is increasingly placed on households through access to debt (Crouch 2009; Krippner 2011). The original Keynesianism relates to the **welfare state**, while the privatised Keynesianism is associated with **neoliberalism**.

This regime emerged in response to the constraints imposed by inflationary pressures, public debt limits, and political resistance to high levels of government spending. As Crouch (2009) argues, governments adopted policies that actively promoted private borrowing to compensate for stagnating wages and the retrenchment of welfare states. Through instruments such as deregulated mortgage markets and consumer credit expansion, households were encouraged to take on debt to sustain consumption, effectively replacing public stimulus with private financial leverage.

> Under original Keynesianism it was governments that took on debt to stimulate the economy. **Under the privatised form individuals, particularly poor ones, took on that role by incurring debt on the market**. The main motors were the near-constant rise in the value of owner-occupied houses and apartments alongside an extraordinary growth in markets in risk. (Crouch 2008)

The rise of privatized Keynesianism is inseparable from broader processes of financialization. Households increasingly rely on asset inflation—particularly in housing and stock markets—to supplement stagnant incomes and build wealth. As Krippner (2011) notes, this represents a shift in the locus of economic dynamism from production to finance. What emerges is an asset-based welfare system, in which social security is tied to the performance of financial markets rather than to redistributive public programs. This process reinforces socioeconomic inequality, as those with greater access to credit and assets benefit disproportionately.

The 2008 global financial crisis revealed the structural fragility of privatized Keynesianism. When asset bubbles burst and credit flows contracted, household balance sheets collapsed, exposing the unsustainability of a growth model based on debt-financed consumption. Streeck (2014, 2016) describes this trajectory as part of a broader sequence in which capitalist economies have relied successively on inflation, public debt, and private debt to defer systemic crises. The failure of privatized Keynesianism to provide durable economic stability underscores the need to rethink the institutional foundations of demand management in post-industrial economies.

In sum, privatized Keynesianism represents a defining feature of neoliberal economic governance, wherein the function of macroeconomic stabilization is outsourced to financial markets and private households. While it has enabled the maintenance of consumption levels in an era of wage stagnation and fiscal austerity, it has done so at the cost of rising inequality, financial vulnerability, and a diminished role for democratic control over the economy. Understanding this model is essential for evaluating the possibilities and limits of economic policy in contemporary capitalism—and for envisioning alternatives that prioritize social equity and macroeconomic resilience.

4.2 Smart Cities and Privatized Keynesianism

Most of the cities that embarked on smart city experiments have also been captured by the regulatory mechanisms of neoliberalism. The relationship between smart cities and privatized Keynesianism highlights how contemporary urban transformations are embedded in the logics of financialization and neoliberal governance. Smart cities—urban environments enhanced through the integration of digital technologies—are often promoted as efficient and sustainable responses to urban challenges. However, their implementation tends to align with market-oriented growth models, where private investment, platform-based services, and digital consumption replace traditional public welfare mechanisms. Rather than representing a new paradigm of public infrastructure, smart cities often mirror the principles of privatized Keynesianism by shifting the burden of economic stimulation from state-led investment to private credit and consumption (Birch 2020; Cardullo et al. 2021).

From a positive perspective, smart cities offer opportunities for urban innovation, environmental sustainability, and improved resource efficiency. Data-driven solutions can enhance public transportation, reduce energy consumption, and foster new channels for civic engagement. In this view, smart cities represent a techno-infrastructural upgrade of Keynesian public works, with a key difference: investments are predominantly private and focused on the development of urban digital markets. As such, they are seen as drivers of economic growth in post-industrial economies, where digital infrastructure replaces the industrial base as a site of value creation.

However, from the standpoint of privatized Keynesianism, smart city development is often sustained by debt-driven growth, gentrification, and the financialization of urban space. Public-private partnerships, concession models, and municipal bond issuance are frequently used to finance digital infrastructure. These mechanisms privatize profits while socializing risks, shifting responsibility to local governments and reinforcing socio-spatial inequalities. The emphasis on digital solutions also contributes to the commodification of urban life, where access to services increasingly depends on one's ability to engage in data-generating and credit-based consumption (Kitchin et al. 2015; Sadowski 2020).

Moreover, the techno-solutionist discourse surrounding smart cities can obscure structural problems such as income inequality, housing access, and digital exclusion. The logic of privatized Keynesianism manifests in the way technological consumption—smartphones, mobility apps, digital surveillance—is positioned as a remedy for complex social issues. Citizens are reframed as consumers of urban digital services, and public goods are replaced with subscription-based, credit-mediated alternatives. This approach reinforces neoliberal patterns of individualized risk management and undermines democratic control over urban planning and service provision (Datta 2018).

In conclusion, the intersection between smart cities and privatized Keynesianism reveals both potential benefits and critical drawbacks. While these cities may generate innovation and economic dynamism, they often do so by reproducing neoliberal tendencies that replace public welfare with commodified services. Recognizing the underlying financial and political structures of smart city development is essential for imagining more equitable and democratically governed urban futures.

4.3 Surveillance Capitalism, Smart Cities, and Privatized Keynesianism: A Critical Perspective

The intersection between surveillance capitalism, smart cities, and privatized Keynesianism reveals complex dynamics in contemporary urban governance. Smart cities use digital technologies to optimize urban services by collecting vast amounts of citizen data. This model often aligns with surveillance capitalism, in which private companies monetize personal data to predict and influence behavior, turning human experience into raw material for predictive products (Zuboff 2019). Rather than public infrastructure expanding citizens' rights, smart cities can become arenas for data extraction and behavioral manipulation.

On the other hand, the concept of privatized Keynesianism describes the shift in economic stimulus responsibility from the public to the private sector, encouraging consumption through credit expansion and financialization. In smart cities, this logic materializes when private investments in digital infrastructure replace direct public investments, promoting economic growth based on technological platforms and data services (Sadowski 2020). This often results in an overreliance on public-private partnerships, where profits are privatized while risks are socialized (Cardullo et al. 2021).

This dependence on private investment in smart city infrastructure can lead to the prioritization of corporate interests over public welfare. Extensive data collection and surveillance may compromise citizen privacy and reinforce social inequalities, especially when access to urban services becomes conditioned on digital participation. Furthermore, the emphasis on technology as a solution can overlook structural challenges like income inequality and housing insecurity (Schuilenburg and Palli 2022; Söderström et al. 2014). Although technology is at the core of the smart city concept and is a necessary condition, it is not sufficient, because the application of **technology is not enough to create public value for citizens**.

In this context, surveillance capitalism becomes an operational layer of smart cities, reinforcing the commodification of daily life. As citizens move through urban spaces, their behaviors and preferences are tracked by sensors, apps, and surveillance systems, feeding massive datasets controlled by private firms. These data are monetized not for public service optimization, but for commercial targeting, predictive analytics, and consumption stimulation—thus aligning with the logic of privatized Keynesianism, which seeks to sustain economic growth via private consumption rather than public redistribution (Zuboff 2019; Sadowski 2020). This highlights how many smart city initiatives have become intertwined with **neoliberal policies**, where **privatization** and **market-driven approaches** shape the governance and development of urban technologies and services.

Moreover, smart cities may deepen existing social inequalities. Access to digital public services often depends on device ownership and digital literacy, excluding marginalized populations. The drive for "efficiency" and "innovation" can legitimize the use of intrusive surveillance technologies in the name of public safety, disproportionately targeting racialized and vulnerable communities (Kitchin and Dodge 2017; Monahan 2022). This creates cities that are not only unequal, but also governed through opaque and undemocratic technological infrastructures.

In response, scholars and activists have called for alternative smart city models—ones that prioritize digital justice, democratic control, and public ownership of urban data. Concepts such as data commons and platform cooperativism offer strategies to reclaim digital infrastructure for collective benefit rather than private profit. Reorienting smart city development toward social equity and public accountability could preserve the innovative potential of digital technology while avoiding the extractive logic of surveillance capitalism and privatized Keynesianism (Morozov and Bria 2020).

In summary, although smart cities promise efficiency, sustainability, and innovation, their implementation within the frameworks of surveillance capitalism and privatized Keynesianism presents significant risks. Without democratic oversight and a strong commitment to equity, these cities may reproduce neoliberal logics of privatization, surveillance, and social exclusion. Recognizing and confronting these dynamics is essential for building inclusive, democratic, and just urban futures.

5 The Pivotal Role of Nature

The conceptualization of humanized smart cities needs a critical interrogation of technological narratives through the lenses of regeneration, proximity, and care. Central to this discourse is the pivotal role of nature as a mediating entity between urban environments, technological advancements, and ethical responsibilities. Nature functions as a dialogical bridge, harmonizing the human aspiration for an automated future with the intrinsic needs of territories, land, and ecosystems. The sustainability of future urban landscapes is inextricably linked to the regeneration of natural systems, which underpins the conditions essential for life and its continuous development.

Recent scholarly contributions underscore the imperative of integrating ecological principles within urban infrastructures to foster resilience and systemic balance. Zari (2019) emphasizes that regenerative urban design must move beyond sustainability toward actively restoring ecological processes. This vision necessitates a paradigm shift where cities are not merely seen as technologically advanced spaces but as living systems that participate in broader ecosystems. The notion of biomimicry, wherein cities emulate nature's self-sustaining mechanisms, suggests that the future of urban design must prioritize bio-integrated structures that regenerate rather than exploit the environment.

In parallel, the concept of the 'Internet of Nature,' as proposed by Galle et al. (2019), advocates for a reconfiguration of digital infrastructures to include nature as an intelligent actor in urban systems. By embedding ecological intelligence within smart city frameworks, technology can be leveraged to enhance environmental stewardship rather than merely optimize human convenience. This approach repositions the role of digital innovation from one of extraction to one of co-existence, where data-driven systems facilitate regenerative practices by monitoring biodiversity, water cycles, and urban microclimates.

The notion of proximity, both spatial (or functional) and relational, is another critical dimension in theorizing a humanized smart city. Bollier and Helfrich (2019) argue that commons-based urbanism, where natural and digital infrastructures are governed collectively, fosters a participatory ethos that counteracts the top-down imposition of technological systems. This vision aligns with Latour's (2020) call for a redefinition of human-environment relations, in which cities become loci of care that extend beyond human-centered priorities to encompass ecological well-being. A city structured around care inherently integrates nature as a co-actor in governance models, ensuring that the automated future does not bypass environmental justice.

This perspective necessitates rethinking the ethical frameworks underlying smart city development. Haraway (2016) contends that the Anthropocene's challenges require a departure from hierarchical governance structures toward multispecies and multi-entity networks of care. In this context, nature is not a passive resource to be managed but an agent that actively shapes urban evolution. A regenerative smart city, therefore, does not merely deploy technology for efficiency but constructs new epistemologies of urban living, where technological progress is measured by its ability to sustain and restore life-supporting systems.

A humanized smart city is not merely an automated, data-driven environment but a system in which nature mediates the interactions between technology, urban planning, and care ethics. The integration of regenerative design, ecological intelligence, and participatory governance ensures that urban futures are not detached from the fundamental conditions that sustain life. By positioning nature as the central dialogical bridge, urban development can navigate the fine balance between automation and ecological integrity, ensuring that technological progress coexists with the imperative of planetary regeneration.

6 For a City that Really Cares: A Dialogue Between Smart and Care, Where Technology Meets Nature and Humanism

A critical engagement with the smart city paradigm is essential to uncover the power structures, exclusions, and sociotechnical imaginaries underlying its implementation. Although often presented as neutral or inherently progressive, smart city technologies are shaped by specific political and economic agendas. Without a reflexive approach, these urban innovations risk deepening inequalities, reproducing surveillance regimes, and replacing democratic participation with data-driven governance. As several scholars have pointed out, the techno-solutionist orientation of smart urbanism can obscure the structural causes of systemic urban challenges such as poverty, housing insecurity, and digital exclusion.

One of the most pressing concerns surrounding smart cities is the rise of algorithmic exclusion. As data-driven systems increasingly mediate access to urban services—from mobility and housing to policing and welfare—algorithmic biases can reproduce and even intensify existing social inequalities. Marginalized populations are often rendered invisible or misrepresented within datasets, leading to discriminatory outcomes in automated decision-making processes. Moreover, the opacity of proprietary algorithms and the lack of democratic oversight raise urgent questions about accountability, transparency, and the right to the city in the digital age (cf. O'Neill 2016). A step ahead in this path should be to establish contracts with private companies that provide these services emphasizing **free software** and **open-source subscriptions**.

This accentuates the need for a transition from private or corporate **data ownership** models to ones that **ensure community-driven governance of digital resources**. By creating structures that allow for collective control and management of data, cities can better protect the rights of citizens, promote transparency, and ensure that data generated in the public domain benefits the community as a whole, rather than being exploited for commercial gains.

Smart cities also face critical challenges in relation to sustainability. While often promoted as environmentally friendly through the use of sensors, data analytics, and energy-efficient infrastructure, smart city initiatives frequently prioritize technocentric "green fixes" that sidestep deeper questions about urban consumption, extractivism, and ecological justice. The material footprint of digital infrastructures—servers, devices, rare earth minerals—contributes to global environmental degradation and reproduces unequal geographies of resource extraction and waste disposal. A truly sustainable urban future demands more than smart technologies; it requires structural transformations in how cities produce, distribute, and consume resources.

In this context, the concept of the caring city emerges as a relevant counterpoint. Grounded in the political economy of care, the caring city values interdependence, relationality, and the prioritization of human and ecological well-being over technological efficiency or economic growth. Instead of seeing residents as data points or consumers of urban services, this approach recognizes them as agents embedded in networks of mutual support and collective responsibility. It calls for a reorientation of urban planning toward social justice, inclusion, and lived experience (cf. Concilio and Rizzo 2016).

The notion of the *city that cares*, as articulated by Ezio Manzini, offers a transformative lens through which to rethink the development and implementation of smart city technologies. Rather than centering innovation solely on efficiency, optimization, and control, Manzini's framework calls for a shift toward fostering networks of care, empathy, and mutual support. In this vision, cities are not merely systems to be managed, but ecosystems of human and more-than-human relationships to be cultivated. Smart technologies, when aligned with this paradigm, can serve as enablers of distributed social innovation—tools that amplify citizens' capacities to organize, collaborate, and co-create solutions grounded in local needs and collective well-being.

Furthermore, the notion of the city that cares invites us to rethink the role of the state, the commons, and ethics in urban governance. Ostrom (1990) defined commons as shared resources—such as forests, fisheries, or water systems—that are collectively used and managed by communities through locally crafted rules and institutions. Contrary to the idea of the "tragedy of the commons," where individuals are expected to act selfishly and deplete the resource, Ostrom demonstrated that collective action can lead to sustainable management. The core paradox involves aligning individual interests with collective well-being: while each person may benefit from overusing the resource, such actions ultimately harm the group and even the individual in the long term. The author found that when users have trust, communication, and a sense of mutual responsibility, they often devise effective systems of self-governance. Her work challenged top-down and purely market-based solutions, showing that communities themselves can be the best stewards of common resources.

This perspective highlights the centrality of care infrastructures—such as healthcare, education, housing, and accessible public spaces—as foundational to urban life. In contrast to the commodified logic of smart urbanism, the city that cares foregrounds public investment, democratic participation, and solidarity as structuring principles. It advocates resisting the automation of care and instead revaluing the labor—both paid and unpaid—that sustains urban life.

This approach stands in stark contrast to the dominant logic of many smart city projects, which often privilege centralized data infrastructures and corporate-led governance. Manzini's concept urges a redefinition of the "smart" in smart cities—not as technological sophistication per se, but as the collective intelligence of a community capable of caring for itself and others. It emphasizes small-scale, situated practices that can scale by replication rather than standardization, nurturing what he calls "cosmopolitan localism." Within this framework, smart city technologies must be open, accessible, and responsive to the lived experiences of residents—supporting forms of urban life where care is not an afterthought, but a central organizing principle.

Bringing together a critique of the smart city with the ethics of care is not merely an academic exercise, but a political and normative imperative. In the face of intersecting urban crises—from climate change to social fragmentation—it becomes urgent to imagine and build alternative urban futures. The smart city must be reimagined not as a technological fix, but as a contested space where values such as justice, care, and democracy can be reclaimed and re-embedded in urban governance. Only through this critical and imaginative effort can we move toward cities that are not only smart, but truly humane.

A truly caring city is one that places human and ecological well-being at the center of urban governance, infrastructure, and daily life. This means recognizing care not as a private or residual activity, but as a fundamental social function that must be supported through public investment, inclusive planning, and participatory processes. A city that cares actively designs its policies and spaces to reduce precarity and vulnerability—ensuring access to affordable housing, quality healthcare, education, and safe public spaces. It acknowledges the uneven distribution of care burdens, especially along lines of gender, race, and class, and seeks to redistribute care work in just and equitable ways. In doing so, it confronts the structural inequalities that often remain hidden behind the rhetoric of technological progress and urban innovation. A caring democracy is one in which citizens and institutions take up the responsibilities of care, not only in private life but also in the public realm, shaping policies and environments that enable all people to live and flourish (Tronto 2013).

The concept of the *city that cares* offers a powerful ethical counterpoint to the prevailing smart city model, particularly when viewed through the lens of Michael Sandel's critique of market society. Sandel (2012) warns against the unchecked expansion of market logic into all domains of life, arguing that when public goods are commodified, essential civic values—such as solidarity, dignity, and the common good—are undermined. Many smart city initiatives, driven by private platforms and subscription-based services, risk reproducing precisely this dynamic: public services become digital commodities, and citizens are reframed as consumers in algorithmically managed urban environments. In contrast, the caring city resists this commodification by emphasizing relational infrastructures, shared responsibilities, and the intrinsic value of care. It asks not only what technologies are efficient or profitable, but what kinds of urban life we want to sustain—and who benefits from them. By integrating Sandel's moral perspective, the city that cares reclaims the political and ethical dimensions of urban governance, grounding innovation in justice, inclusion, and human dignity.

Finally, smart cities are increasingly embedded in a broader socio-political context of surveillance capitalism and post-capitalist imaginaries (cf. Zuboff 2019). Surveillance technologies, often justified under the guise of efficiency and security, normalize the collection of vast amounts of personal data by both state and corporate actors. This dynamic contributes to a mode of governance that prioritizes control, predictability, and profit over democratic deliberation and collective agency. In opposite, our approach ensures that cities or organizations using these services are not locked into proprietary systems, and it **promotes innovation** by allowing **continuous improvement from a wide pool of contributors**. This approach encourages **flexible and adaptive urban innovation**. By testing various solutions in smaller, controlled environments, cities can identify which projects effectively serve the public interest and discard those that fail to provide tangible benefits.

This **iterative process promotes efficient resource use** and prevents the large-scale implementation of ineffective or unpopular projects.

Yet, amid these tensions, smart cities also hold the potential to become laboratories for post-capitalist experimentation—spaces where cooperative economies, platform commons, and digital rights can be foregrounded. To accomplish this potential, however, it is essential to resist the reduction of urban life to data flows and instead reclaim technology as a tool for emancipation and care.

Beyond infrastructure and policy, a city that truly cares also fosters an ethic of relationality and interdependence among its inhabitants. It encourages everyday practices of mutual aid, solidarity, and empathy, cultivating a civic culture where people feel seen, valued, and supported. In this sense, care becomes both a principle of design and a practice of citizenship. Technologies and data systems can play a role in this process—but only if they are deployed in ways that reinforce, rather than replace, human connections and collective responsibility. In this sense, technology can be a powerful enabler. The caring city is not merely a more compassionate version of the smart city; it is a fundamentally different urban ideal—one where innovation serves life, not the market, and where the future is imagined through the lens of justice, dignity, and care for all.

References

Allam Z, Bibri SE, Chabaud D, Moreno C (2022) The '15-Minute City' concept can shape a net-zero urban future. Humlties Soc Sci Commun 9(1):1–5

Birch K (2020) Technoscience rent: toward a theory of rentiership in technoscientific capitalism. Sci Technol Hum Values 47(3):405–429. https://doi.org/10.1177/0162243919829567

Bollier D, Helfrich S (2019) Free, fair, and alive: the insurgent power of the commons. New Society Publishers, Gabriola Island, Canada

British Standards Institute (2014) Smart city framework. Guide to establishing strategies for smart cities and communities. PAS 181:2014. BSI, London

Caragliu A, Del Bo CF (2018) Smart innovative cities: the impact of smart city policies on urban innovation. Technol Forecast Soc Change. https://doi.org/10.1016/j.techfore.2018.07.022

Cardullo P, Di Feliciantonio C, Kitchin R (2021) The right to the smart city. Emerald Publishing, London

Concilio G, Rizzo F (2016) Human smart cities: rethinking the interplay between design and planning. Springer, Switzerland

Crouch C (2008) What will follow the demise of privatised Keynesianism? Polit Q 79(4):476–487

Crouch C (2009) Privatized Keynesianism: an unacknowledged policy regime. Br J Polit Int Rel 11(3):382–399. https://doi.org/10.1111/j.1467-856X.2009.00377.x

Cugurullo F (2019) The origin of the smart city imaginary: from the dawn of modernity to the eclipse of reason. In: The Routledge companion to urban imaginaries. Routledge, Oxon

Dameri RP (2017) Smart city implementation. Springer, Heidelberg

Datta A (2018) The digital turn in postcolonial urbanism: smart citizenship in the making of India's 100 smart cities. Trans Inst Br Geogr 43: 405–419. https://doi.org/10.1111/tran.12225

Espejo R (2014) Cybernetics of governance: the Cybersyn project 1971–1973. In: Social systems and design. Springer, New York, pp 71–90

Galati SR (2018) Funding a smart city: from concept to actuality. In: Smart cities. Springer, Cham, pp 17–39

Galle NJ, Nitoslawski SA, Pilla F (2019) The Internet of Nature: how taking nature online can shape urban ecosystems. Anthropocene Rev 6(3):279–287. https://doi.org/10.1177/2053019619877103
Groys B (2022) Philosophy of care. Verso, London
Halegoua GR (2020) Smart cities. The MIT Press, London
Haraway DJ (2016) Staying with the trouble: making kin in the Chthulucene. Duke University Press. https://doi.org/10.2307/j.ctv11cw25q
IBM (2013) Smarter cities. Available at: https://web.archive.org/web/20151031211926/ibm.com/smarterplanet/ie/en/smarter_cities/overview/
IBM (2014) IBM smarter cities. Available at: https://web.archive.org/web/20190514085323/ibm.com/smarterplanet/us/en/smarter_cities/overview/. Accessed on 18 March 2025
ICF—Intelligent Communities Forum (2025). List of all intelligent communities. Available at: https://www.intelligentcommunity.org/intelligent_communities_list. Accessed on 18 March 2025
Kitchin R, Dodge M (2017) The (in)Security of smart cities: vulnerabilities, risks, mitigation, and prevention. J Urban Technol 26(2):47–65. https://doi.org/10.1080/10630732.2017.1408002
Kitchin R et al (2015) Smart cities and the politics of urban data. Routledge, London
Krippner GR (2011) Capitalizing on crisis: the political origins of the rise of finance. Harvard University Press, Cambridge
Kozłowski W, Suwar K (2021) Smart city: definitions, dimensions, and initiatives. Eur Res Stud J XXIV(3):509–520
Lamperti G (2022) Papanek's green imperative: an ethically responsible future for design. Available at: https://www.salonemilano.it/en/articles/papaneks-green-imperative-ethically-responsible-future-design#:~:text=A%20green%2C%20ethical%20approach%20that,consequences%20of%20a%20new%20product. Accessed on 22 March 2025
Latour B (2020) Down to earth: politics in the new climatic regime. Polity Press, Cambridge
Lee JH, Phaal R, Lee S (2013) An integrated service-device-technology roadmap for smart city development. Technol Forecast Soc Chang 80(2):286–306
Manzini E (2019) Politics of the everyday. Bloomsbury Publishing, London
Manzini E (2022) Liveable proximity: ideas for the city that cares. Egea, Milano
Monahan T (2022) Crisis vision: race and the cultural production of surveillance. Duke University Press, Durham
Morozov E, Bria F (2018) Rethinking the smart city. Democratizing urban technology 2. Rosa Luxemburg Stiftung, New York. Available at: https://sachsen.rosalux.de/fileadmin/rls_uploads/pdfs/sonst_publikationen/rethinking_the_smart_city.pdf. Acessed on 29 July 2025
Moura F, Abreu e Silva J (2023) Smart cities: definitions, evolution of the concept and examples of initiatives. In: Leal Filho W et al (eds) Industry, innovation and infrastructure. Encyclopedia of the UN sustainable development goals. https://doi.org/10.1007/978-3-319-71059-4_6-1
Nam T, Pardo TA (2011) Conceptualizing smart city with dimensions of technology, people, and institutions. In: Proceedings of the 12th annual international digital government research conference: digital government innovation in challenging times, pp 282–291
O'Neill C (2016) Weapons of math destruction: how big data increases inequality and threatens democracy. Crown, New York
Ostrom E (1990) Governing the commons: the evolution of institutions for collective action. Cambridge University Press, New York
Papanek V (1995) The green imperative: ecology and ethics in design and architecture. Thames & Hudson, New York
Reichental J (2020) Smart cities for dummies. Wiley, New Jersey
Sadowski J (2020) Too smart: how digital capitalism is extracting data, controlling our lives, and taking over the world. MIT Press
Sandel MJ (2012) What money can't buy: the moral limits of markets. Farar, Starus and Giroux, New York

Schuilenburg M, Palli B (2022) Smart city imaginaries looking beyond the techno-utopian vision. In: Schuinlemburg M, Peeters R (eds) The algorithmic society: technology, power, and knowledge. Routledge, London

Simon HA (1996) The sciences of the artificial, 3rd edn. MIT Press, London

Söderström O, Paasche T, Klauser F (2014) Smart cities as corporate storytelling, city. Anal Urban Trends Cult Theory Policy Action 18(3):307–320. https://doi.org/10.1080/13604813.2014.906716

Streeck W (2014) Buying time: the delayed crisis of democratic capitalism. Verso Books, London

Streeck W (2016) How will capitalism end? Essays on falling system. Verso Books, London

Tronto J (2013) Caring democracy: markets, equality and justice. NYU Press, New York

UNESCO (2021) DDS: De Digitale Stad/The Digital City. Available at: https://www.unesco.org/en/memory-world/dds-de-digitale-stad/digital-city. Accessed on 18 March 2025

Yovanof GS, Hazapis GN (2009) An architectural framework and enabling wireless technologies for digital cities & intelligent urban environments. Wirel Pers Commun 49(3):445–463. Available at: https://doi.org/10.1007/s11277-009-9693-4. Accessed on 18 March 2025

Zari MP (2019) Regenerative urban design and ecosystem biomimicry. Routledge, Abingdon. https://doi.org/10.4324/9780429505737

Zuboff S (2019) The age of surveillance capitalism: the fight for a human future at the new frontier of power. PublicAffairs, New York

Fashion Design and Product Design and Innovation

Navigating by Sight. Rethinking Fashion Education from Historical Archives to Artificial Intelligences Through Collaborative Practices

Paolo Franzo, Ines Tolic, Flavia Piancazzo, Andrea Quartu, and Denis Domenichetti

Abstract Fashion education needs to be rethought in light of the contemporary cultural, technological and social context. With the advent of AI and the uncontrolled production of images, one of the main challenges is to recover the historical dimension of fashion and to better exploit archival materials. To achieve this, we need to identify collaborative practices that bring together various disciplines and approaches, generating knowledge and sharing values, while producing culturally and socially sustainable products and narratives. To reflect on this, this paper analyses the experience of the workshop "Weaving Archives: Activating Textile Archives Through Artificial Intelligence", promoted by the Master's Degree in Fashion System Design of the Department of Architecture—University of Florence, in collaboration with the Master's Degree in Fashion Studies of the Department of Arts—University of Bologna and the Fashion Research Italy Foundation in Bologna. The aim is to demonstrate how collaborative workshops provide valuable opportunities for training future professionals to balance the global reach of fashion with its complex local implications. This endeavour seeks to chart alternative paths that illuminate the intricate expanse of the modern fashion landscape.

Keywords Fashion education · Collaborative design practices · Fashion archives · AI · Cultural sensitivity

P. Franzo (✉)
Università degli Studi di Firenze, Florence, Italy
e-mail: paolo.franzo@unifi.it

I. Tolic · F. Piancazzo
Università degli Studi di Bologna, Bologna, Italy

A. Quartu · D. Domenichetti
Università della Campania Luigi Vanvitelli, Caserta, Italy

D. Raposo et al. (eds.), *Perspectives on Design IV*, Springer Series in Design and Innovation 59, https://doi.org/10.1007/978-3-032-00733-9_14

1 Introduction: Fashion Education

To explain what he meant when he spoke of sustainability, Ezio Manzini used the image of a sailboat: a sophisticated, light vessel that is able to make the best use of both technology and the skills of its crew "to sail with the winds and currents, correcting its course moment by moment, without losing sight of its objectives" (Manzini 1997, p. 44). Borrowing his metaphor and referring it to industrial society, if we were to compare fashion to a boat we would think of an oil tanker instead of a sailing boat, belching smoke and noisily navigating the oceans without regard to people, seas and climates. To be sustainable in the most current and broadest sense, fashion today should be reinterpreted in light of the values that characterise contemporary society. At the same time, it is necessary to train a new generation of professionals, i.e. sailors capable of navigating and operating successfully in the most complex of cultural and creative industries. In our view education is the key to initiating a revision that over the long run can produce systemic changes and transform fashion into a fleet of agile sailboats. Yet, precisely because of the speed of the changes taking place, those working in fashion education do not have enough tools to chart a course towards this goal. The title of this paper—"Navigating by Sight"—suggests the need to explore innovative educational pathways that use collaborative practices to pave the way for inherently inclusive and sustainable fashions.

The ecological and digital transition process, accelerated at the national and European level by Next Generation EU project funds, is part of a global context already pervaded by epochal socio-cultural changes. We are witnessing a "relational turn", i.e. an increasingly widespread interpretation of contemporary contexts and phenomena through an ecological lens, and this "relational" perspective (Guattari 2005) is stimulating approaches and methodologies capable of shaking up the paradigms inherited from the past century. University fashion education, characterised by a close connection between research and teaching (Lupano and Vaccari 2014), thus reflects current changes, modulating in contexts made up of a plurality of subjects and objects through continuous contact between students, teachers, researchers, professionals and companies (Franzo 2021). Within these temporary communities, which are created during a course or workshop, students hone their critical skills by approaching topics that are often highly complex. These teaching moments thus become rhizomatic spaces (Deleuze and Guattari 1988) where knowledge and the protagonists of these temporary communities intertwine without hierarchies. This formative dynamic was skilfully described by Gilles Clément through the image of an "edifice of knowledge [that] does not correspond to a pyramid whose summit is occupied by masters and sages, but to a rough copy of the cosmos, in which particles are enriched by contact" (Clément 2008, p. 76). The condition of continuous negotiation and hybridisation of knowledge and skills is thus translated into an interdisciplinary path of co-creation whose aim is the exchange of signs, images, narratives, forms and languages. To summarise, applying these reflections to the contemporary context, the aim is to train fashion designers and all those working in the fashion industry so that they can understand the changes taking place, thus stimulating the creation of updated value imaginaries.

This approach was the basis for the organisation of the workshop "Weaving Archives: Activating Textile Archives Through Artificial Intelligence" (Bologna and Rimini, 6–8 November 2024). Promoted by the Master's Degree in Fashion System Design of the Department of Architecture—University of Florence, the workshop was held in collaboration with the Master's Degree in Fashion Studies of the Department of Arts—University of Bologna and the Fashion Research Italy Foundation (F.FRI). The aim of the workshop was to verify the interactions between artificial intelligence and archival materials from a practical point of view and in a controlled setting, and was one of a growing number of initiatives in Italy and abroad actively reflecting on fashion studies education. As this essay wants to demonstrate, "Weaving Archives" and similar initiatives represent valuable opportunities to educate future professionals to balance the global pervasiveness of fashion with its complex local implications. In fact, by distancing itself from a merely aesthetic and functional point of view, and also moving away from quantitative and economic assessments, the workshop was intended to be a space for reasoning about the dialogue between technology, sustainability, culture and politics (Bhardwaj and Fairhurst 2010). With the dual aim of addressing the contemporary and imagining possible futures, during "Weaving Archives" students and teachers applied a transdisciplinary approach to explore the relationships between the physical and digital dimensions, the synergies between theoretical and design aspects, the relationship between archives and innovation, and the original dialogue between human creativity and artificial intelligence, verifying some of the many new vectors necessary for today's fashions.

2 Spaces, Times and Fashions

Recent reflections on recognised practices in the area of cultural sensitivity have gained prominence in fashion studies (Green and Kaiser 2017) and have emphasised the need to adopt a sensitive, respectful approach to cultural influences, highlighting the challenges the sector faces (Piancazzo 2025). Prominent among these is the topic of cultural appropriation, which consists of the misuse of other people's cultural forms and elements, and which in recent times has led to important discussions on the responsibilities of fashion in twenty first century society. The latter have also been widely accepted and explored in Italy by academics, who are interested in better understanding the social implications arising from insensitive or ethically problematic designs, garments and communication choices. A striking example of this are the communication crises that have involved many brands and professionals (Cantoni et al. 2020, p. 121). Notable cases include the offensive commercial produced by Domenico Dolce and Stefano Gabbana in 2018 (Huang and Janssens 2019), the headgear made by Gucci that replicated the turbans of the Sikh community (Noris and Cantoni 2022, pp. 6–93), and the controversies sparked by celebrities "donning the robes" of certain religious figures, which triggered media scandals and raised accusations of cultural insensitivity (Piancazzo 2023a). These cases often arise from distorted representations that perpetuate long-established stereotypes in the industry (Piancazzo 2025) and overlook the cultural and social complexities of different spatial and temporal contexts.

In the face of these critical issues, there has been a growing reflection on the role of fashion and the need to recognise its cultural and geographical roots. In other words, we can now say that there is a collective realisation at the global level that fashion originates in specific historical and geographical contexts, and that therefore, as Simona Segre Reinach has suggested, it would first of all be more appropriate to use the term in the plural and speak of fashions rather than fashion (Segre Reinach 2019). Time and space are two inescapable references to identify the place of each of these fashions in the endless, stormy ocean of contemporaneity. However, as coordinates have long been identified according to Euro-American parameters, knowledge of fashions is extremely limited (Ahmed 2022). According to Cupples (2019, p. 2), the scientific output of Western universities still reflects Eurocentric epistemologies (Ahmed 2022), presenting them as objective. In order to chart new courses and thus generate new narratives, there is a need now more than ever to decentralise, decolonise and deconstruct Eurocentric assumptions, proposing points of view that are historically aware and geographically inclusive (Cheang and Suterwalla 2020), bearing in mind that, as pointed out by Loomba (2006), colonial methods and imagery vary in time and space.

This reasoning was explored in November 2024 by the authors of this paper thanks to the "Weaving Archives" workshop. The Fashion Research Italy Foundation (F.FRI) in Bologna, which collaborated with the workshop, is a fashion research and educational centre and has an archive group containing 30,000 textile designs on paper and fabric, the result of the activity of Renzo Brandone's converter, Silkin (D'Angelo 2019). Produced between the late 1970s and 2013, this archival material provides insight into much more than just the company's entrepreneurial trajectory. In fact, thanks to the recollections of the most renowned national and international graphic design studios, including Giuseppe Spadacini, D.R., Farkas Ortensi, Nuova Idea and Studio Lanati (Fashion Research Italy 2025), which for several decades have collaborated with the most important fashion brands, including Aspesi, Iceberg, La Perla, Les Copains, Loewe, Max Mara, Moschino, Prada, Ralph Lauren and Versace (Fashion Research Italy 2018, p. 21), the archive allows us to retrace the routes that led Silkin and its creatives to explore territories considered original in their day. Among them, two continents stand out for the number of references: Africa and Asia (Fig. 1).

The "Weaving Archives" project is based on a circumscribed selection of materials that, while representing a very small part of the documents in the Brandone collection, seemed capable of producing reflections on the meaning of cultural relations and thus take the discourse on fashion far beyond clothing (Ling and van Dartel 2019). Operating in this way, the working group tried to orient the research and teaching towards a transcultural approach capable on the one hand of connecting people, cultures and everyday practices, and on the other inserting them into global historical dynamics (Ling 2019). The search for "exotic" motifs as a symbol of

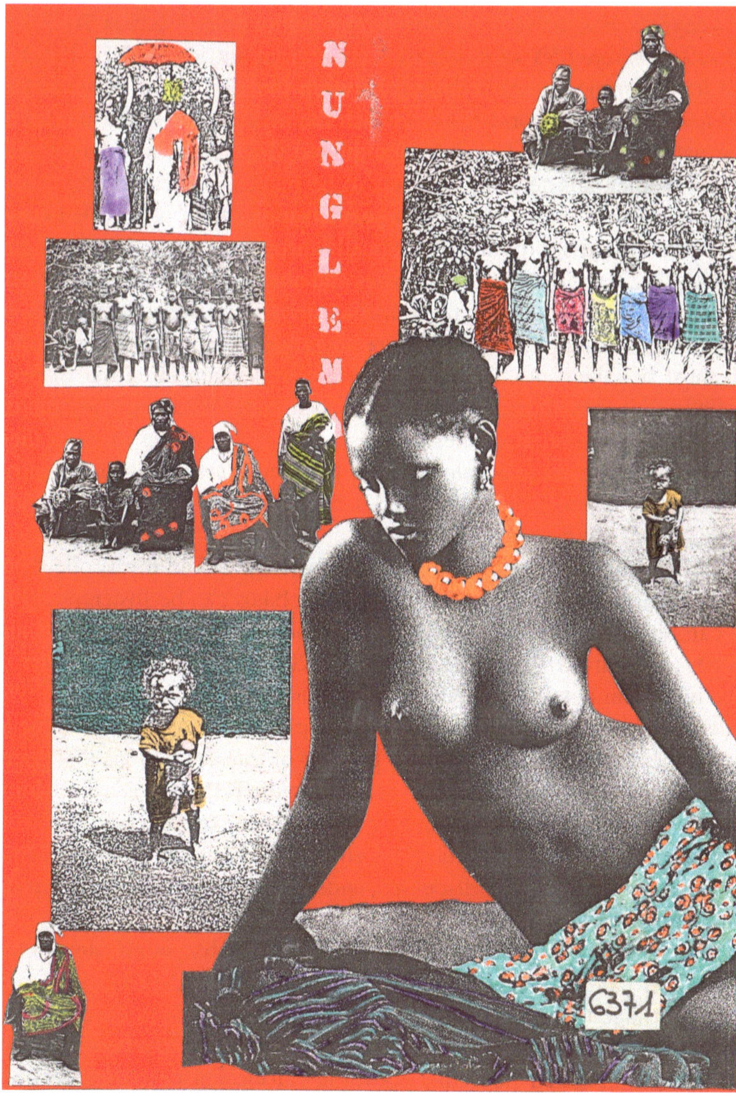

Fig. 1 Fondazione Fashion Research Italy, drawing 6371, 1987, from the Renzo Brandone archive

"novelty" starting in the last century (Kawamura and de Jong 2022, pp. 51–52) over time produced some designs and textiles that are considered extremely original, but, when analysed through contemporary cultural sensibilities (Craik 2018, pp. 97–118), highlight issues that the working group made up of teachers and students from different geographical backgrounds sought to tackle. In fact, the profile of the participants at "Weaving Archives" reflected that of the students enrolled in the two

Master's degrees, benefiting from the internationality of the Fashion Studies department (UniBo), where 85% of its students come from outside the EU. The drawings selected for the project were produced between the 1970s and 1980s and share a colonial and colonising worldview, in which the myth of the white man as ruler of both the natural and animal worlds is amplified (Fig. 2). In a different selection, indigenous women are depicted as fruits to be picked, objects to be possessed or animals to be tamed, reflecting a hyper-sexualisation of the other (De Petris 2005, p. 270). Local customs, presented uncritically through a "white lens" (Hogans and McAndrews 2023, p. 501), fuel stereotypes and alter images and meanings, making a true understanding of complex phenomena impossible. These documents, while a small part of the Brandone collection, are nevertheless a constituent part of it, and as such must be taken into account in Silkin's arc, which on closer inspection is intertwined as much with the history of products Made in Italy as with that of Italy. Deconstructing the content of these representations, confining them in their own space and time, invites us to develop a capacity for critical reading, thus contributing to a process of decolonising fashion and decentralising its histories, an endeavour that today is more necessary than ever.

By highlighting the complexities of fashion heritage and archival collections such as the Renzo Brandone Archive Group, broad debates can be fostered and fashion studies can be steered towards more inclusive positions on the one hand and new approaches to research on the other (Calanca 2020). At a time when Western universities are increasingly talking about cancel culture (Norris 2023, p. 146), "Weaving Archives" chose to use documents that today appear controversial to say the least in order to "understand social change as a product of transformations that occur over time and influence the practices and phenomena studied" (Magaudda and Mora 2019, p. 3). In an environment dedicated to learning, to peer exchange and "protected" in a way that only a research institution can ensure, the contents of archival materials have been brought to light to be discussed, understood and subsequently transformed through participatory cultural and creative practices.

The Brandone archive group, as much as any other public or private archive, now more than ever can no longer be understood only as a place for the preservation of cultural heritage (Pompa 2020, p. 33), nor as a mere tool to enhance corporate identity and longevity (Riviezzo et al. 2016), being reduced to a heritage marketing tool (Mosca 2017). Rather, the archive constitutes an ideal place to share and therefore to teach fashion, as it represents a space offering "a convergence between material, immaterial, natural and digital aspects according to a dynamic vision of culture that continuously produces and reproduces its expressions" (Tufarelli 2022, p. 17). In other words, these are spaces that foster both tangible and intangible aspects of cultural heritage and related practices (Harrison 2010, pp. 8–9), operating on three main levels: cognitive, symbolic and social. On the one hand, these spaces produce and sediment knowledge, holding meanings that represent symbols of shared traditions and identities, and on the other they act on processes of communication, organisation and distribution (Tufarelli 2022, p. 36). In this perspective, the materials of the Renzo Brandone Archive Group selected for the "Weaving Archives" project have been "activated" in their complexity and by virtue of it, promoting critical and

Fig. 2 Fondazione Fashion Research Italy, drawing 6655, 1975–1986, from the Renzo Brandone archive

design reflections capable on the one hand of spotlighting the historical dimension of the heritage as a whole, and on the other the respect for the cultural geographies from which they draw inspiration. In this sense, the "Weaving Archives" project aimed to promote a more ethical and sensitive approach to cultural diversity (Piancazzo 2023b), allowing the complexities and contradictions of the many fashions that sail our seas to be addressed.

3 Languages, Images and Data

Contemporary fashion, and consequently fashion education, require a balance between issues of cultural sensitivity and the exploration of developments related to digital technologies, in particular the widespread and easy accessibility of visual and textual generation tools through artificial intelligence. As is well known, fashion and other cultural and creative industries are going through a phase of significant transformation induced by the "digital turn". This technological and digital transformation dissolves the boundaries between the material object and the imaginary realm (Han 2023), evolving into a hybrid system supported by complex networks of human and non-human actors. Today, artificial intelligences, protagonists of a post-human vision of fashion (Smelik 2019), are revolutionising the system in which human and non-human, organic and artificial connect in a single system of relations (Franzo et al. 2024). The data exploited by AI, derived from text, images and video, together with the algorithms that process them, are intrinsically involved in a process that results in new ways of dis-organising, de-stratifying and de-territorialising fashion. This is why it seems particularly relevant to "test" AI with a system such as that of an archival group with diametrically opposed characteristics. Starting from this premise, the "Weaving Archives" workshop was conceived to explore the possible routes of reactivating archive materials in collaboration with AI, bringing the documents and the narratives they bear into contemporary cultural sensitivity. The difficult interaction between past and present was compounded by the complexities of working in groups, which involved students from different geographical areas and thus characterised by varied cultural sensitivities.

During the first phase, each group was assigned a textile image from the Renzo Brandone Archive Group and asked to critically analyse it (Fig. 3). Supported by text-to-text generative artificial intelligence tools, materials, colours, shapes and patterns were described, delving into the historical, cultural, material and immaterial aspects of textiles and identifying possible critical issues of meaning and representation. The decision to enter only textual data into the AI software at this stage and not directly the image provided derived not only from the need to respect the copyright and to use the images provided by F.FRI in a responsible manner, but also from the need not to feed the algorithm with data and information that would risk reiterating stereotyped and, as such, outdated narratives. The students were then asked to develop a project with a common goal: to redesign the textile designs by transforming the critical aspects of the archive material into innovative elements. In particular, using text to image and image to video generative artificial intelligence tools, each group designed and implemented an advertising campaign to promote the redesigned fabrics, which convey updated values.

The decision to focus the workshop on the development of an advertising campaign was motivated by the realisation that in the field of design and fashion advertising still plays a crucial role not only for the promotion of products, but also as a means of building brand identity and nurturing the imagination of consumers. As shown by the numerous campaigns published in Vogue Italia since the first editions in

Fig. 3 Fondazione Fashion Research Italy, drawing 6407, 1986, from the Renzo Brandone archive

1964, Italian textile companies presented the innovation of their products by collaborating with various international fashion brands, highlighting the fundamental role of the textile industry in the national economy of the twentieth century. While in the past such partnerships had proved to be an effective promotional tool, in the contemporary context historical analysis and semiotic criticism have highlighted the profound social implications of certain marketing rationales. Over time, the pervasiveness of promotional campaigns and messages has shaped social, cultural and identity issues related to bodies, sexualising them, commodifying them, perpetuating gender stereotypes and reinforcing often unattainable aesthetic standards (Wolf 1991). On the other hand, such campaigns contributed—albeit indirectly—to the colonial imagery typical of the era, reinforcing the stereotypes (Faloppa 2022) that influenced society (Berger 1972; Bordo 1993; Evans 2003; Solomon et al. 2006). The visual synthesis imposed by communication through images proved to be the ideal tool for "Weaving Archives" to address the theme of social, economic and cultural change, while at the same time exploring the possibility of developing ethical, inclusive and sustainable narratives from the most controversial characteristics of the images (Fig. 4).

Fig. 4 Fondazione Fashion Research Italy, drawing 7263, 1987, from the Renzo Brandone archive

In this perspective, artificial intelligences can be seen as an actor, enabling them to generate, visualise and interact with new design possibilities, becoming an interlocutor with which to exchange ideas, visual references, opinions and feedback, and to elicit empathy. Not without difficulty, AI seems to push the fashion designer to a quantum leap: no longer a solitary monad, but rather the central node of an autopoietic system, where it performs the task of organising and activating data through conversational and generative processes. To do this, it is crucial to test the robustness of AI algorithms and understand their mechanisms, integrating ethical and value considerations into the co-creation process. The creation of advertising campaigns by means of artificial intelligence systems, both textual and visual, as in the case of the "Weaving Archives" workshop, fostered a critical understanding of the potentials and limitations of these technologies, prompting participants to an inclusive and sensitive use of the tools available for creativity. In this way there was a shift towards the creation of new value images capable of responding to the needs of a contemporary, more attentive and aware public (Floridi 2018).

In this sense the role of AI is not to replace human creativity, but rather to amplify it, acting as partners in the creative process, opening up new opportunities for co-design between humans and machines (Manovich 2020). In fact, in academia the use of AI to generate textures, patterns or visual concepts offers students new expressive possibilities. However, these technologies require training that meshes technical skills with a solid critical and creative foundation. Indeed, the language and prompts used in AI play a crucial role in the creation of new content. As mentioned above, in fashion and design courses students can use prompts to explore alternative scenarios, create complex narratives and develop visual outputs that challenge conventions. Such approaches are supported by a growing literature, which highlights how controls can act as catalysts for innovative design processes (Dale 2020). The use of prompts enables the generation of ideas that would not be able to emerge using traditional methods, thus promoting experimentation that is as free of preconceptions as possible. Bender and Koller (2020) emphasise the importance of language, understood as a tool capable of constructing shared contexts, as a fundamental aspect in the development of creative processes.

In the contemporary landscape, the relationship with data acquires new complexities, as the data itself becomes an autonomous entity that invites intimate interactions and manifests distinctive characteristics. Data acts as a catalyst for processes of discovery, fostering the construction of connections that go beyond traditional socio-cultural constructs. This evolution pushes human beings—and designers in particular—beyond the boundaries of their ontological nature, deconstructing the anthropocentric paradigm and giving rise to new relational modes, not only between individuals but also between people and data. These reflections make a reconsideration of previous rationales imperative, placing the role and competences of the fashion designer and the professionals involved at the centre of the debate (Vaccari 2012). By collaborating with emerging knowledge and technologies, the designer becomes the explorer who, armed with their own sail, ventures out to discover new territories and new conceptions of making and understanding fashion.

Fashion thus emerges as a set of dynamic and collaborative design spaces where a variety of interpretations and perspectives can be explored and included, allowing stories and visions to evolve and take on different forms, transcending a single meaning or traditional vision. In this global ocean, the data is not reduced to a mere aesthetic or symbolic encoding, but constitutes a linguistic act capable of operating on different sensory and interpretative levels (Nachtigall 2021). According to the observations that emerged from the workshop, the information presented and the cultural and design choices of fashion designers—our "sailors"—can play a decisive role in shaping the perception of users, re-orienting emotions and generating a renewed sensitivity.

4 Conclusions: Towards New Fashions

Recalling the theories of Latour (1999) applied to fashion design by Joanne Entwistle, a reconsideration of the concept of actor emerges, which is not limited to identifying a human subject endowed with consciousness, but includes any entity capable of acting or generating action (Entwistle 2015). This approach is part of a vision of fashion as an intrinsic hybrid between culture and nature: it represents the end product of complex interventions in the natural world and is deeply connected to global networks of actors, both human and non-human. Consequently, fashion should be interpreted as a heterogeneous system composed of a wide range of interconnected actors (Entwistle and Slater 2014; Sommerlund 2008). Nowadays, while on the one hand researchers aim to review the history of fashions through an inclusive approach, on the other hand we are witnessing the emergence of new actors who are redefining the boundaries of the possible as we write. Artificial intelligences offer tools that can transform the way fashion designers can conceive and realise their designs, generating images in real time and hypothesising predictions of new trends.

Considering the contemporary landscape of the emergence of AI in fashion, it is possible to outline two key points that are also useful for rethinking fashion education. The former concerns artificially produced visual narratives, the latter the practices through which such creations are realised.

In light of the reflections that have emerged in this paper, we feel it is important to emphasise that the education of fashion designers requires a multidisciplinary approach that is both critical and historical, and capable of integrating tangible and intangible aspects. From this point of view, exploring the interactions between archival materials and artificial intelligence offers itself not only as an opportunity to generate new images, but also and above all as a test bed to verify the correctness of our educational approach. In the case of "Weaving Archives", the use of materials with a well-defined spatio-temporal imprint made it possible to control and direct the creation of new visual products, producing results in line with the most current debates on cultural sensibility. The archive was thus never intended as a mere container of images and visual stimuli, but as a repository of critically valued and historically structured knowledge.

Second, the growing importance of co-production and co-designing practices in fashion must be emphasised. These concepts find resonance in the design theories of Manzini (2015) who defines co-design as "a social conversation in which different actors interact in different ways, ranging from collaboration to conflict" (Manzini 2015, pp. 4, 49). Fashions today are thus understood as the results of a social conversation, which requires the contribution of different actors to take place. Thanks to such diversity, or rather through a socially and culturally inclusive approach, designers' practices can challenge traditional canons and push towards fashions that, while retaining their specificities, contribute to the definition of images with an intrinsically global value.

In conclusion, the new vectors of the many fashions that make up today's system call for a profound change in teaching methods before professional practice, aiming at inclusive collaboration and conscious creativity. In the "Weaving Archives" workshop the students shared critical approaches using technological tools in a system that enhanced the skills of each and promoted a creative, conscious approach to archive materials. In this process, recognising the importance of emotional positions in teaching and learning has become a key aspect. Only through "learning to unlearn", i.e. by freeing oneself from the paradigms of thought imposed by educational and cultural traditions (Cheang and Suterwalla 2020), will it be possible to open oneself up to new perspectives and charting new courses, navigating towards alternative horizons, aware of how challenging the contemporary currents in the field of fashions can be.

Acknowledgements The workshop "Weaving Archives: Activating Textile Archives Through Artificial Intelligence" was funded by the Department of Architecture, University of Florence, through the 2024 call for national and international educational initiatives.

References

Ahmed T (2022) What is education in fashion? Reflecting on the coloniality of design techniques in the fashion design educational process. Int J Fash Stud 9:401–412. https://doi.org/10.1386/infs_00079_7

Bender EM, Koller A (2020) Climbing towards NLU: on meaning, form, and understanding in the age of data. In: Proceedings of the 58th annual meeting of the Association for Computational Linguistics, pp 5185–5198. https://doi.org/10.18653/v1/2020.acl-main.463

Berger J (1972) Ways of seeing. Penguin Books, London

Bhardwaj V, Fairhurst A (2010) Fast fashion: response to changes in the fashion industry. Int Rev Retail Distrib Consum Res 20(1):165–173. https://doi.org/10.1080/09593960903498300

Bordo S (1993) Unbearable weight: feminism, western culture, and the body. University of California Press, Oakland

Calanca D (2020) Archivi digitali della moda e patrimonio culturale tra descrizione e integrazione. ZoneModa J 10(2):11–25. https://doi.org/10.6092/issn.2611-0563/11793

Cantoni L, Cominelli F, Kalbaska N, Ornati M, Sádaba T, SanMiguel P (2020) Fashion communication research: a way ahead. Stud Commun Sci 20(1):121–125. https://doi.org/10.24434/j.scoms.2020.01.011

Cheang S, Suterwalla S (2020) Decolonizing the curriculum? Transformation, emotion, and positionality in teaching. Fash Theory 24(6):879–900. https://doi.org/10.1080/1362704X.2020.1800989

Clément G (2008) Il giardiniere planetario. 22 Publishing, Milano

Craik J (2018) Exotic narratives in fashion: the impact of motifs of exotica on fashion design and fashionable identities. In: Jansen MA, Craik J (eds) Modern fashion traditions: negotiating tradition and modernity through fashion. Bloomsbury Academic, London, pp 97–118. https://doi.org/10.5040/9781474229524.ch-005

Cupples J (2019) Introduction: coloniality resurgent, coloniality interrupted. In: Cupples J, Grosfoguel R (eds) Unsettling Eurocentrism in the Westernized University. Routledge, Abingdon, pp 1–22

D'Angelo C (2019) Il ruolo degli archivi della moda nella «formazione» per le piccole e medie imprese. Econ Cult 1:129–134. https://doi.org/10.1446/93085

Dale R (2020) GPT-3: what's it good for? Nat Lang Eng 27(1):113–118. https://doi.org/10.1017/S1351324920000601

De Petris S (2005) Tra “agency” e differenze. Percorsi del femminismo postcoloniale. Studi Culturali 2:259–290. https://doi.org/10.1405/20855

Deleuze G, Guattari F (1988) A thousand plateaus: capitalism and schizophrenia. Bloomsbury Publishing, London

Entwistle J (2015) Bruno Latour: actor-network-theory and fashion. In: Rocamora A, Smelik A (eds) Thinking through fashion: a guide to key theorists. I.B. Tauris, London, pp 269–284

Entwistle J, Slater D (2014) Reassembling the cultural: fashion models, brands and the meaning of “culture” after ANT. J Cult Econ 7(2):1611–1677. https://doi.org/10.1080/17530350.2013.783501

Evans C (2003) Fashion at the edge: spectacle, modernity and deathliness. Yale University Press, New Haven

Faloppa F (2022) Sbiancare un etiope: la costruzione di un immaginario razzista. De agostini - Utet Libri, Milano

Fashion Research Italy (2018) Polo d’eccellenza a servizio delle manifatture moda. https://www.erchives.it/erchives/download/FRI%20-%20Corporate.pdf

Fashion Research Italy (2025) Fondo Renzo Brandone. https://www.ffri.it/archivio-moda-online/fondo-renzo-brandone/#page-1

Floridi L (2018) Soft ethics and the governance of the digital. Philos Technol 31(1):1–8. https://doi.org/10.1007/s13347-018-0303-9

Franzo P (2021) Il portfolio come progetto. Una ricerca nei corsi di laurea in moda. Bembo Officina Editoriale, Venezia

Franzo P, Quartu A, Cianfanelli E (2024) Co-designing fashion with AI in the phygital scenario. Convergences. J Res Arts Educ 17(34):81–92. https://doi.org/10.53681/c1514225187514391s.34.269

Green DN, Kaiser SB (2017) Fashion and appropriation. Fash Style Popul Cult 4(2):145–150. https://doi.org/10.1386/fspc.4.2.145_2

Guattari F (2005) The three ecologies. Bloomsbury Publishing, London

Han B (2023) Le non cose. Come abbiamo smesso di vivere il reale. Super ET, Bologna

Harrison R (2010) What is heritage? In: Harrison R (ed) Understanding the politics of heritage. Manchester University Press, Manchester, pp 5–42

Hogans K, McAndrews L (2023) Away from violence toward justice: a content analysis of cultural appropriation claims from 2013–2020. Fash Pract 15(3). https://doi.org/10.1080/17569370.2022.2118333

Huang Q, Janssens A (2019) Come mangiare un cannolo con le bacchette: the contested field of luxury Fashion in China, a case study of the 2018 Dolce & Gabbana Advertising Incident. ZoneModa J 9(2):123–140. https://doi.org/10.6092/issn.2611-0563/9970

Kawamura Y, de Jong JWM (2022) Cultural appropriation in fashion and entertainment. Bloomsbury Publishing, London

Latour B (1999) On recalling ANT. Sociol Rev 47(1):15–25. https://doi.org/10.1111/j.1467-954X.1999.tb03480.x

Ling W (2019) Global fashion as a tool in and out of the ethnographic. In: Curating fashion in and out of the ethnographic, conference presentation, Tropenmuseum, Amsterdam

Ling W, van Dartel D (2019) Global fashion as a tool in the ethnographic museum. ZoneModa J 9(2):71–88. https://doi.org/10.6092/issn.2611-0563/10048

Loomba A (2006) Colonialismo/postcolonialismo. Meltemi, Roma

Lupano M, Vaccari A (2014) Insegnare il design della moda. Aracne Editrice, Roma

Magaudda P, Mora E (2019) The contamination of practices: how practice theories matter in multiple domains. Sociologica 13(3):1–10. https://doi.org/10.6092/issn.1971-8853/10269

Manovich L (2020) AI aesthetics. Strelka Press, Moscow

Manzini E (1997) Progettare la sostenibilità, Leapfrog: anticipazioni di un futuro possibile. Domus 789:41–47

Manzini E (2015) Design, when everybody designs: an introduction to design for social innovation. MIT Press, Cambridge

Mosca F (2017) Heritage di prodotto e di marca: modelli teorici e strumenti operativi di marketing per le imprese nei mercati globali del lusso. Franco Angeli, Milano

Nachtigall T (2021) Data as a material for fashion. Eburon Academic Publishers, Utrecht

Noris A, Cantoni L (2022) Digital fashion communication: an (inter) cultural perspective. Brill

Norris P (2023) Cancel culture: myth or reality? Polit Stud 71(1):145–174. https://doi.org/10.1177/00323217211037023

Piancazzo F (2023a) Celebrities dressed like a goddess: admiration, cultural appropriation and disrespect. Film Fash Consump 12(2):231–246. https://doi.org/10.1386/ffc_00064_1

Piancazzo F (2023b) Developments of cultural appropriation in fashion: an in-progress research. In: Sabatini N, Sádaba T, Tosi A, Neri V, Cantoni L (eds) Fashion communication in the digital age. FACTUM 2023. Springer proceedings in business and economics. Springer, Cham, pp 136–143. https://doi.org/10.1007/978-3-031-38541-4_13

Piancazzo F (2025) Moda e sensibilità culturale. Tra appropriazione e ibridazioni. Collana: Cultura, moda e società, Mimesis Edizioni, Sesto San Giovanni

Pompa C (2020) La memoria 'estesa' della moda. Come valorizzare l'heritage aziendale con la tecnologia. ZoneModa J 10(2):27–41. https://doi.org/10.6092/issn.2611-0563/11911

Riviezzo A, Garofano A, Napolitano RM (2016) "Il tempo è lo specchio dell'eternità". Strategie e strumenti di heritage marketing nelle imprese longeve italiane. Il capitale culturale 13:497–523. https://doi.org/10.13138/2039-2362/1360

Segre Reinach S (2019) Un mondo di mode. Laterza, Bari

Smelik A (2019) New materialism: a theoretical framework for fashion in the age of technological innovation. Int J Fash Stud 5(1):33–54. https://doi.org/10.1386/infs.5.1.33_1

Solomon MR, Bamossy G, Askegaard S, Hogg MK (2006) Consumer behavior: a European perspective. Pearson Education, London

Sommerlund J (2008) Meditations in fashion. J Cult Econ 1(2):165–180. https://doi.org/10.1080/17530350802243578

Tufarelli M (2022) Design heritage e cultura digitale. Scenari per il progetto nell'archivio diffuso. Firenze University Press, Firenze

Vaccari A (2012) La moda nei discorsi dei designer. Clueb, Bologna

Wolf N (1991) The beauty myth: how images of beauty are used against women. HarperCollins, New York

“Physical and Intangible: Rethinking How to Teach Fashion Design in the Age of Immateriality”

Giovanni Maria Conti

Abstract The following chapter investigates the various avenues that the teaching of Fashion Design is now called upon to explore in a context in which virtual and augmented reality, as well as the metaverse, call into question the very meaning that the term ‘reality’ now holds within the sector. The vast array of questions that will arise on the roles of design, trained professionals, and even the relationship between manual work and mechanical work when hybridised—which has always been a feature of Italian products—will be considered not with a view to finding an answer, but as a means of bringing our focus back to education and training and, specifically, to the fundamental role that they play in constructing the kind of useful, versatile knowledge base and skill set that are becoming increasingly crucial in the professional world. Whilst fashion has always been the stage for an intertwining of bodies, arts, techniques, politics, approaches and materials, this is now truer than ever thanks to new technologies, which allow us to work in the virtual sphere as if it were our own reality, making it important to analyse and consider where we now stand as designers. In this day and age, the concept of remanufacturing (Barucco et al. in Remanufacturing Italy. L’Italia nell’epoca della postproduzione [Remanufacturing Italy: Italy in the Age of Post-Production], Venice, Mimesis, p 11 (2020))—a metaphor that encapsulates the current era of post-production in which products are manufactured from objects that are already in circulation, rather than from raw materials—has some clear points of affinity with the concept of the virtual, suggesting the possibility that traditional production systems could be reconfigured through these new technologies. Finally, we will consider the idea of ‘Made in Italy’: an international metabrand, an intricate and multifaceted heritage of techniques and products, approaches and materials that is yet another creative workshop that fashion designers must engage with. Indeed, it is these very designers who, faced with the paradigm shift being brought about by a wave of new technologies, are paving the way for a new development model in which the concept of creativity is no longer linked exclusively to the creation of products and systems, but instead to designing the memory of the future (Ibid. p.11.).

G. M. Conti (✉)
Design Department, Fashion Design Degree Programme, Politecnico di Milano, Milan, Italy
e-mail: giovanni.conti@polimi.it

D. Raposo et al. (eds.), *Perspectives on Design IV*, Springer Series in Design and Innovation 59, https://doi.org/10.1007/978-3-032-00733-9_15

Keywords Fashion design · Training · Real-virtual · Remanufacturing · Made in Italy

1 Introduction

We exist in an age in which changes to our living conditions are the order of the day; everything we know—or think we know—the rules and dynamics of is constantly being challenged and overturned, ultimately leading us to a series of necessary reflections: what values do we convey through the things we design? What is our role as designers within an ever-changing context?

Is the common concept of fashion still centred around consumerism, beauty, luxury, elegance and distinction? Does it still exist in the collective imagination as a 'magical world' of sorts, considered by many to be ephemeral, in which the image constitutes the sole fundamental element?

Or, conversely, is fashion now a more diverse and multifaceted landscape that is once again placing people at its heart, now no longer merely as consumers, but instead as part of a community for which we design not only products but, increasingly in the coming years, new ways of interacting with objects?

During the FARB research project at the Politecnico di Milano entitled 'Con_Text: Design-Oriented Solutions to Connect Young Creative Talents with the Knitwear and Textile Manufacturing Sector',[1] an opportunity arose to observe first-hand how in the Italian textile and clothing sector, a growing number of debates and new considerations are emerging regarding the role of the fashion designer in a context in which the product is increasingly obfuscated, clouded and obscured in favour of systems and services tasked with modifying, enhancing and redefining the meaning of the product itself.

> Companies such as Loro Piana need highly competent interfacing people capable of engaging with both sides: people who can understand the creative aspect of design and grasp where the brand wants to go with the product and what they are looking for, but also the constraints associated with the technical world underpinning this process. Enrico Ciccone, Head of Yarns B.U. Sales, Loro Piana.[2]

It is in this context that there is an emerging interest in the concept of 'post-growth fashion', in which the role of the fashion designer is to design a fresh set of values and in which prosperity and wellbeing are not exclusively linked to a purely economic concept, but to the blossoming of a collective sense of wellbeing; in this designer's role, our collective wellbeing becomes clear in the relationships that exist between us and other people, between us and our objects, and between us, our objects and the planet. Design is once again thrust into a central role, not only in terms of the design of individual product systems, but also in the construction of unprecedented

---

[1] https://www.context.polimi.it/

[2] Report from the Con_Text research project, available for download from https://www.context.polimi.it/wp-content/uploads/2023/06/Report-doppia-pagina.pdf, p. 16.

systems of storytelling and communication, in the identification of novel business models, or in the reclamation—through a new lens—of approaches to production and consumption from a bygone era.

The paradigm of Industry 4.0,[3] as it currently exists, was never fit for purpose in the context of a climate crisis and a planetary emergency, nor did it incorporate measures to address our deep-rooted social tensions. Industry 5.0, meanwhile, represents an evolutionary step up from Industry 4.0, placing a strong emphasis on human-centricity, sustainability and resilience; as such, it becomes the duty of university education to contribute to a fresh vision, oriented towards a new way of planning business models, insofar as they are the value chains in which collaboration with companies and industry promotes new skills and approaches to research and innovation. Technology must be used to facilitate human work, rather than to replace it, thus striking an ideal balance between mechanical efficiency and human creativity.

As for the field of fashion specifically, if it is to survive, it must reclaim its original function and cater first and foremost to the needs of the public, of everyday people in real life; universities play an active role in promoting this new vision, including from an industrial perspective, by preparing students to contribute to a future in which technology and humanity coexist in a harmonious, productive way.

1.1 'Made in Italy' as a Metaphor for Cultural Manufacturing

A passionate belief in a job well done, along with an in-depth knowledge of materials and production techniques, are the elements that have contributed most significantly to the quality of Italian-made products. Stefano Micelli, in his book *Futuro Artigiano* [*Artisanal Future*],[4] the tacit forms of knowledge linked to artisanal production—which have long constituted a competitive edge for Italian-made products on the international scene—are not only the levers that now demonstrate the product's excellence, but also highlight the artisans themselves, giving them a strategic role in innovation processes; indeed, although Gucci is an international brand, it still keeps its production chain based in Italy—specifically in Tuscany—because it is there that the cultural and operational seeds associated with handbag production planted over a century ago have sprouted and blossomed, allowing its unique manufacturing culture to take root in the area.

Renowned anthropologist Desmond Morris once said that "art is what makes the ordinary extraordinary, in order to amuse the brain. Science is what makes

[3] Industry 5.0: A Transformative Vision for Europe ESIR Policy Brief No. 3, various authors, 2022: 5.

[4] Micelli (2011)

the complex simple, in order to understand existence."[5] Design, meanwhile, has always transformed otherwise unremarkable everyday objects into products capable of feeding the imagination and inspiring an emotional response in consumers. In Italy alone, professionals such as Achille Castiglioni, Marco Zanuso, Aldo Rossi, Vico Magistretti and many more besides redefined the aesthetic redesign of everyday items through their work. The desire of these designers was to create products that would fly in the face of the tendency towards standardisation caused by mass production, offering consumers the opportunity to stand out from the crowd through the beauty of the object itself. At the heart of the new approach proposed by these designers were the artistic sensibilities of many Italian entrepreneurs who decided to take a risk by producing products with unusual shapes and colours, setting them apart from everything else available on the market at the time. Whilst this focus on the aesthetics of a product initially developed in the homeware and furnishing sectors, it quickly spread to other sectors of Italian-made products; Italian fashion underwent a rapid evolution in the '70 s and '80 s at the hands of designers such as Gianni Versace, Giorgio Armani, Gianfranco Ferré and Missoni who, by devoting themselves to prêt-à-porter offerings, focused their efforts on creating clothing that was more removed from the elitism of French haute couture, as well as more in tune with modern lifestyles and changes in society.

With this in mind, the importance of the link between culture and manufacturing—or rather, Italian manufacturing culture—cannot be overstated, insofar as it sheds light on the 'Made in Italy' concept's proven ability to provide a fresh and updated reinterpretation of the expertise inherent in traditional production. Seen from this perspective, one might say that manufacturing not only produces high-quality products, but indeed produces culture: a sort of cultural manufacturing that feeds on and develops cultural content.

The spread and growth of the Internet has triggered the development of a new generation of platform-based companies that have proven themselves capable of paving fresh and original avenues towards growth, breaking away from the paths established by the major companies that preceded them. In this sense, the concept of the virtual has a great deal in common with the idea of remanufacturing,[6] positing the potential for traditional production systems to be reconfigured through the capabilities of new technologies, with each individual component of the supply chain being promoted and incorporated into new production networks.

Several examples such as ArtKnit, Hands on Design and Up To You[7] demonstrate that, for what we can redefine as 'circular industries', new professionals must be trained in a way that ensures that they can effectively combine technical innovation

[5] Interview with Desmond Morris conducted by Paolo Foschini and published in Il Corriere della Sera on 20/09/2014, http://archiviostorico.corriere.it/2014/ottobre/20/DESMOND_MORRIS_CREA_SCIMMIA_NON_co_0_20141020_05896b7a-581b-11-e4-870b-5a57e987836f.shtml.

[6] Conti and Franzo (2020).

[7] Further details of the cases mentioned here are available in the chapter by Conti, G.M., Franzo, P. entitled Distretti produttivi virtuali. La transizione del Made in Italy nella moda [Virtual Production Districts: the Transition of 'Made in Italy' Fashion], ibid. p.132, 135, 137.

with environmental and social awareness within the framework of a highly complex system.

The evolution of the Internet and consumption habits also poses two new challenges to Italian companies. The first has to do with communication: in other words, the goal is now not simply to open one's own online sales channel or make the product available through one of the many existing multibrand portals, but rather to understand that each market can have its own platforms. The most interesting example of this comes from observing sales dynamics in China, where all users use WeChat rather than other types of platforms to purchase all manner of goods.

The second is technological in nature: although we are used to wearing clothes made of physical fabric and materials, an increasing number of brands are investing in virtual reality, in terms of both fashion items—that is, actual digital clothes or accessories that can be worn by avatars and virtual alter egos—and the places where these avatars can flaunt their virtual wardrobes, parallel realities where users can take part in events and meet new people.

This naturally leads us to a reflection on the world of digital and virtual fashion, on the very fabric of the clothes of the future: will it be dematerialised and immaterial? Further to this, will the body itself be completely dematerialised—an avatar, nothing more than a digital being—or will the physical body be enhanced by what it wears in the real world?

1.2 *Fashion Designers in the Age of Immateriality*

> In short, it will be crucial to fully reconsider - with the help of these very same technologies and new materials - the idea of a fashion created for human beings, so that we can be comfortable in our own bodies, which are not all modelled on a single, stereotyped ideal, but rather come with different ages, sizes, abilities, colours, skin types and fragilities.[8]

What does the world of fashion offer today and what are its new horizons, such as digital and virtual fashion? What will we be wearing in the not-too-distant future? How will the world of fashion change and what new forms of expression will materials take?

Clothing has always been the weapon and the flag of seduction; it is a physical product with a certain weight, shape and scent, made of materials in one or more colours. But in our modern age, we could say that the product of fashion is no longer simply clothing and all the various reflections that have gone along with it, but instead something new, something that we cannot yet see or, if it exists, perceive. Because there is no question about it: we are faced with a complex paradigm shift in which clothes, bodies, and methods of production and consumption are changing, and the logics and narratives of the past are no longer capable of fully grasping what we, as designers and instructors, are witnessing and perhaps even participating in.

---

[8] Calefato (2021).

The impact generated by the Internet on traditional companies has mainly affected their way of interacting with the demand of their end users; shared virtual concepts released online—on platforms open for designers to submit their ideas, such as in the case of Hands on Design[9]—are proposed to artisans who decide which ones to produce depending on their own skills and affinities, without ever necessarily finding out the name of the designer who created them. With this approach, the work of designers and artisans alike is guided by the platform in establishing a coherent image that links the various objects.[10]

During the COVID-19 pandemic, interest in the virtual world exploded in an unprecedented way. An ever-growing number of users decided to explore new ways of dressing up their digital alter egos, turning to new companies such as The Fabricant.

The Fabricant is a virtual fashion brand founded in 2018 by Kerry Murphy and Amber Jae Slooten.[11] The brand designs garments that exist only digitally, created with the use of 3D software, in a way that allows them to move in exactly the same ways as if they were being worn in the real world. Anyone who purchases their clothes can have their avatar wear them on social media, in game environments, and in all manner of virtual worlds. Slooten, a digital fashion designer and co-founder of The Fabricant, said: "*How do we want to represent ourselves within the virtual space? If we can be anything, will we still want to be ourselves?*", espousing the view that thanks to virtual fashion, people are free to live out their fantasies online.[12] The Fabricant hopes to make the world of digital fashion more accessible to fashion lovers, in contrast with physical fashion, which is all too often the preserve of industry insiders.

Creative, interactive digital spaces are a natural evolution of how people use technology, reflecting the growing amount of time that consumers spend online.

Robert Triefus, the former marketing director of Gucci, said: "*[People are] probably underestimating the value placed on people wanting to express themselves in a virtual world with a virtual product, [through] a virtual person*".[13] Indeed, it was Gucci that created the Gucci Garden, a space within the Roblox gaming platform that has attracted an impressive 19 million visitors, designed as a means of using the gaming platform to forge new communities within virtual worlds.

In 2022, the metaverse celebrated the first edition of Metaverse Fashion Week, a virtual fashion event in which various prestigious brands presented their collections on the virtual catwalks of Decentraland: a virtual social world that is the first decentralised metaverse, as well as the only open-source one. Within the Decentraland platform—which can be accessed through a browser or via a desktop client—users can create, experiment with and monetise content and applications, as well as socialise and take part in a vast range of community-led events held on a daily basis. Etro not only opened its first pop-up store in the Luxury Fashion District of

---

[9] Op. cit., (2020), p.137.

[10] Op. cit., (2020), p.137.

[11] Crook (2021)

[12] Crook (2021)

[13] Balchandani et al. (2022)

Decentraland, but also used it to present a collection celebrating the new Liquid Paisley motif: "a new, contemporary interpretation of one of the brand's most iconic codes in a vibrant, fresh and joyful colour palette, featuring lively shades such as candy pink, emerald green and sunshine yellow".[14] The outfits were worn by 20 avatars representing the genderless idea central to the collection whilst at the same time conveying a message of inclusivity and positivity to all users. Even the new pattern, inspired by the fluid spirit of freedom and harmony espoused by the brand, breaks down barriers and predefined norms to translate the idea of fashion into an inclusive vision that is perfectly in keeping with the metaverse as a dimension: open and accessible to all.

DressX was created in 2020 as the brainchild of Daria Shapovalova, Natalia Modenova and Julie Krasnienko, who wanted to revolutionise the fashion industry by harnessing the power of innovation in the digital sphere. The first step was allowing users to digitally try on the clothes shown in the photos on the website, and the service immediately attracted a huge audience, largely consisting of Millennials and Gen Zers in search of fashion experiences that were engaging, digital and, crucially, mindful of sustainability. It is exceptionally easy and intuitive to get started: a series of outfits designed by various digital fashion designers are available on the brand's platform, which can be browsed just like any other online retail platform.

The user chooses a garment and buys it, uploads a photo of themselves, and then the team receives the materials and creates a photo of the user wearing their chosen outfit. In just a few quick, easy steps, the user has purchased a new garment that they can wear digitally and show off on various social media platforms, obtaining their digital content in the space of 24 h. DressX has declared this array of opportunities to enrich one's virtual wardrobe—dubbed the 'metacloset'—a success amongst digital natives who love keeping up with the latest fashion trends. What's more, in 2024, they introduced a generative artificial intelligence tool that allows their audience to create customised digital fashion outfits using text or images as a reference point and inspiration; the tool then dresses them virtually in a matter of seconds, further blurring the already slim line between users' imagination and creativity and the world of virtual reality.

In short, the landscape is shifting and changing, so fashion designers must be aware that their work concerns an object that can be virtual as well as physical. This evolution in fashion product design will increasingly affect how it is taught; it is not a matter of promoting education and training that leads to competition between specialised, disruptive innovations, but rather one of adopting collaborative learning practices that involve sharing and exploring possible alternative paths[15]; indeed, the value of a designer's contribution is rapidly evolving from shaping future scenarios

[14] Various authors, (2022), La sottile linea tra reale e digitale, la moda sfila nel metaverso [The Thin Line Between Real and Digital: Fashion Takes to the Catwalk in the Metaverse], ansa.it, last accessed on 25.01.25 at 00.41.

[15] Badalucco and Cristofoli Ghirardello et al. (2020).

to educating people, citizens, consumers and users on the practices of imagining futures and the various processes and dynamics that go along with that.[16]

As such, technology will be the bridge between tangible and intangible reality, increasingly blurring the line between the two, with not only products, but also services taking on a growing importance for the promotion of high-quality goods; and even further to this, where the integration of the range of product manufacturing services available stitches together the real and the virtual.

2 Concluding Remarks

With technology being integrated into fashion, the experimental approach taken by brands aims to redefine the business models of traditional fashion—a cycle of production and consumption—by creating more transparency, efficiency and fairness. This is the case with The Dematerialised: the new technology that has been integrated into the fashion world has allowed the brand to create a digital fashion marketplace with fairness, accessibility and transparency as the cornerstones of its ethos. The brand considers itself responsible for the creation of new scenarios and worlds within the metaverse: inclusive, exceptionally innovative spaces in which a vast array of different identities and personalities can emerge, each one free to express their vision and creativity. Karinna Nobbs, CEO of The Dematerialised, told Vogue that digital fashion is a new, limitless way of experiencing fashion and expanding creativity, and as we are no longer limited by forces such as gravity or our physical location, we can experiment more freely with materials, textures and shapes. Indeed, it even allows us to evolve our identity, not only in digital environments such as gaming platforms or metaverse spaces, but also temporarily in real life thanks to augmented reality.

Getting dressed is a symbolic act: in so doing, we prepare ourselves to face the world, to interact with our respective communities, all with the aim of living a fulfilling life. Although one may believe that they dress exclusively to cater to their own personal tastes, Lipovetsky (2007) suggests that even the very act of choosing what to wear involves an implicit engagement with the wider community that surrounds each of us. If this has been true up to now—and perhaps still is—everything that is happening in the virtual world seems to raise a question concerning the body: who actually wears a garment these days? A physical body or an intangible body? The social nature of clothing derives from the fact that interpersonal communication, being intrinsically linked to the body, habitually makes use of clothing as a tool. In light of this, our first fundamental point becomes clear: without the body, we cannot communicate and, when it comes to teaching fashion design, we cannot disregard the fact of having a body to cover, to clothe. Clothing and the body are inextricably linked, and therefore any and all concrete reflections on clothing must necessarily consider the body as an integral part thereof.

[16] Celaschi et al. (2014).

The way a garment suits someone is not only about its relationship with the body, but also its relationship with the social context within which it is 'displayed'—'performed', even. As such, clothing the body means investing a certain audience with the responsibility of attributing meaning to it and, consequently, interacting with us in a certain way (Volonté 2003). This brings us to our second fundamental point, another concern in the discipline of fashion design, namely that it is almost impossible to attribute a specific meaning to a garment in isolation; we must always consider the clothed body as a whole within a certain social context and group or community.

The fundamental principle that underpins fashion never changes: the desire to appear a certain way, to attract other people's attention, to position oneself as an object of interest—or, alternatively, of disgust—remains the same as it ever was. Even those who object to fashion, positioning themselves outside of it for ethical or ideological reasons, nonetheless adopt a lifestyle that is in itself a form of exposition and differentiation. We cannot understate the importance of clothing as a key tool in expressing identity, both personal and collective. Indeed, clothing functions as an integral and even essential part of the human being—an additional organ, almost, without which social life would be unthinkable.

Due to the complexities hidden within the relationship between the body, its clothing and the space in which they communicate and interact with others, it would be rather challenging to define how this relationship may evolve in the near future, especially in light of technological development, which is a factor that may further influence this relationship, completely transforming it and perhaps even rewriting it into the virtual dimension. We have seen over the course of this in-depth exploration of the matter how fashion is increasingly using these channels to send new messages, or at least to convey them in unprecedented ways.

This brings with it a shift in the interactions between brands and users, but also between consumers themselves, who use virtual channels to communicate their personality and creativity in ways that can diverge significantly from their real-world selves, precisely because intangible realities place no limits on creativity. New digital media and the new worlds being created have the power to completely shatter the paradigm of communication as we have always known it, though whether this is for the better or for the worse remains to be seen.

The idea of wearing the intangible and expressing oneself through what is ultimately no more than a collection of data can either result in entirely new, more inclusive messages, creating communities that are accessible to an ever-growing audience, or alternatively foster utopian and wholly unrealistic canons of expression, increasing the frustration of their users once they are forced to return to the real world.

> What we need now is to rediscover our humanity and the otherness that inhabits us and the relationship with the other that constitutes us and our various spirits that link us to our land, to other cultures, to the other - be it animal, vegetable, material, machine or network - and to the intelligent, technological system that we ourselves have created. At that point, ethics

becomes the way we live and exist in a place that is shared with others, and culture is our ethical responsibility towards the community and how we care for that place.[17]

References

Badalucco L, Cristofoli Ghirardello P (2020) Nuovi modelli rigenerativi di produzione e consumo [New Regenerative Models of Production and Consumption], in Buracco MA, Bulegato F, Vaccari A, Remanufacturing Italy. L'Italia nell'epoca della postproduzione [Remanufacturing Italy: Italy in the Age of Post-Production], Venice, Mimesis, p 29

Balchandani A, Berg A, Hedrich S, Ekeløf Jensen J, Le Merle L, Rölkens F (2022) How the fashion industry can get into a metaverse mindset, in McKinsey and BOF Fashion Report 2022, mckinsey.com. Last accessed on 27.01.25 at 16.05

Barucco MA, Bulegato F, Vaccari A (2020) Remanufacturing Italy: un'introduzione [Remanufacturing Italy: An Introduction] in Buracco MA, Bulegato F, Vaccari A, Remanufacturing Italy. L'Italia nell'epoca della postproduzione [Remanufacturing Italy: Italy in the Age of Post-Production], Venice, Mimesis, p 11

Celaschi F, Celi M, Formia E (2014) Quando il design incontra il futuro. When des meets future. Plan des Technol 3:21–29

Conti and Franzo (2020) Distretti produttivi virtuali. La transizione del Made in Italy nella moda [Virtual production districts: the transition of 'Made in Italy' Fashion], in Buracco MA, Bulegato F, Vaccari A, Remanufacturing Italy. L'Italia nell'epoca della postproduzione [Remanufacturing Italy: Italy in the Age of Post-Production], Venice, Mimesis, p 126

Calefato P (2021) La moda come sistema di segni nel mondo iperconnesso [Fashion as a system of signs in a hyperconnected world], Cuaderno No 127, Centro de Estudios en Diseño y Comunicación, Facultad de Diseño y Comunicación, Universidad de Palermo, Buenos Aires, Argentina, ISSN Online: 1853–3523. https://dspace.palermo.edu/ojs/index.php/cdc/article/view/4580

Crook L (2021) Amber Slooten reveals her vision for a more inclusive fashion industry in a live Dezeen 15 interview, dezeen.com. Last accessed on 08.01.25 at 20.30

Fiorani E (2021) Scintille di umanità [Sparks of Humanity], Lupetti, Milan, p 183

Micelli S (2011) Futuro artigiano. L'innovazione nelle mani degli italiani [Artisanal Future: Innovation in the Hands of Italians], Venice, Marsilio

Lipovetsky G (2007) Una felicità paradossale. Sulla società dell'iperconsumo, Raffaello Cortina Editore, Milano

Volonté P (2003) La creatività diffusa. Culture e mestieri della moda oggi, FrancoAngeli, Milano

Giovanni Maria Conti is not only the Scientific Coordinator of the KnitDesign Research Group at the Design Department of the Politecnico di Milano, but also an expert collaborator of the Istituto Italo-Latino Americano - IILA in the Pymes Forum for cooperative projects relating to textiles and fashion.

He is a member of the editorial board of the Lupetti - Editori di Comunicazione publishing house, and in 2020, he received an Honourable Mention at the XXVI Compasso d'Oro awards for his 'DigiKnit' research project.

He is a member of the Advisory Board of CiAUD, the Centre for Research into Architecture, Urbanism and Design at the Lisbon School of Architecture, as well as a member of LeNS - the International Learning Network of on Sustainability.

[17] Fiorani et al. (2021).

Technological Innovations in the Production of Knitted Uppers for Sneakers

Regina Aparecida Sanches, Adriana Yumi Sato Duarte, Francisca Dantas Mendes, Maurício de Campos Araújo, Camilla Borelli, and Maria Goreti Issac

Abstract Footwear, initially designed solely for foot protection, has evolved throughout history, gaining prominence and acquiring varied characteristics. The evolution of technology has led to the discovery of new materials and production processes, turning the piece into a fashion item consumed by all the public. The main objective of this article is to present the new technologies to produce seamless knitted uppers and compare these new technologies with the traditional processes for the production of fabric sneakers. The methodology used was a narrative literature review. Sneakers with knitted uppers produced using the new technologies are more sustainable, have a shorter production time, consume less energy, and occupy less space within footwear industries, increasing the competitiveness of companies.

Keywords Knitted uppers · Flat knitting machines · Circular knitting machines · Warp knitting machines

1 Introduction

The footwear industry is one of the oldest in the world. In ancient times and prehistoric times, humans already used rudimentary forms of footwear, covering their feet with animal skins to move during walks and protect themselves from harsh environmental conditions, crafting these items as a primitive form of business (Tomassini 2011). Later, footwear began to be produced by artisans (shoemakers and their assistants) in their small workshops (UNIDO 2007).

R. A. Sanches (✉) · F. D. Mendes · M. de Campos Araújo · C. Borelli
University of São Paulo, São Paulo, Brazil
e-mail: regina.sanches@usp.br

A. Y. S. Duarte
São Paulo State University, Bauru, Brazil

M. G. Issac
NAPI SUSTEXMODA, Research Support Center of the University of São Paulo, São Paulo, Brazil

D. Raposo et al. (eds.), *Perspectives on Design IV*, Springer Series in Design and Innovation 59, https://doi.org/10.1007/978-3-032-00733-9_16

The formalization of the footwear industry came with the Industrial Revolution. The first factories and machines used exclusively for footwear production emerged in the United States and Europe in the 1870s, causing a growth in shoe production, which began to be produced on a large scale at a lower cost (Silva and Lepikso 2019).

After World War II until the end of the 1960s, footwear production was centralized in wealthier countries, such as Japan, which showed high production performance.

However, from 1970 onwards, it was observed that the production model created by developed countries, due to production costs, could not maintain the productivity gains achieved in the past (da Costa and Fligenspan 2013).

With easily accessible technology, the footwear industry can be found in different regions and countries. Aiming to reduce costs, footwear producers seek regions that offer lower wages and abundant labor, placing this industry in permanent migration, not only within the same country but even crossing continental barriers (Guidolin et al. 2010).

When the capacity of a country or territory to offer competitive conditions is exhausted, or another location with superior cost advantages emerges, the industry promptly moves to the new region. In the first major migration wave, the transfer of footwear production to the so-called Asian Tigers (South Korea, Taiwan, and Hong Kong) and Brazil was observed. In the search for cheaper labor, another migration movement of the footwear industry occurred to other, still underdeveloped, Asian countries, such as the Philippines, Indonesia, Thailand, and China (Tomassini 2011).

Since the Industrial Revolution, there has been a great evolution in design, production processes, and materials used for footwear manufacturing. However, the human component has not been fully replaced by equipment automation; part of the footwear manufacturing processes still has characteristics that resemble the activities performed by the first shoe artisans and their assistants in their small establishments (Perera et al. 2009).

In summary, the traditional process of producing uppers consists of cutting various materials and then sewing the cut parts to assemble the upper, generating a large amount of waste in the cutting stage and using a lot of labor to perform this operation (Duarte and Sanches 2022).

In 2012, Nike and Adidas revolutionized the market with the launch of sneakers with seamless flat knitted uppers, marketed under the brands Flyknit and Primeknit, respectively (Troynikov and Watson 2015).

Currently, many footwear industries produce their sneakers using seamless uppers. In addition to the production of uppers in flat knitting, the industry offers the following technologies for the production of seamless knitted uppers: large-diameter and small-diameter circular knitting machines, and raschel machines. These new technologies produce uppers ready to be assembled to the soles.

The objective of this article is to present the new technologies to produce seamless knitted uppers and compare those with the traditional processes to produce fabric sneakers. The methodology used was a literature review, which consisted of reviewing the literature related to the topic addressed. Books, journals, articles, internet sites, and other sources were used.

2 Footwear Industry

2.1 Global Scenario

The global production of the sector totaled 22.6 billion pairs in 2023. In 2022, the four largest global footwear producers are countries located in Asia—China, India, Vietnam, and Indonesia—which accounted for 76% of global footwear production. Brazil is the fifth-largest footwear producer in the world, the largest outside of Asia. China is the main producer, responsible for more than half of the world's footwear production (Abicalçdos 2024).

According to the same source, global footwear production is being impacted by the fragmentation and reorganization of international geopolitics, as well as the relocation of intra-Asia footwear production. In the last decade (2013–2022), there has been exponential growth in footwear production in Asian countries: Vietnam, Turkey, Cambodia, and Indonesia. During the same period, Chinese footwear production recorded a decline. Even so, Chinese footwear production is five times the total production of India, the second-largest producer.

According to Abicalçdos (2021), factors such as input costs, labor prices, taxes and charges, export support, among other factors, reduce the competitiveness of companies located outside Asia.

2.2 Productive Sector

The development in the textile sector is accelerated by the automotive, aerospace, and mechanical engineering industries, civil engineering, agriculture, footwear, sports, and medicine. The knitting sector also follows this trend, where the interaction between designers, creators, and raw material suppliers, knitting machine manufacturers, and direct users of knitted products dictate strategies that aim at the production of innovative, multifunctional, and seamless products. These strategies include measures for the gradual transition to the circular economy (Troynikov and Watson 2015).

Footwear production, for the analysis of the productive sector, can be divided by (Abicalçdos 2021): the predominant material: plastic, rubber, synthetic laminate, textile, leather, and other materials (Abicalçdos 2024); by gender: female, male, children, and unidentified; and (A-ZEN 2022) by type of use: slippers, casual and formal, sports, safety, and orthopedic (Abicalçdos 2024). According to the same source, among the shoes classified by type of use are sneakers, which are subdivided into sports and casual.

Casual sneakers are inspired by the older models of sneakers (da Costa 2013), from the early twentieth century, which were relegated for a period of the last century, brought back in 1993 and maintained in the market to this day.

The sports sneakers currently found are inspired by the diversity of existing sports modalities. Thus, given the technology employed, each model has different materials and characteristics, aiming to achieve different objectives, such as comfort, lightness, design, durability, and other characteristics (Marcondes Silva 2023).

Sports Footwear. According to Turner (2019), the development of sports footwear was primarily encouraged by physical recreation and sports practiced at the time, the late nineteenth century, especially tennis, football, and cricket. These sneakers could be used habitually not only by athletes but also by the general population, and such products gave an appearance of modern men and women. According to Herrmann (2017), the textile and clothing chain consumes, globally, around 98 million tons of non-renewable resources per year, and only 13% of the materials consumed in the production of clothing are reused.

Currently, sneakers are increasingly specific; it is easy to find sneakers for everyday use, running shoes for long or short distances, walking shoes for flat or rugged terrain, sneakers with various soles, sneakers with uppers in leather or synthetic, artificial or natural fabrics, sneakers in various colors, sneakers with various cushioning systems, and several other characteristics that aim to serve the public belonging to a pre-determined niche (Nike 2021).

Main Parts of a Sneaker. Figure 1 shows the four main parts of a sports sneaker.

The upper has the function of providing more firmness and protection to the upper part of the foot. The insole has the function of providing firmness when walking and support to the medial longitudinal arch. The midsole is an intermediate layer placed between the insole and the sole, with aesthetic and cushioning functions. The outsole, which is the part that is in direct contact with the ground, plays a fundamental role in the quality and grip of the shoe (Sanches et al. 2024).

Fig. 1 Main parts that make up a sports shoe. *Source* Adapted from Dinato (2018, p. 9)

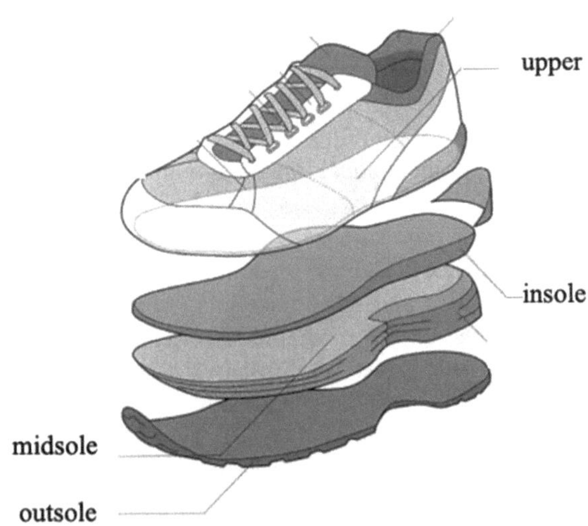

2.3 Technologies Used in the Production of Textile Uppers

The footwear chain operates in a linear system production, in which significant amounts of non-renewable resources are extracted from nature and used as raw materials for the production of shoes (Duarte et al. 2020).

According to Cuden (2022) and Sanches et al. (2021), the production of uppers in textile materials can be carried out by the conventional process, which is traditionally used by footwear companies and includes the following activities: (a) cutting the parts and (b) joining these cut parts, or by seamless technology, which is simplifying the production flow of uppers, with the production of seamless knitted uppers, ready to be used by the footwear industry.

Making Uppers: Traditional Process. According to Duarte and Sanches (2022), the production process of manufacturing conventional fabric sneakers begins in the textile chain, with the extraction/production of the textile fiber, yarn production, obtaining woven fabric, knitted fabric, or nonwoven fabric, textile finishing, and continues to the footwear industry for the production of the sneaker, the final products are delivered to retail stores, where they are purchased by consumers, and ends with disposal by consumers after the end of the products' useful life. Figure 2 illustrates the main stages of the production process of traditional fabric sneakers.

Environmental impacts caused by this production chain are present in all stages, from the extraction/production of fibers to the disposal of the final product by consumers. During the stages of the textile sneaker production process, high consumption of water and energy, the use of toxic chemicals, the disposal of solid waste, and the pollution of effluents are observed (Shirvanimoghaddam et al. 2020).

Since the Industrial Revolution, in the footwear industry, shoes have been produced using pieces of different materials that are cut and sewn to make the uppers. Then, the uppers are glued or sewn to the sole to form a single piece. Part of the production process is carried out with the use of machines, but many steps are still carried out by hand, requiring several skilled workers to join the parts of the sneaker. Furthermore, the raw materials come from different places, requiring a large logistics operation for the manufacture of shoes (Cunha 2017). The conventional uppers are composed of about 20 components (Ozdemir et al. 2020).

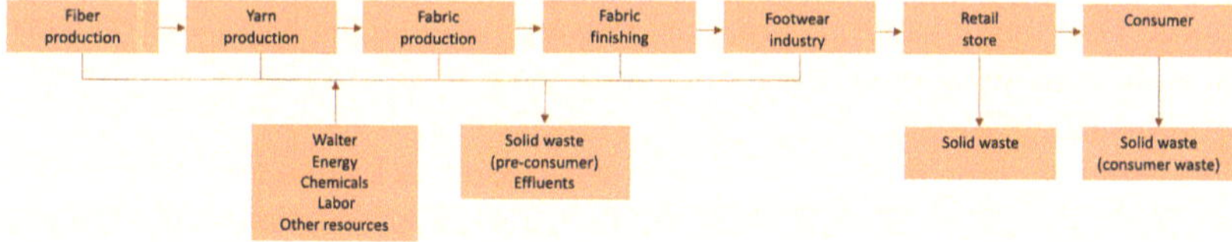

Fig. 2 Fabric footwear production chain (from the textile industry to the final consumer)—traditional process. *Source* Adapted from Duarte et al. (2020)

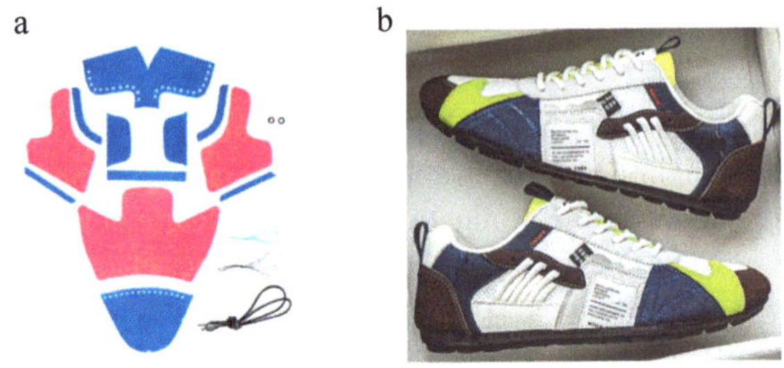

Fig. 3 Conventional upper components and the lines used to join the parts (**a**) and sneakers produced with uppers made from various materials (**b**). *Source* (**a**) Ozdemir et al. (2020), *Source* (**b**) www.temu.com

In the traditional process, the uppers are produced using knitted fabrics manufactured by warp knitting as raw materials. This technology requires a stage of preparing warp rolls before the production of the knits, and to make the production process viable, it is necessary to produce large quantities of fabrics (Cuden 2022). Figure 3 shows the components of a traditional shoe (3a), and a pair of sneakers produced with uppers made by the traditional process (3b).

Currently, there is a greater focus on the recycling of raw materials, knit waste, and the development of more sustainable processes. In this same vein, manufacturers of knitting machines have created equipment that offers more sustainable production techniques, with fewer negative impacts on humans and the environment, and a significant reduction in waste generation. The new technological processes were developed to save production time, space, energy, labor, and reduce the disposal of raw materials (Cuden 2022).

According to Troynikov and Watson (2015), the elimination of seams and the ability to seamlessly incorporate various knit structures in the upper offer greater flexibility, reduced weight, foot conformity, and better fit of the shoe to the foot, providing the necessary support and flexibility, providing thermophysiological requirements to the foot, such as ventilation, thermal transfer, and breathability.

The process of making seamless uppers can be carried out in both weft knitting and warp knitting. According to Cuden (2022) and Sanches et al. (2021), weft knitting uses the following technologies: flat knitting machines, large-diameter circular knitting machines, and small-diameter circular knitting machines. And warp knitting uses raschel machines. Figure 4 shows the technologies to produce seamless knitted uppers.

According to Sanches et al. (2021), weft knitting is characterized by the production of knit loops in the transverse direction of the fabric, i.e., each row of knit is produced by a single yarn (Fig. 5a), and in warp knitting, the loops are produced in the longitudinal direction of the fabric, i.e., each row of knit is produced by several warp yarns (Fig. 5b).

The manufacturing process of sneakers with seamless knitted uppers begins in the textile chain where the stages of textile fiber extraction/production, yarn production, textile finishing of the yarns, production of knitted uppers are carried out, it continues to the footwear industry where the manufacture of sneakers is carried out, which are sold to retail stores, then purchased by consumers, and at the end of their useful life,

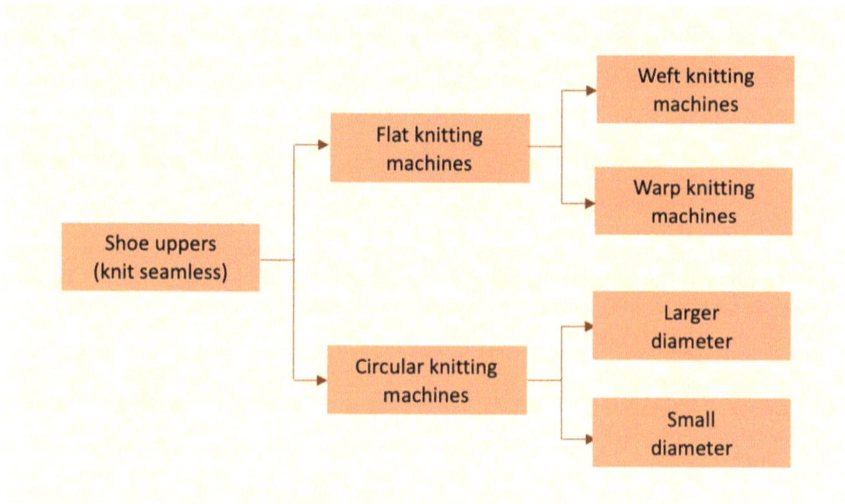

Fig. 4 New technologies for the production of seamless mesh uppers. *Source* Sanches et al. (2025)

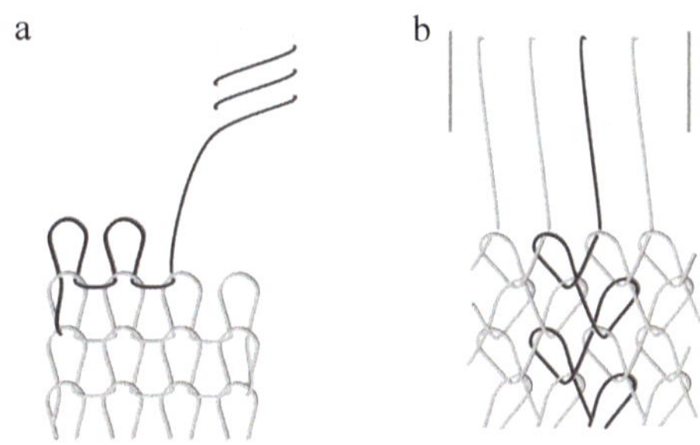

Fig. 5 Schematic representation of fabric produced in weft knitting. *Source* www.karlmayer.com (2021)

they are discarded (Duarte and Sanches 2022). Figure 6 illustrates the main stages of the production process of sneakers with seamless knitted uppers.

Flat Knitting Machines (V-bed). The trend for next-generation knitted footwear began in 2012, with the production of uppers on flat knitting machines. The equipment allows to produce fully modern shoe uppers, with a compact structure and integrated loops for lacing. Knitted sports shoes have a sock-like fit, are lightweight, comfortable, and provide sufficient support for the foot. Sports shoes with knitted uppers are used both for professional sports and for leisure (Kaziur et al. 2022). Figure 7 shows an upper and the sneaker produced by this technology.

Small Diameter Circular Knitting Machines (sock-shoe). In recent decades, the adaptation of sock machines, due to the decreasing consumption of pantyhose in the

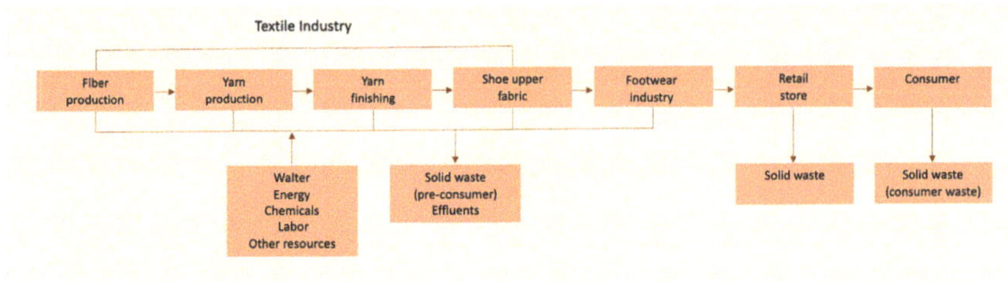

Fig. 6 Production chain of textile footwear (from the textile industry to the final consumer)—new technologies for upper production. *Source* Sanches et al. (2025)

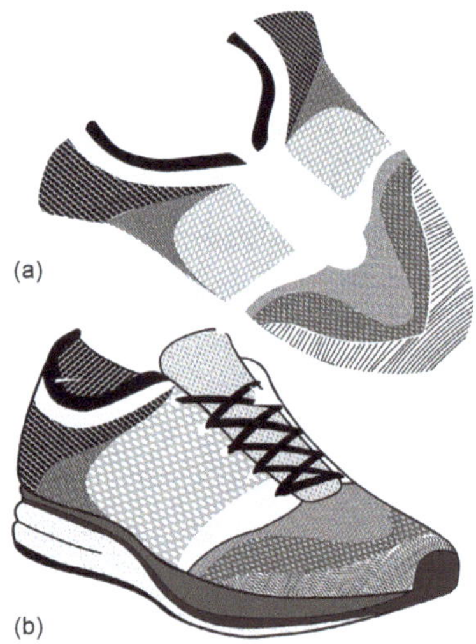

Fig. 7 Upper produced in flat knitting (**a**) and sneaker (**b**). *Source* Troynikov and Watson (2015)

1970s, enabled the development of seamless machines to produce apparel and, more recently, for the production of socks for the manufacture of the so-called "sock-shoe." The machine produces uppers for a wide range of applications, from technical and sports footwear to casual and fashion footwear (Cuden 2022).

Small diameter machines have design systems and additional technologies, such as printing or embroidery, making it possible to produce shapes and colors that were not feasible in the past. During the production process, the machine can create 3D areas, eyelets for laces in a single piece on the upper part of the upper.

An important benefit is the ability to knit numerous intarsia patterns for any type of seamless upper. It also allows mapping different areas depending on the type of yarn, including breathable mesh areas with cushioning components and/or transfer loops and 3D areas (Santoni 2020).

The knitted uppers made on sock machines are knitted in a single 3D piece and are ready to be fitted onto the mold for the finishing process (Cuden 2022). The production process is available in any color or design; the flexible technology allows the creation of any type of knit stitches, as well as the rapid change of style, model,

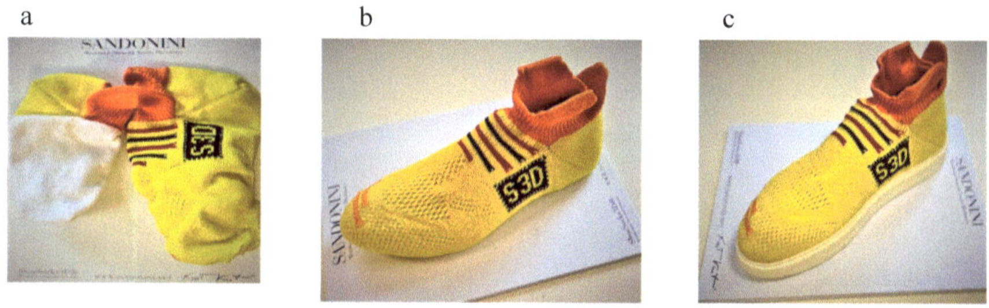

Fig. 8 Production process steps of 3D "sock-shoes". *Source* www.ShoeSocks3D.com (2020)

and size, according to customer needs. The uppers can be made in Jacquard knitted fabric with various patterns, with a logo, and in various colors (Cunha 2017).

Just like those produced on flat knitting machines, they are delivered to the footwear industry ready to be attached to the soles. Figure 8 illustrates the main stages of the production process: (8a) Upper production; (8b) Fixing on the form and finishing; and (8c) sneaker.

Large Diameter Circular Knitting Machines. Large diameter circular knitting machines produce various multicolored Jacquard structures and are equipped with easy-to-program software, which facilitates the production of both smaller and larger batches with the desired color combination (Mayer and Cie 2020).

Circular knitted uppers are used in the production of both sports and leisure sneakers. Unlike flat knitting and small diameter circular machines, these machines produce fabrics with the uppers knitted side by side. Figure 9a illustrates the fabric with the uppers on a large diameter circular knitting machine and 9b an example of sneakers produced with uppers using this technology.

Warp Knitting Machines (double-bar raschel machine). Warp knitting machines, like large diameter circular machines, produce fabrics with multiple uppers knitted

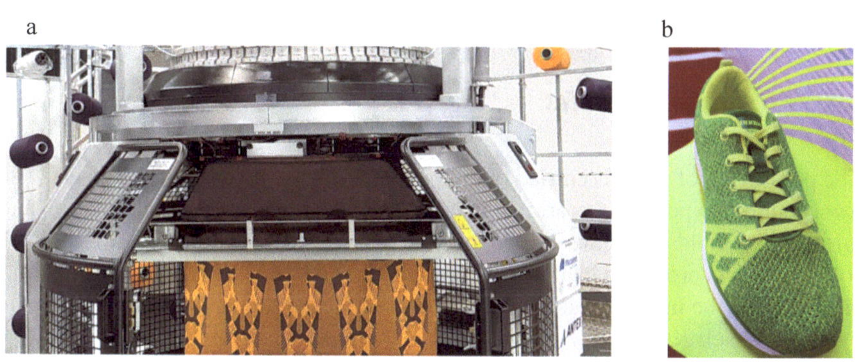

Fig. 9 **a** Uppers produced on large diameter circular knitting machines. *Source* www.outatex-egypt.com (2018) and **b** final product—sneakers. *Source* www.mayercie.com

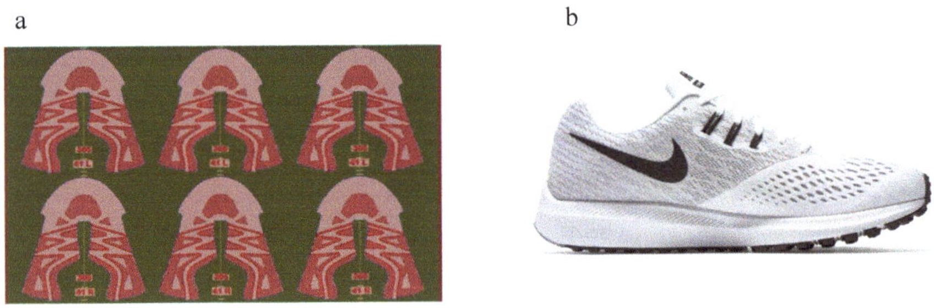

Fig. 10 **a** Fabrics with uppers in warp knitting and **b** produced sneaker. *Source* A-ZEN Academy (2022)

side by side. Warp knitting machines produce innovative and high-performance fabrics for footwear.

The uppers are manufactured with multicolored effects with integrated functional regions and complex designs. To create a colored shoe fabric, the yarns from the multicolored warp roll are knitted on both sides of the fabric (Curen 2022). The uppers produced by this technology are used in the production of sports and leisure sneakers.

The raschel machines produce 4D-Knit structures with 3D designs with relief on both sides. The 4D knit structure is not a classic mesh fabric with monofilaments with different shrinkages, as the space between the inner and outer surfaces of the upper is filled with a bulky yarn. In this way, bulky fabrics are created, which can be produced with either raw or colored yarns, in various patterns (Textile Today 2021).

Double-bar raschel machines have a special configuration to produce these colored uppers. The upper is knitted with a combination of specific colors and patterns. The bars are assembled so that one color of yarn will be knitted on the inside of the uppers and the other colors on the outside (A-ZEN 2022). Figure 10 shows the fabric with the uppers produced by warp knitting (10a) and a sneaker produced with warp uppers (10b).

2.4 Production of Soles

The production of soles is carried out in parallel with the production of uppers. The most common types of soles are: (a) injected sole: the selected material is heated and injected into molds that have the shape and size of the sole (Onodera 2016); (b) EVA sole: the sole is produced by cutting its profile from a block of EVA (Kick 2017); (c) molded sole: the upper and sole parts are formed separately and molded in a single operation (Kick 2017).

2.5 *Assembly of Sneakers*

The assembly sector aims to join the uppers to the soles, giving the sneaker its final shape. The assembly consists of stretching and molding the upper material onto the last (assembly component that functions as a mold with the desired silhouette volume and configuration for the model) and joining the lower part of the upper with the insole.

This joining can be done with tacks, adhesives, or seams. The most common method of joining the sole to the upper is by adhesive, which is selected according to the type of sole and according to the adhesive manufacturer's guidelines (Marcondes Silva 2023). The assembly operation is divided into three stages: preparation of the material fixing process, assembly of the upper on the sole, and joining of the upper to the sole.

3 Production Flow of Fabric Sneakers in the Footwear Industry

3.1 *Traditional Process*

In the footwear industry, the traditional sneaker production process occurs in seven main stages: modeling, cutting the upper parts, sewing, assembly, finishing, verification, and packaging. During the production process, there is consumption of energy, labor, and other resources; large quantities of solid waste are discarded by companies, and this waste is sent to landfills or incinerated (Ellen Macarthur Foundation 2017).

Modeling is the beginning of everything; in this phase, the creation and development of the shoe to be manufactured takes place. It is during this first stage that the product can be adapted for its production, according to material specifications, available machinery capacity, and costs involved (Marcondes Silva 2022).

The upper, like conventional sportswear, is usually composed of various materials to meet the physiological requirements of the product, such as heat transfer, vapor transfer, and moisture management. And at the same time, they must be rigid and resistant to provide support and flexibility to the final product (Troynikov and Watson 2015). The various raw materials that form the upper and other components are joined by seams and glues. According to Textilemagazine (2019), about 25% of the raw materials used by the footwear sector are discarded, in the form of material scraps, as waste.

Then, during assembly, the upper is joined to the sole by means of a new seam, by gluing or pressing. Heel, toe caps, and insoles are added in sequence (Faustino 2023). While finishing, the shoe is removed from the last and goes through the final details: lining placement, cleaning, among others. Then, quality control is carried out when the shoes are all ready to be analyzed and packaged (Sanches et al. 2024).

3.2 New Technologies

Some companies in the footwear sector outsource part of their production process, such as cutting, sewing, pre-fabrication, embroidery, and others; the outsourced companies are called ateliers and carry out part of the production process without being the holders of the final product (Abicalçados 2024).

Flat Knitting Machines (V-bed). In the production process of flat knitted uppers, the footwear industry receives the ready-made upper (3D) from the knitting mill, in the size and shape to be assembled on the sole (Sanches et al. 2024). Flat knitting machines take about 10 min to knit an upper and knit three uppers simultaneously (Krinner 2023).

According to Basu and Gupta (2019), uppers knitted on flat knitting machines do not generate waste, as they use the exact amount of yarn to produce each upper, and when the knitting process is finished, the yarn is cut, and the remaining yarn that is wound on the cone is used to produce another upper.

The footwear industry receives the upper in the size and shape to be assembled on the sole; the production process reduces production costs, time, and does not generate waste of raw material by the footwear industry.

The main stages to produce flat knitted uppers, carried out by the textile industry, are yarn manufacturing, yarn dyeing, and knitting of seamless mesh uppers. The footwear industry receives the ready-made uppers and carries out the stages of assembly, quality control, and packaging of the sneakers, which follow the same production flow described for sneakers produced by the traditional process (Sanches et al. 2024).

Small Diameter Circular Knitting Machines (sock-shoe). Small diameter circular knitting machines produce the entire body of the shoe in the shape of a sock (3D) in three minutes. The 3D knitted upper is made in one piece, ready to be assembled onto any type of sole by the footwear industry (Sandonini 2017).

The equipment reduces production costs, time, and does not generate waste.

The cutting stage is eliminated by the footwear industry, there is no waste of raw material, the number of work phases is reduced, and the finishing process is shortened.

The main stages for the production of small diameter circular knitted uppers, carried out by the textile industry, are: yarn manufacturing, yarn dyeing, knitting of seamless knitted uppers; the footwear industry receives the ready-made uppers and carries out the stages of assembly, quality control, and packaging of the sneakers, which follow the same production flow described for sneakers produced by the traditional process (Sanches et al. 2024).

Large Diameter Circular Knitting Machines. The productivity of large diameter circular knitting machines is, depending on the type of structure used in knitting, ten to twenty times higher than the productivity of flat knitting machines, but the process produces material waste, as the uppers are knitted side by side, leaving fabric around the knitted uppers that need to be cut (Cunha 2018). According to

Textilemagazine (2019), about 10% of the raw materials used are discarded as waste, in the form of material scraps, by the footwear sector.

These machines have a higher productivity; the energy consumption per unit produced on a large diameter circular machine is, to produce similar products, about five times lower than on a flat knitting machine (Cunha 2017).

The main stages for the production of large diameter circular knitted uppers, carried out by the textile industry, are: yarn manufacturing, yarn dyeing, knitting of fabrics with seamless knitted uppers; the footwear industry receives the fabrics and carries out the stages of cutting and closing the uppers, then the assembly, quality control, and packaging operations of the sneakers are carried out, which follow the same production flow described for sneakers produced by the traditional process (Sanches et al. 2024).

Double-Bar Raschel Warp Knitting Machines. Warp knitting machines require a preparation stage for knitting, which is the making of warp rolls, making this technology viable only to produce large quantities of uppers (Krinner 2023). The uppers are knitted in the fabric, side by side; the excess fabric between the uppers needs to be cut. This technology wastes about 10% of the raw materials in the cutting sector of footwear industries, which are discarded as waste, in the form of material scraps.

The production of uppers on raschel machines can reach 25 m/h, with a daily production of about 3000 pairs of uppers (A-ZEN Academy 2022). As these machines have very high productivity, the energy consumption per unit produced is lower than in other technologies.

The main stages for the production of uppers on raschel machines, carried out by the textile industry, are: yarn manufacturing, yarn dyeing, yarn warping, knitting of fabrics with seamless knitted uppers; the footwear industry receives the fabrics and carries out the stages of cutting and closing the uppers, then the assembly, quality control, and packaging operations of the sneakers are carried out, which follow the same production flow described for sneakers produced by the traditional process (Sanches et al. 2024).

4 Final Considerations

The conventional textile footwear production chain is long, starting in the textile industry with all the stages to produce warp knitted fabrics.

The production chain of seamless knitted textile footwear is shorter when compared to the traditional process, starting in the textile industry with the production of seamless uppers by one of the technologies: flat knitting, large diameter circular knitting, small diameter circular knitting, or warp knitting.

The different technologies used in the production of knitted uppers produce different textile products, but there are also products that can be made by more than one technology, as is the case with uppers for sports shoes. Knitting mills typically select the technology based on the needs of their customers.

Given the above, it is possible to affirm that the technology of producing seamless uppers will contribute to reducing the environmental impact caused by the traditional footwear industry, by the disposal of textile solid waste in the environment, in addition to reducing the number of stages in the production process, contributing to increase the competitiveness of companies, reducing their costs with taxes and charges, increasing production, and reducing human errors.

Although the production process of seamless knitted uppers has eliminated or reduced the cutting and sewing operations of the upper parts, considerably reducing the workflow, the assembly stage is not fully automated. The attachment of the uppers to the soles is carried out manually, it is the same as that carried out by the traditional process of producing uppers, each upper part of the sneakers needs to be heated individually after being knitted and sewn to the sole. This operation requires many operations and a lot of labor.

References

Abicalçados – Associação Brasileira das Indústrias de Calçados (2021) Relatório anual de atividades. Novo Hamburgo. https://www.abicalcados.com.br. Last acessed on 25 Feb 2025

Abicalçados – Associação Brasileira das Indústrias de Calçados (2024) Relatório anual de atividades. Novo Hamburgo. https://www.abicalcados.com.br. Last accessed on 12 March 2025

A-ZEN (2022) Applictions of shoe textile. https://www.azentex.com/sports-shoes.html. Last accessed on 12 March 2025

Basu SK, Gupta M (2019) Knitted footwear technology—an approach towards sustainability. J Emerg Technol Innov Res (JETIR) 6:1069–1073

Cuden AP (2022) Recent developments in knitting technology. In: Maity S, Rana S, Pandit P, Singha K (eds) Advanced knitting technology. Elsevier

Cunha R (2017) Máquinas de tricô e impressão 3D estão revolucionando a fabricação de calçados. https://www.stylourbano.com.br/maquinas-de-trico-e-impressao-3d-estao-revolucionando-a-fabricacao-de-calcados. Last accessed on 2 March 2025

Cunha R (2018) Com seu design híbrido, o tênis meia é a maior tendência na indústria de calçados. https://www.stylourbano.com.br/com-seu-design-hibrido-o-tenis-meia-e-a-maior-tendencia-na-industria-de-calcados. Last accessed on 8 March 2025

da Costa CR (2013) Uma análise nos principais aspectos da construção calçadista para o desenvolvimento de um solado protótipo feito a partir de fibra de coco. Dissertação de Mestrado, Universidades de São Paulo, São Paulo

da Costa AB, Fligenspan FB (2013) Odeslocamento d e empresas decalçadosparaonordeste-brasileiro. Editora da UFRGS, Porto Alegre

Dinato RC (2018) A influência de diferentes sistemas de amortecimento do calçado esportivo na economia de corrida e no desempenho. Tese de doutorado. Escola de Educação Física e Esporte da Universidade de São Paulo, São Paulo

Duarte A, Sanches RA (2022) Proposta de fabricação de cabedal de malha utilizando o modelo de economia circular. Revista de Ensino em Artes, Moda e Design, v. 6

Duarte AYS, Sanches RA, de Lima FS, de Queiróz RS (2020) Technological innovations in the production of sportwear: from conventional production to industry 4.0. Product Manag Dev 18:19–25

Ellen Macarthur Foundation (2017) A new textiles economy: redesigning fashion's future. https://emf.thirdlight.com/file/24/uiwtaHvud8YIG_uiSTauTIJH74/A%20New%20Textiles%20Economy%3A%20Redesigning%20fashion%E2%80%99s%20future.pdf. Last accessed on 25 Feb 2025
Faustino RC (2023) Investigação dos processos produtivos dos tênis esportivos e proposta de desenvolvimento de um modelo com cabedal em malha sem costura. Dissertação de Mestrado, Universidade de São Paulo
Guidolin S, Costa A, Rocha E (2010) Indústria calçadista e estratégias de fortalecimento da competitividade. BNDES Setorial. n. 31
Herrmann S (2017) A new textiles economy: redesigning fashion's future. https://www.ellenmacarthurfoundation.org/assets/downloads/publications/A-New-Textiles-Economy_Full-Report.pdf. Accessed on 17 Feb 2025
Kaziur P, Mikołajczyk Z, Kłonowska M, Woźniak B (2022) Design methodology and technology of textile footwear materials. https://doi.org/10.3390/ma15165720. Last accessed on 29 Jan 2025
Kick PGS (2017) Projeto SOU: Tênis personalizado e customizável através das tecnologias de Prototipagem Rápida. Trabalho de Conclusão de Curso. Bacharelado em Design de Produtos, Universidade Estadual Paulista
Krinner S (2023) The influence of knitting technologies within e-textiles. In: Smart clothes and wearable technology. Elsevier
Marcondes Silva (2022) Estudo comparativo do processo de fabricação dos cabedais pelo processo convencional e cabedais sem costura produzidos em máquina retilínea. Dissertação de Mestrado. Escola de Artes, Ciências e Humanidades. São Paulo, Universidade de São Paulo
Marcondes Silva A (2023) Estudo comparativo do processo de fabricação dos cabedais pelo processo convencional e cabedais sem costura produzidos em máquina retilínea. Dissertação de Mestrado. Escola de Artes, Ciências e Humanidades. Universidade de São Paulo, São Paulo
Mayer and Cie (2020) https://www.mayercie.com. Last accessed on 20 Feb 2025
Nike (2021) Componente dos tênis. https://www.nike.com.br. Last accessed on 21 March 2025
Onodera AN (2016) Influência das características mecânicas da entressola e da estrutura do cabedal de calçados esportivos na concepção do conforto e na biomecânica de corrida. Tese do Doutorado, Faculdade de Medicina. Universidade de São Paulo, São Paulo
Ozdemir M, Cascini G, Verlinden J (2020) A mass personalization framework for knitted footwear. In: Trabalho apresentado no 9th international conference on massa customization and personalization—Community of Europe
Perera L, Kimura H, Kerr R (2009) Estratégias de internacionalização da indústria de calçados: conhecendo o consumidor americano. In: Simpósio de Administração da Produção, Logística e Operações Internacionais – SIMPOI, 13, São Paulo, 2009. Anais. São Paulo, FGV
Sanches RA, Duarte AYS, Sbordone MA, Ranzo P (2021) Tecnologia da malharia: processos e principais produtos. Modapalavra e-periódico 14(32):51–72
Sanches RA, Faustino RC, Duarte AYS (2024) Processos produtivos dos tênis esportivos de malha sem costura e proposta de desenvolvimento de um modelo com cabedal tricotado em máquina retilínea. Cuadernos del Centro de Estudios de Diseño y Comunicación 1:79–102
Sanches RA, Duarte AYS, Mendes FD, Araújo MC, Borelli C, Issac MG (2025) Fotos dos arquivos pessoais das autoras
Sandonini (2017) https://sandonini.net. Last accessed on 15 March 2025
Santoni (2020) http://www.santoni.com/en/products/footwear/mec-mor-cmp. Last accessed on 25 Feb 2025
Shirvanimoghaddam K, Motamed B, Ramakrishnac S, Naebe M (2020) Death by waste: fashion and textile circular economy case. Sci Total Environ 718. https://doi.org/10.1016/j.scitotenv.2020.137317. Last accessed on 28 Feb 2025
Silva ESS, Lepikso HA (2019) Abordagem de uma indústria 4.0 para personalização da produção de calçados. Simpósio Internacional de Inovação e Tecnologia. Salvador (B.A.), 9–11 outubro

Textile Today (2021) Karl Mayer introduced new 4D-KNIT warp knitted textiles. https://www.textiletoday.com.bd/karl-mayer-introduced-new-4d-knit-warp-knitted-textiles. Last acessed on 27 Feb 2025

The Textile Magazine (2019) Circular knitting machines: most ideal to make shoe uppers. https://www.indiantextilemagazine.in/circular-knitting-machines-most-ideal-to-make-shoe-uppers. Last accessed on 8 March 2025

Tomassini RAS (2011) A História Interessa – path dependence e a Indústria Calçadista. Dissertação de Mestrado, Programa de Pós- Graduação em Administração de Empresas da PUC- Rio, Rio de Janeiro

Troynikov O, Watson C (2015) Knitting technology for seamless sportswear. In: Shishoo R (ed) Textiles for sportswear. Woodhead Publishing Limited and The Textile Institute, pp 95–117

Turner T (2019) The sports shoe: a history from field to fashion

UNIDO - Organização do Desenvolvimento Industrial das Nações Unidas (2007) A resposta mundial ao rápido crescimento do suprimento de calçados do sul da Ásia. Painel Industrial do couro e produtos do Couro, 16, Gramado, Brasil

A Project to Create Value with Local Wool in Fashion Design Products: A Social, Environment and Economic Sustainable Model

Madalena Pereira, Carlos Ferrão, António Quaresma, and Tiago Quaresma

Abstract Throughout history, the interior region in Portugal as characterized by the exodus of people who sought better living conditions in other locations and communities. Nevertheless, some of the shepherds of Serra da Estrela persist, resulting in a love for the profession and a legacy of generations. Knowing the harsh reality of a profession, guardians of sheep of the Bordaleira breed, it is necessary to guarantee an economically viable and consequent response that would create favorable conditions for them to continue the profession and ensure the desired quality of life in the community. Along with cheese, there was an urgent need to find a solution for wool region, which in many situations is already a waste, without value. Grupo Valor do Tempo, leader of the project, in partnership with associations in the region Ancose, and researchers from academia, developed a project to solve the problem: valorization of wool from Serra de Estrela sheep and, consequently, impact on the quality of life in the community of shepherds existing in the region of Serra de Estrela. The project, developed in a concept of Economic, Social and Environmentally Sustainability dimensions, involving Fashion Design and product development, using companies and artisans in the region and Portugal for the viability and valorization the wool from Bordaleira sheep. It is developed with a project methodology, which guarantees the valorization of wool directly for the shepherds of Serra de Estrela between 1000 and 2000%, and thus being able to contribute to improving the living conditions of the shepherds and reducing wool waste in the region. Finally, the validation and communication of the project culminated in the opening a pop-up store—Serra de Estrela DOP and a creation a Serra de Estrela sheep interpretation centre.

M. Pereira
FibEnTech, University of Beira Interior (UBI), Covilhã, Portugal

M. Pereira (✉)
UNIDCOM/IADE-UE, University Europeia, Lisbon, Portugal
e-mail: madalena.pereira@ubi.pt

C. Ferrão
Artesões da Serra de Estrela, Seia, Portugal

A. Quaresma · T. Quaresma
Grupo Valor do Tempo and Serra de Estrela Sheep Interpretation Center, Seia, Portugal

D. Raposo et al. (eds.), *Perspectives on Design IV*, Springer Series in Design and Innovation 59, https://doi.org/10.1007/978-3-032-00733-9_17

Keywords Sustainable project · Wool · Fashion design project · Serra de Estrela · Social economy

1 Introduction

The importance of sustainable development objectives (ODS) on the planet at a global level has been one of the issues that scientists (Daly 1994; Visser 2017), politicians such as Al Gore, Ajay Banga, Bill Gates, Gro Harlem Brundtland, organizations United Nations (UN), International Labour Organization (ILO), European Commission, and others and ONG's society: Green Peace, MacArthur Foundation, Fashion Revolution, develop and care about ODS, with are currently increasingly concerned about, such as the management of existing resources without compromising the resources of future generations, and the way in which this resource management affects society and human life on the planet. Beginning in the 1960s, this concern only gained greater attention with the Brundtland report (WCED 1987), prepared by the Commission on Environment and Development and approved by the General Assembly of the United Nations, aimed to present future options for economic development that responds to emerging environmental concerns, ensuring social justice and inclusion, later adopted by the ON General Assembly, the 2030 Agenda is a universal and ambitious action plan that aims to promote sustainable development for People, Planet, Prosperity, promoting Peace and Partnerships: the 5 "P". In addition, there is also the Kyoto Protocol and the Paris Agreement in this direction.

The efforts of the ON with the introduction of the sustainable development goals (ODS) and more recently European Commission with the emergence of the European Green Deal (European Commission 2024) which establishes sustainable development goals for 2030 and 2050. However, its application has not been easy.

Elkington in 1994 (Elkington and Rowlands 1999), identify sustainable development in three dimensions: environmental, social and economic, also called triple button, it is about thinking about the generation of value in an organization, based on these three dimensions.

But the costs associated with sustainable practices have prevented their application and evolution as desired. On the other hand, the costs associated with its implementation and offering products to the market are high.

Eco-Branding strategies, identified by a differentiation strategy based on environmental and social attributes targeting market niches, have been an example applied by countries since the beginning of the century (Orsato 2006; Das et al. 2024). But consumers need to clearly understand the benefit and reason for their purchase. Credible information is the second prerequisite for a product with a differentiating strategy; and the third attribute is associated with an environmental differentiation of the product and involves the barrier to imitation. Micceli (2011), De Marchi et al. (2013), Bettiol et al. (2019), also identified Eco-Branding strategies as a form of differentiation through products that differentiate market positioning and have environmental characteristics, namely recyclable raw materials and artisanal production.

Barucco et al. (2020), currently characterizes, called post-production, a strategy for Italy of remanufacturing: "the term remanufacturing indicates the regeneration of a product obtained by replacing worn parts with new elements or recovered elements from pre-existing objects" says the author. The transformation includes also the transformation of the discourses and narratives that circulate around it and contribute to renewing its definition. Raposo et al. (2020), considered the identity promotes sustainability and respect for people's quality life and conclude "the visual identity continues to function as a cultural interface of the territory in a sustainable logic and not business requirements".

The project developed in this research aims to consider the relationship between economic, social and environmental sustainability and using its close relationship in the territory as inspiration and with the creation of the pop-up store in homage to the shepherds of Serra de Estrela.

2 Circular Economy and Sustainable Development

2.1 Definitions

Nowadays there is still some confusion in the definition of sustainable development and circular economy. The most common is to consider sustainable development as the main objective of the circular economy but reconciling social and economic sustainability is still debatable. The circular economy depends on an alliance between the various stakeholders in the chain. The study on the circular economy concludes that only 8.6% of the planet is circular, and that it is a means of achieving sustainable development in the economic, social and environmental components. Kirchherr et al. (2023), after a survey of 221 definitions associated with sustainability define sustainability in a broad view of the system as:

> The circular economy is a regenerative economic system which necessitates a paradigm shift to replace the 'end of life' concept with reducing, alternatively reusing, recycling, and recovering materials throughout the supply chain, with the aim to promote value maintenance and sustainable development, creating environmental quality, economic development, and social equity, to the benefit of current and future generations. It is enabled by an alliance of stakeholders (industry, consumers, policymakers, academia) and their technological innovations and capabilities.

Bartola and Colombi (2024) address the three dimensions of sustainability: economic, environmental and sociocultural and their triple transition through design with focus in product and process innovation; value chain positive transformation, business model innovation and supply chain positive transformation (see Fig. 1). The author refers to "as we navigate the intricate landscape of transforming the fashion system sustainably, it becomes imperative to acknowledge and grapple with the inherent challenges".

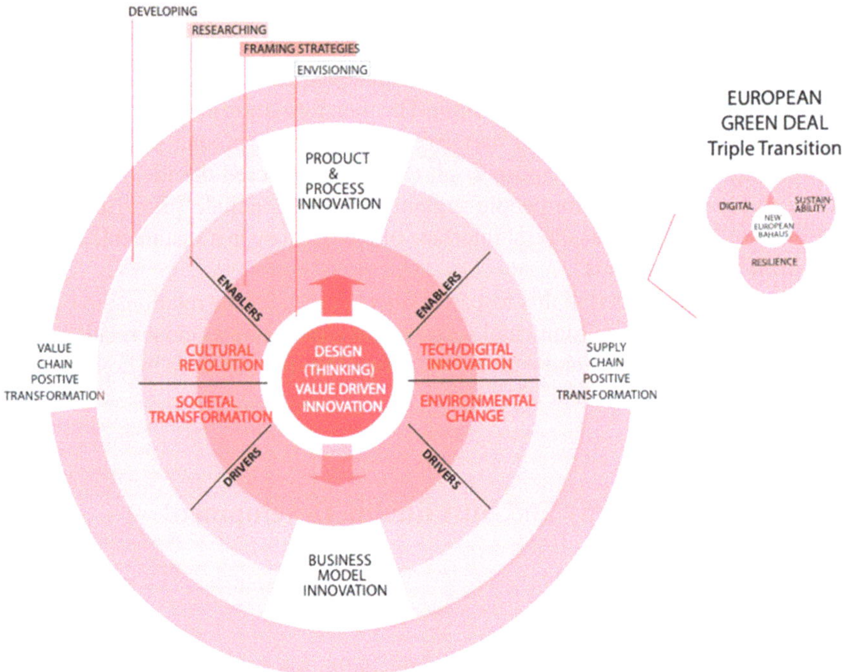

Fig. 1 Triple transition through design. *Source* Bertola and Colombi (2024)

For this transformation, consumer behaviours and attitudes also need to change more and more for a new green consumer (Vignali et al. 2019; Shiel et al. 2020).

2.2 *Wool for Sustainable Development*

In history of wool Calado (2008), says "wool as a raw material takes us back to our origins and constitutes one of the most vivid affirmations of man's relationship with nature".

All natural fibers have characteristics that synthetic fibers do not have, one of which is biodegradability at the end of their life, causing much less harm to the environment than synthetic fibers. But other resources are used in its production and currently there is discussion about the transformation of processes in terms of energy, land use, CO_2 emissions, water use, using regenerative agriculture, reuse of raw materials using recycling process or use waste with new solutions for the market (Wang et al. 2024; Walter et al. 2019; Pereira et al. 2024). The production of natural fiber: wool and recycling wool, is in 2023 1.134.033 tonnes, with 6% are recycling and 94% virgin wool (Pereira et al. 2025). In Portugal wool production are only 6.218 (Textile Exchange 2025) tonnes and involves not only the textile sector and artisans but also the food sector in Portugal. The Portuguese Serra de Estrela wool from sheep that provide milk for different types of cheese (the most famous are Serra de Estrela cheese), and meat is not always considered in the present market as a value-added product for fashion industry. According to Calado (2008), in the territory between the Douro and Serra de Estrela, the Mondegueira breed of sheep stands out, characterized by its high milk quality, which, adapted to the environment, presents a unique rustic characteristic, transported to the churra-type wool fiber adapted to the artisanal and industrial production of artifacts such as blankets and carpets. The fibers are 20–22 cm long and have a diameter of 36.4–40.4 μm. This problem occurs in both Portugal and Italy, where the criteria for fibre fineness and respective properties do not reach the famous Australian Merino wools, with values of 17–23 μm, and designed for fine fibre (Calado 2008; Instituto Nacional de Estatísticas 2024; Woolmark 2025). Most recent studies such as Ferreira et al. (2024), Gaddi and Mastrolonardo (2024), Lanz et al. (2024), and Puppim and Broega (2021), develop studies to find solutions through technology, design and business models that allow them to contribute to the use of wool, which is considered no used in value market and also waste, and extend the life of the material through recycling and the creation of value-added products. In this way they will be able to contribute to sustainable development.

3 New Model Business for Prosperity

3.1 Social Economy and Social Innovation

Social innovation is one of the areas that has sparked interest among several researchers, and which aims to improve the living conditions of human beings and their communities by creating models of social relations and cooperation. Manzini (Munari and De Vasconcelos 1981) proposes that design should go beyond creating products, acting as a facilitator of social change. He defines social innovation as new ideas that meet social needs and create new relationships or collaborations. For these social changes to materialize, it is necessary to develop and apply new business models that have an economic impact throughout the chain, from the production of raw materials to a more conscious consumer who changes behaviours.

Stiglitz (2009, 2018, 2020) advocates a global concept of social economy and say "Inequality is a choice, not of the poorest, but of our political systems (…) the way we structure the economy, the tax systems, the labor laws, the way we conduct monetary policy," he said. These choices influence "market power, the bargaining power of different groups", justifying inequalities in income and opportunities.

The principles of the social economy revolve around the appreciation of human beings, solidarity, democratic participation and sustainability, above individual profit. They guide the functioning of cooperatives, associations, social enterprises and other organizations that put people at the center of the economy. Several economists, designers, sociologists and philosophers have developed this transdisciplinary area of social economy, such as Elinor Ostrom, Ezio Manzini, Paul Singer, Jean-Louis Laville; and others however, its application and growing evolution today has not been easy. Where are we failing and how can we ignore the international context of war that increasingly limits its evolution? The connections with associations, ONG's, organizations with companies, society and new era of consumers and human needs are fundamental.

4 Project

4.1 Methodology

Considered a design methodology existent for project product design: Munari and De Vasconcelos (1981), Bonsiepe et al. (1984), Bomfim (1995), Baxter (2018), and Lobach (2001), these methods can be basically defined in 4 phases: preparation, generation, evaluation and implementation. The choice of the appropriate method for each of the projects developed or its adaptation depends on several variables.

Munari methods, it began with the definition of the problem. Munari and De Vasconcelos (1981): "Designing is easy when you know what to do. Everything becomes easy when you know how to proceed to reach the solution to a problem" (Munari, p. 12). What we intend here is to present a possible way of conceiving, since "he designs method is nothing more than a series of necessary operations, arranged in logic, dictated by experience. Your goal is to achieve the best result with the least effort" (Munari and De Vasconcelos 1981, p. 20). Munari and De Vasconcelos (1981) proposes the following project phases: (1) Problem; (2) Problem definition; (3) Components of the problem; (4) Data collection; (5) Data analysis; (6) Creativity; (7) Materials and technologies; (8) Experimentation; (9) Models; (10) Verification; (11) Solution and (12) Assessment.

Bonsiepe (2012) approaches as "methodology" a set of questions to regarding the problem: What, Why and How that gives rise to new products. Bonsiepe (2012) mentions that in the current historical phase where the development of sustainable design projects requires interdisciplinary work, not limited to environmental but also social aspects.

In project fashion design products Jones (2005) are the most used in method project, but without include social aspects.

Most recently, Norman (2023) says "the key to change is human behaviour, meaning, sustainability, and humanity-centeredness. Emphasize quality of life, not monetary rewards; restructure how we live to better protect the environment; and focus on all of humanity." We need to give a sign for the fashion product, or only a design product for all life. Norman defends the methodology user-centered design. The methodology combines the adaptation of several mentioned here, but give a meaningful, sustainable and centred in shepherds of Serra de Estrela, emphasize quality of life of them.

This study adopts a qualitative and exploratory methodology grounded in design research, participatory observation, and applied project development. The methodological approach is structured around the principles of user-centred design (Bonsiepe 2012) and project-based problem-solving (Stiglitz 2009, 2018), with integration of social design concepts (Gaddi and Mastrolonardo 2024). The project follows a five-phase framework adapted for fashion and product design development: (1) Diagnosis and Contextualisation: Direct field observation of shepherds' living conditions and wool management. Interviews with shepherds (Ancose members), artisans, and local stakeholders. Review of historical, cultural and economic data about wool production in Serra da Estrela. (2) Problem Definition and Opportunity Mapping: Identification of systemic issues related to wool devaluation. Analysis to explore potential market repositioning. Co-creation sessions with community members to validate the design challenge. (3) Ideation and Concept Development: Brainstorming and material exploration workshops. Development of eco-branding narratives and symbolic value propositions. Preliminary sketches, prototypes, and storyboards. (4) Product Development and Prototyping: Use of natural fibres from Bordaleira sheep. Collaboration with regional textile industries and artisans for jacquard weaving. Creation of limited-edition products (e.g., numbered cushions), integrating identity, functionality and storytelling. (5) Validation and Impact Assessment: Product testing with user groups: assessing perception pre/post awareness of value chain. Public engagement via pop-up store with storytelling installations and regional promotion.

4.2 Objectives and Research Question

The main objective of this project is to improve the living conditions of the shepherds of Serra de Estrela by revalorise Bordaleira Serra de Estrela wool that is almost worthless waste, and today that does not contribute to improving their living conditions or the society in which they live.

As a secondary objective, we intend to development a fashion design project find a solution that fits into the concept of the Valor do Tempo Group, with authenticity, with purpose, embedded in an eco-branding strategy and that is valued in society. The following question is thus posed: How can design-led strategies enhance the social, cultural, environmental, and economic value of Serra de Estrela wool, and contribute to the revitalisation of rural communities in Serra da Estrela?

4.3 The Project Leader and Partners

The Valor do Tempo group is the project leader and defines its strategy as follows: "For more than two decades we have dedicated ourselves to working on the best that Portugal has to offer the world. We are fascinated by the history of our country and the multiple stories it must tell. We believe in a human economy, centred on human and based on a sharing model. It is from there that true Added Value is achieved, which is nothing more than ever than the result of the Soul placed in the projects."

Also involved in this project was Ancose, as the National Association of Sheep Breeders in Serra Da Estrela, including "famous people" such as the guardian shepherds, participation of the Serra da Estrela artisan's association and scientific organization Beira Interior University.

4.4 Results and Analysis

During the project development phase and adapted from the methodologies and methods referred to in point 4.1, five phases presented in Table 1 are considered. In the first research phase, followed by Identify and understand the fundamental Problems, Ideation and solution, Development of the solution and Validation, communication with society.

During the research, direct observations of the living conditions of the shepherds were carried out, identifying components of the problem and collecting information

Table 1 Project phases

Phases	
Research	Research into projects, markets, and scientific work in the area of using wool waste, application of wool with fineness above 25 µm (coarse wool). Social, environmental and economic sustainability strategies and projects
Identify and understand the fundamental problems	Low economic value of the raw materials of Serra de Estrela Bordaleira sheep, which consequently does not contribute to the prosperity of the shepherds' living conditions. Wool with low quality characteristics for the classic existing market, which in many situations is considered a waste and does not contribute to the environmental, economic and social sustainability of the existing shepherding community
Ideation and solution	Explore product solutions (aesthetics, functionality and processes), services, business model, communication that value Bordaleira Serra de Estrela wool in the three dimensions of sustainability: economic, social and environmental
Development the solution	Develop the solution, using environmentally, socially and economically sustainable techniques and processes. Use of natural wool colors, regional and national industry and the collaboration of artisans from the Serra de Estrela region
Validation, communication with society	Validation before a group of users, and the way in which the project will be communicated: pop-up store paying homage to shepherds (human-centered), transmitting social economy values with authenticity, eco-design, environmental advantages and connection with nature

Fig. 2 Images of inspiration of shepherds of Serra da Estrela, in the past and present. *Source* Own, Grupo Valor do Tempo (2022) and Calado (2008)

from those involved. Additionally, scientific and historical research was carried out into the evolution of the use of wool, techniques and traditions in the region.

The use of wool to produce Burel was identified as a successful case of its application and avoiding waste, but the amount paid to shepherds remains below what is necessary improves the living conditions of shepherds in various periods, also caused by markets that are still to be consolidated.

After observation, images were collected that could communicate the values of the region, the community and its relationship with nature and history (see Fig. 2).

In the problem identification phase, and together with the shepherds and Ancose, a five-year protocol was established for the amount to be paid to the shepherds for each kilo of wool, to contribute to a social economy and prosperity. An added value of the product of 1000% was established, in relation to the current value. In this way, the product or service solution is an important but secondary point of the identified problem. It was necessary to contribute to the valorization of the wool of the Bordaleira sheep of the shepherds of Serra de Estrela by 1000%. In the current era, a kilogram of wool was often paid at a symbolic value that would not be enough to cover the costs of shearing the wool, which as we know is mandatory to avoid the death of animals in the hottest seasons.

In the idea's exploration phase, several products and different ways of communicating them were identified together with the values of the associated brand—Grupo

Valor do Tempo. Figure 3 present an example of the exploratory work of ideas in terms of product and communication of the value of wool from the region.

After the idea exploration phase and based on the constraints of techniques and processes, it was decided to develop a jacquard fabric with the sheep's identity and in the natural colors of the wool, for an exclusive, limited-edition product. Each cushion contained 1 kg of wool equivalent to the average provided by each sheep, which would be numbered and dated. When developing the product, we also thought about extending the product's lifespan by creating a cover that can be removed from the base product, highlighting the interior with wool from Serra de Estrela sheep: unique, exclusive, authentic and sustainable (see Fig. 4).

Fig. 3 Exploration of ideas, products and communication. *Source* Own

Fig. 4 Fashion design product use wool region valorization and sustaibable. *Source* Own

It was also decided that different projects would be developed each year after validation and feedback each year.

For the validation phase, a test was carried out with a small group of users to analyze the perception that the product gave in terms of economic, social and environmental sustainability without providing any information. In a second phase, a set of preliminary information was provided before its analysis: the objective of the project, the product, the reason for exclusivity and authenticity and its association with eco-branding. The result obtained from the research group was different with and without information. The group with information about the cause of this project give more value to the product.

Also, for the validation phase, the idea of visual merchandising for the pop-up store was developed, and it was important to pay tribute to the shepherds of Serra de Estrela Each of the pastors associated with Ancose had a tribute with their name engraved on one of the walls of the pop-up store (see Fig. 5).

In resume, the implementation of the design project led to a series of measurable and observed outcomes across social, economic, and environmental dimensions:

1. Economic Impact: The project enabled a price increase for local wool, from an average of €0.20/kg to over €2.00/kg for selected finished products, through value-added transformation and storytelling-based marketing. Collaborations with regional textile mills and artisans generated local income, while the production of limited series items (e.g., artisanal cushions) demonstrated the commercial viability of design-led valorisation.
2. Social Engagement: Direct involvement of shepherds, local associations, and regional stakeholders facilitated knowledge exchange and collective ownership of the outcomes. The co-creation process strengthened local identity and pride, particularly by reviving narratives around pastoral culture and material heritage.
3. Environmental Value: By utilising natural, biodegradable fibres and promoting local production, the project significantly reduced the ecological footprint compared to conventional fashion supply chains. The reuse of waste wool contributes to circular economy goals, while the limited, locally produced items align with principles of slow design.

Fig. 5 Pop-up store with final solution of fashion design product and names of shepherds Serra da Estrela region. *Source* Own

5 Conclusions

This project confirms that design with social economy can act as a strategic tool to reconfigure value chains, transforming undervalued materials into culturally significant and economically viable products and human-centered. The participatory, design-led model developed in the Serra da Estrela region demonstrates that rural development and sustainable fashion are not mutually exclusive but can reinforce one another when aligned through systemic thinking. By engaging local communities in all phases of the project, from ideation to validation, this approach promotes inclusive innovation and contributes to the regeneration of peripheral territories. This project may also be interpreted as an act of symbolic human justice. Through the inscription of the shepherds' names on the walls of the pop-up store designed to present the products, the Valor do Tempo Group transcends the realm of tribute, instituting a new paradigm of collective authorship in which the originators of the material content are formally acknowledged, publicly celebrated, and reintegrated into the chain of meaning and value creation. Future work should focus on expanding the model to other projects in next years, with objective for development a sustainable model in cultural, social, economic and environmental, conducting longitudinal impact assessments, and integrating more robust quantitative evaluation tools. Grupo Valor do Tempo also developed an interpretation Center for the Serra de Estrela Sheep, also known as Bordaleira and as an indigenous Portuguese breed, which stands out for the quality of its milk for the production of Serra de Estrela cheese and meat; in this case, wool is a by-product in the twenty-first century due to its characteristics and changes in the market for the production of wool articles, which seek extremely fine and soft-to-the-touch wool, which in the case of Bordaleira wool is characterized by a fine cross-type, slightly wavy, and a soft or rough touch. With the aim of continuing the project to enhance the value of Bordaleira Serra da Estrela wool and its shepherds. In the future, new projects will be developed to apply and enhance the value of wool from Serra de Estrela sheep. Always focused and centered on the need to improve the quality of life of shepherds and contribute to the economic, social and environmental sustainability of the Serra de Estrela region.

Acknowledgments Thanks to Grupo Valor do Tempo, which financed the project and provided a practical example of social, economic and environmental sustainability, improving the living conditions of the shepherds of Serra de Estrela https://ovalordotempo.pt/marcas/centro-interpretativo-ovelha-da-serra-da-estrela/.

The authors acknowledge the support of this work through FCT—Foundation for Science and Technology, I.P., within the scope project UIDB/00711/2020 from UNIDCOM/IADE—research unit in Design and Communication.

The authors are very grateful for the support granted by the Research Unit of Fiber Materials and Environmental Technologies (FibEnTech-UBI), through the Project reference UIDB/00195/2020, funded by the Fundação para a Ciência e a Tecnologia, IP/MCTES through national funds (PIDDAC), and DOI: 10.54499/UIDB/00195/2020 (https://doi.org/10.54499/UIDB/00195/2020).

References

Barucco M, Bulegato F, Vaccari A (2020) Remanufacturing Italy: l'Italia nell'epoca della postproduzione. Mimesis, DCP/IUAV

Baxter M (2018) Product design. CRC Press

Bertola P, Colombi C (2024) Can fashion be sustainable? Trajectories of change in organizational, products and processes, and socio-cultural contexts. Sustain Sci Pract Policy 20(1):2312682

Bettiol M, Maria ED, Micelli S (2019) New craft. Saper fare, tecnologia e design per una produzione sostenibile. MD J 7:14–21

Bomfim GA (1995) Metodologia para desenvolvimento de projetos. Editora Universitária/UFPB, João Pessoa

Bonsiepe G (2012) Design como prática de projeto. Editora Blucher

Bonsiepe G et al (1984) Metodologia experimental: desenho industrial. CNPq/Coordenação Editorial, Brasília

Calado E (2008) Rota da Lã Translana. Edição Museu de Lanifícios da Universidade da Beira Interior

Daly HE (1994) For the common good: redirecting the economy toward community, the environment, and a sustainable future, no. 73. Beacon Press

Das T, Patnaik A, Mapatro SK (2024) The impact of eco-brand expansion strategy on brand engagement concerning the fashion industry: a bibliometric review. In: Future of management: embracing sustainability, diversity, and inclusivity, pp 24–37

De Marchi V, Maria ED, Micelli S (2013) Environmental strategies, upgrading and competitive advantage in global value chains. Bus Strateg Environ 22(1):62–72

Elkington J, Rowlands IH (1999) Cannibals with forks: the triple bottom line of 21st century business. Altern J 25(4):42

European Commission (2024) European green deal. Available at https://commission.europa.eu/strategy-and-policy/priorities-2019-2024/european-green-deal_en. Last accessed on 23/04/28

Ferreira S et al (2024) Development of recycled yarns as a contribution to sustainability in the fashion pipeline and consumption. In: International fashion and design congress. Springer Nature Switzerland, Cham

Gaddi R, Mastrolonardo L (2024) Micro-reti locali per la transizione verde della filiera della lana.-local micro-networks for green transition of the wool supply chain. AGATHÓN 15:344–353

Instituto Nacional de Estatísticas (2024). https://www.ine.pt/xportal/xmain?xpgid=ine_main&xpid=INE. Last accessed 2024/12/18

Jones SJ (2005) Fashion design: manual do estilista. Editora Gustavo Gil, Barcelona

Kirchherr J, Yang NHN, Schulze-Spüntrup F, Heerink MJ, Hartley K (2023) Conceptualizing the circular economy (revisited): an analysis of 221 definitions. Resour Conserv Recycl 194

Lanz IE et al (2024) A mapping of textile waste recycling technologies in Europe and Spain. Text 4(3):359–390

Lobach B (2001) Design industrial: bases para configuração dos produtos industriais. Tradução Freddy Van camp, Edgard Blücher, Rio de Janeiro

Micelli S (2011) Futuro artigiano: l'innovazione nelle mani degli italiani. Marsilio Editori SpA

Munari B, De Vasconcelos JM (1981) Das coisas nascem coisas

Norman DA (2023) Design for a better world: meaningful, sustainable, humanity centered. MIT Press

Orsato RJ (2006) Competitive environmental strategies: when does it pay to be green? Calif Manag Rev 48(2):127–143. https://doi.org/10.2307/41166341

Pereira M et al (2024) Sustainability and circularity in textile industry: new approach for development of fashion products with textile pre-consumer waste. In: Meeting of research in music, arts and design. Springer Nature Switzerland, Cham

Pereira M et al (2025) Portugal. In: Wynn M, Wiegand T (eds) Sustainability, the circular economy and digitalisation in the European Textile and Clothing Industry. SDGs and textiles. Springer, Singapore. https://doi.org/10.1007/978-981-97-9116-3_8

Puppim R, Broega AC (2021) An innovative textile product proposal based on sustainability: recycling wastes from the wool industry. In: Machado J et al. (eds) Innovations in mechanical engineering. Springer International Publishing, Heidelberg, Germany
Raposo D et al (2020) Visual identity design as a cultural interface of a territory. In: Advances in ergonomics in design: proceedings of the AHFE 2020. Virtual conference on ergonomics in design, 16–20 July 2020, USA, Springer International Publishing
Shiel C, do Paço A, Alves H (2020) Generativity, sustainable development and green consumer behaviour. J Clean Prod 245:118865
Stiglitz J (2009) Moving beyond market fundamentalism to a more balanced economy. Ann Public Coop Econ 80(3):345–360
Stiglitz J (2018) O fim das desigualdades: sociedades desiguais e como as mudar. Bertrand, Lisboa
Stiglitz JE (2020) Rewriting the rules of the European economy: an agenda for growth and shared prosperity. WW Norton & Company
Textile Exchange (2025). https://textileexchange.org/wool/. Last accessed on 2025/04/23
Vignali G et al (eds) (2019) Technology-driven sustainability: innovation in the fashion supply chain. Springer
Visser W (2017) Our common future ('The brundtland report') world commission on environment and development (1987). In: The top 50 sustainability books. Routledge, pp 52–55
Walter LF et al (2019) A review of the socio-economic advantages of textile recycling. J Clean Prod 218:10–20
Wang X et al (2024) Social sustainability and the impact of social media influencer campaigns on the positive eWOM and purchase intention of youth subcultures. In: The Palgrave handbook of consumerism issues in the apparel industry. Springer Nature Switzerland, Cham, pp 401–420
WCED (1987) Brundtland report. Our common future: report of the world commission on environment and development. Retrieved at: http://www.un-documents.net/wced-ocf(open in a new window)
Woolmark (2025) What are the different types of wool? Retrieved at: https://www.woolmark.com/fibre/what-is-merino-wool/. Last accessed 2025/04/23

Eyes Wide Design: Reflecting and Reformulating Dissemination Methodologies

Liliana Soares, Ermanno Aparo, and Rita Almendra

Abstract This paper expresses the theme of social media doubleness through a double entendre in the title—Eyes Wide Design—and through the choice, for the Stanley Kubrick's movie 'Eyes Wide Shut', whose subtle motifs seems to raise the question whether life is a real or a dream, something real or just virtual. For human being what is significant here is that the events exert such a profound influence on both, the being and his/her environment, that the life of all involved is in continuous change. Thus, the contribution of this study lies in the reflection and reformulation of dissemination methodologies, integrating the phenomenon of social networks, film industry and AI into design research. Methodologically, the study is exploratory and based on field ethnography, establishing the prominence of co-design and the participatory design process to create sustainability, knowledge and innovation. This hypothesis serves as a fundamental intersection where the preservation of manufacturing identity, the development of modern design education, and the application of digital technology converge, shaping new models for culture dissemination.

Keywords Methodology · Sustainability · Design education

1 Introduction

The pandemic has changed the film industry and introduced new economic models of a long-standing tradition to challenge the digital age. Streaming is indispensable following major changes in audience required behavior and content dissemination in online streaming platforms such as, YouTube or Alphabet and prominent players as Netflix, Amazon or Spotify (Ryu et al. 2024). Those global companies and brands have been dominated by powerful domestic platforms that use short-form videos to

---

L. Soares (✉)
Instituto Politécnico de Viana do Castelo, Viana do Castelo, Portugal
e-mail: lsoares@estg.ipvc.com

L. Soares · E. Aparo · R. Almendra
Lisbon School of Architecture, CIAUD, Research Centre for Architecture, Urbanism and Design, Universidade de Lisboa, Rua Sá Nogueira, Polo Universitário do Alto da Ajuda, Lisbon, Portugal

D. Raposo et al. (eds.), *Perspectives on Design IV*, Springer Series in Design and Innovation 59, https://doi.org/10.1007/978-3-032-00733-9_18

enhance millions of individualized experiences not only in the cultural and creative industry, but also in manufacturing industry and public sector.

The rise of short-form video content and social media had redesigned the way culture is communicated, making it more interactive and visually engaging, and expanding their audience. The maturity of digital technologies has further improved digital exemplification and immersive experiences of manufacturing, accelerating their integration with modern technology. These videos are created and shared on various social media platforms and apps, such as TikTok, Instagram Reels and YouTube Shorts. Short video has gained substantial popularity as its format is dynamic, brief and clearly to consume in a brief period. Visual storytelling characterizes this kind of film by the use of editing techniques, popular music, and the ability to quickly convey messages or evoke emotions. These short videos are created by influencers, non-influencers and makers by means of smartphones or other policies with video recording skills (Chen 2023).

In addition, there is the impact of Artificial Intelligence (AI) on culture, society, creativity, representation, and ethics. AI's cultural control is deep, affecting how human being perceive himself/herself, relate with technology, and create his/her identities. Recently, some research had shown that cultural identity can determine responses to AI, proposing that cultural identity shapes how individuals include AI in constructing the self in relation to others and determines the effect of AI on key decision-making processes (Danesi 2024). It seems individualists interpret AI characteristics to transgress on its uniqueness, autonomy, and privacy, while collectivists realize AI as an extension of the self and interpret it to facilitate conformity to consensus (Barnes et al. 2024).

In this process, if language is used as the first global interpretation of the world and, for this reason, it is something that cannot be replaced by anything (Gadamer 2002), how to communicate culture and knowledge in a historical moment people seems with wide shut eyes? Methodologically, in this process, what types of techniques can be created and used? The role of language in the definition of methodology does not involve replacing contents for another, but rather the involvement of several factors, such as, the 17 goals to transform our world (LNCS Homepage 2025a), cultural identity, political, economic, social, technological, environmental and competitive factors to qualify the process in which the interpretation experience is being carried out (Soares 2022).

Supported on this context, the research questions are: If reality is in continuous change and experienced by social media and AI doubleness, how do events influence human beings and the environment identity? What are the focal features of cross-cultural communication comprised in the short videos and AI to promote manufacturing industry and, cultural and creative industry?

2 Research Process

In this paper, the relationship between material life and virtuality is compared with Stanley Kubrick's movie 'Eyes Wide Shut', comparing virtual life with the meaning of dream in the film. Dream was an important tool to understand the study of appearance and reality and to recognize the problem of knowledge (Kubrick 1999). Although the movie turned the public to look beyond appearances and inspect the interiors of stimuli of human behavior, the dream seams to be used as a symbol and an enigma, to encourage the dreamer (characters and/or audience) to gain affective experience instead of knowledge (Ruwe 2002). Therefore, to achieve knowledge dream-motif was highly explored by universal signs and icons.

In design it may means creating multiple opportunities of appearance to communicate a product or service, and tend for it, including events in the process. To choose the appearance means designing images of the product or service, as something ever renewable, rising and sustainable. The designer ponders about images, knowing that it consists of parts in eternal motion, as images that through external perception expects to create sensations and knowledge to the user (Soares and Aparo 2022). Thus, the paper analyze culture as something systemic and sustainable.

3 Methodology

The study is exploratory to produce new knowledge and cultural identity and based on ethnography qualitative methods, using storytelling methodology to reflect and reformulate dissemination methodologies as a form of simplification, complementing information with an engaging and innovative narrative (Soares et al. 2025). To reflect about new dissemination methodologies, it analyses two exploratory and pilot projects, using analogical and digital thinking, sources of artistic culture, and AI as an artistic and experimental typology.

On the one hand, a pilot project was implemented with design students to identify the role of representation and signs in the communication and semiotic process in design education.

On the other hand, the other exploratory case was carried out by design seniors to promote a manufacturing industry and cultural and creative industries.

4 Pilot Project

The rise of short videos and social media has redesigned the way culture is communicated, and innovation happens when there is long-term competitiveness and guarantees, as people buy significances and values and not products (Verganti 2009). In this case, there is design-oriented research, by the identification of 5 W (Who, What, Why,

Where, When), to create new languages. The great levels of aesthetic quality, productivity care, good work and innovation are associated with socio-cultural context of a place. As a storyteller, designer interpreted reality. In addition, it was applied the choice of a new social, and humanist order selecting a universal theme that can involve the uses in the experience of beauty and knowledge (Kant 2007).

Therefore, the authors chose a cultural, artistic and universal theme such as Vincent Van Gogh (1853–1980) artist work: First, assuming taste as a faculty of estimating an object or a mode of representation by means of a delight or aversion from any interest. Second, selecting a universal work of this artist, such as, "The Starry Night" (De sterrennacht), an oil-on-canvas painting, painted by Van Gogh in June 1889.

4.1 The Role of Representation in the Communication and Semiotic Process in Design Education

The first pilot project was carried out in the Draw II course with 31 design students from the Polytechnic Institute of Viana do Castelo, in the North of Portugal (Fig. 1).

Students were invited to draw trees from memory, representing the 4 seasons, using A3 size hard paper, nib pen and ink. Spring, Winter and Summer should include the human figure of an instrumental musician. In Autumn, students were also asked to communicate the language of Vincent Van Gogh (1853–1980) while drawing the tree and a jazz musician.

This work was carried out for 4 hours in classroom with the sound of the jazz standard "Autumn Leaves" composed by Joseph Kosma in 1945. The versions heard were those of Miles Davis (1958), Bill Evans (1959), John Coltrane (1962), Stan Getz (1964), Chet Baker (1976). The presence of a musical piece that talks about Autumn and is interpreted by different songs over time was intended to create in students the ability to act by relating everything, creating connections with other disciplines and knowledge. This environment of variations on a theme served as a motivation for his research to be diversified and transversal.

Students could use digital resources, videos, social networks and the support of AI, searching for references on the internet and in books of works by the great Dutch artist and Cool Jazz period. In this process, it was intended that the student would recognize the artists' universal symbols, conveying a cultural message in their work. It means a creative search focus on Van Gogh and tjazz musicians. Another objective of the workload was to create visual skills in the students to communicate sensations, but also knowledge while drawing.

At the end, only 4 students were able to use symbols from Van Gogh's language and only one student drew a jazz musician. The field ethnography approach allowed a direct and close observation of how young people from generation Z—or the digital generation—use new technological resources. The participation, attraction and practice of social media and AI are total, but they communicate poor, monotonous

Fig. 1 Drawings of a tree in autumn with a musician made by 4 students, using van Gogh's language. Indian ink and nib pen on paper. *Source* The authors

and homogeneous visual culture. There is a fascination with the technique, but a loss of content.

4.2 The Role of Representation in the Communication and in Creative Process Between Design and Business World

The second exploratory project consisted of promoting and disseminating a Portuguese lighting and bathroom accessories company in the north of Portugal. Founded in 1982, the company is engaged in the manufacture and finishing of metallic parts and components for bathroom accessories, decorative and IP rated lighting both for its own brands as well as for its OEM customers (LNCS Homepage 2025b).

The designers wanted to promote a classic bathroom line, where brass is very prominent. In the process, they used AI, asking for a symbolic image that could promote a post for a classic bathroom line. AI gave the possibility to choose the type of drawing style they wanted to represent, proposing, for example: anime, mythology, hyperrealism, Van Gogh, cyberpunk, 2D animation. Considering that the Instagram post is intended to communicate feelings but also knowledge, the designers chose Van Gogh as a universal and transversal solution capable of reaching any user. A company's promotional post must also educate the public and contribute to the dissemination of societal knowledge. The designers chose "The Starry Night" (De sterrennacht), an oil-on-canvas painting, painted in June 1889 that describes a night sky and a village executed in deep blues and brilliant yellows (Howse 2016). This work was chosen not, only because it is considered a touchstone of modern art and one of the most recognizable paintings in Western art but also, because of bright yellow and the wavy lines that define the sky that can symbolizes the gold of the company's brass.

In practical, the designers used manual sketch drawing, crossing it with a drawing created in digital format and using the SHAPR 3D program, Illustrator and Photoshop (Fig. 2).

The proposal was completed and presented to the company manager for review and validation. The cooperative and participatory work process was essential to achieve sustainable and competitive results. The post was published on March 26, accompanied by part of an excerpt from a letter that Vincent Van Gogh wrote to his brother: "if you truly love nature, you will find beauty everywhere" (Thomson 2008).

5 Results

The results achieved revealed that, out of a total of 31 students, 27 students were unable to identify universal symbols of the nineteenth century Western art, such as, the language of Vincent Van Gogh. Additionally, of the 4 students who identified the artist's symbolism, only 1 represented a jazz musician from Cool Jazz period.

This pilot project can also be tested with the other artists and promoted with short-form video content on social media to make it more interactive and visually

Fig. 2 Process of creating a post for a bathroom products company, combining analog and digital resources. *Source* The authors

engaging, to its target. These videos can be shared on social media platforms and apps.

At a later stage, the exploratory project with design students could be confirmed by repeating it with other groups of students from digital generation, creating a questionnaire to understand why it seems they are inept to produce visual culture.

On the other hand, the project developed between senior designers and the bathroom accessories company could be improved at a later stage. This exploratory exercise could be confirmed by repeating it with other paintings from former artists and analyzing the impact of the post on increasing likes, views and followers. The post can be promoted using TikTok, Instagram Reels and YouTube Shorts (Table 1).

Table 1 The experience of beauty and knowledge. *Source* The authors

"The Starry Night" (1889) experience	Users' experience	Cross-communication with social media platforms and apps	Players
1. Taste	**Design Students**	YouTube shorts TikTok Alphabet AI Domestic platforms with short videos	Netflix Amazon Spotify
2. Universal theme	**Designers**		
3. The creation of a rule that is understandable by everyone	**Audience Generation Z**		
4. Transmission of beauty and knowledge	**Audience Generation X**		

6 Conclusions

This paper endorsed Stanley Kubrick's movie 'Eyes Wide Shut', whose sensitive ideas seems to increase the question whether life is just dream or is real. Dream was an important tool to understand the threshold between substance and the multiplies appearances that can represent the essence of a thing, considering events and external factors that define today's reality. Something that considers time, space, causalities and circumstances, and that means pondering culture, social media platforms and apps, communication, inclusivity, environmental impact and sustainability.

The phenomenon of liquidness and unpredictable modernity is accentuated with social media and AI doubleness. Therefore, the authors proved that virtuality communicate symbols and knowledge, based on Immanuel Kant's state 'Experience of Beauty', using social media and Ai as a message, too. It means, choosing works of art that create, not only taste and a sensorial experience in the user, but also, an involvement and practice of comprehension and realization, consuming recognizable works of art of the world. Art can act as satisfying relief, a medium for self-expression, a way to realize the beauty of life. Over art it is possible to produce a narrative of history, embody societal values, and recognize being in the world with an identity.

For design education it means to reformulate dissemination methodologies, integrating the phenomenon of social networks, film industry and AI in teaching practice. It means to assume these social tools in the promotion and recognition of innovative pedagogical practices. This paper confirms the legacy of the culture of drawing.

Cross-cultural communication with comprised in the short videos and used in social media is also a way to promote manufacturing industries. The participatory methods that senior designers used with a business industry created the basis for the development of new techniques of promotion and interaction, linking sensations, culture and knowledge with business values.

The creative process highlights the creation of pedagogical innovation programs, integrating traditional techniques and instruments with innovative and current tools. Considering recent developments, this approach could be an opportunity to explore new ways to communicate a product and, eventually, innovative methods to achieve different markets and, for this reason, an opportunity to create sustainability.

References

Barnes AJ, Zhang Y, Valenzuela A (2024) AI and culture: culturally dependent responses to AI systems. Curr Opinion Psychol 58(1):101838

Chen L (2023) Exploring the impact of short videos on society and culture: an analysis of social dynamics and cultural expression. Pacific Int J 6(3):115–118

Danesi M (2024) AI-generated popular culture: a semiotic perspective. Springer Nature Switzerland, Cham

Gadamer H-G (2002) Verdad y Método II. Sígueme, Slamanca

Howse J (2016) Vincent Van Gogh. Weigl Publishers Incorporated, Alberta

Kant I (2007) Critique of judgement. Forgoten Books, Oxford

Kubrick S (1999) Eyes wide shut. Warner, New York

LNCS Homepage (2025a). https://www.un.org/sustainabledevelopment/. Last accessed 21 Mar 2025

LNCS Homepage (2025b). https://www.furnor.com/_company. Last accessed 02 Apr 2025

Ruwe C (2002) Symbols in Stanley Kubrick's movie 'Eyes Wide Shut'. GRIN Verlag, Munich

Ryu S, Xiang Y, Zhang W (2024) Guide to digital innovation in the cultural and creative industry. SAGE Publications, London

Soares L (2022) The design of interpretation: a contribution to design. Our Knowledge Publishing

Soares L, Aparo E (2022) The concept of edge as design's power to create building's surface. In: Advances in design and digital communication II. DIGICOM 2021. Springer series in design and innovation, vol 19. Springer, Cham

Soares L, Aparo E, Almendra R (2025) Storytelling in design. In: Digital age cultural show travels on network. In springer series in design and innovation. Springer Nature, Switzerland, pp 443–453

Thomson R (2008) Vincent Van Gogh: The Starry Night. MoMa: Museum of Modern Art, New York

Verganti R (2009) Design driven innovation: changing the rules of competition by radically innovating what things mean. Harvard Business School, Cambridge

Reimagining Design in the Anthropocene: Integrating Bioneurodesign for Human and Planetary Well-Being

Maria João Félix

Abstract The Anthropocene confronts industrial design with unprecedented demands, reconciling human needs with ecological fragility, while attending to cultural plurality and ethical responsibility. This chapter examines how Bioneurodesign, an emerging approach rooted in neuroscience, human-centred thinking, and systemic awareness, can reframe design as an instrument for planetary well-being. Rather than replicating conventional models or merely proposing incremental improvements, the chapter advocates for a bioneuroethical shift. In this perspective, design becomes a medium for sensemaking, a way to process complexity, regenerate relationships across species, and respond meaningfully to uncertainty. Drawing on interdisciplinary sources, including cognitive science, posthuman ethics, and environmental philosophy, the chapter articulates a renewed theoretical foundation for design. Central to this inquiry is the idea that design operates between inner cognitive ecologies and outer material systems, shaping how individuals and societies interpret and inhabit the world. The proposal includes a manifesto that positions industrial design as an agent of coherence, emotional resilience, and adaptive transition. Ultimately, the chapter calls for a reimagining of design not as a tool for optimisation, but as a cultural and cognitive force, capable of mediating futures that are ecologically integrated, ethically grounded, and deeply human.

Keywords Bioneurodesign · Cognitive ecologies · Neuroethics · Industrial design · Sensemaking · Regenerative futures

M. J. Félix (✉)
CIAUD, Research Centre for Architecture, Urbanism and Design, Lisbon School of Architecture, Universidade de Lisboa, Lisboa, Portugal
e-mail: mfelix@ipca.pt

D. Raposo et al. (eds.), *Perspectives on Design IV*, Springer Series in Design and Innovation 59, https://doi.org/10.1007/978-3-032-00733-9_19

1 Introduction

1.1 Design in an Age of Fracture

The first decades of the twenty-first century have been marked by a profound sense of disruption. Environmental instability, social fragmentation, and epistemological crises converge in what is now widely referred to as the Anthropocene, a term that signals not only a new geological epoch, but also a rupture in human self-understanding and planetary relations (Crutzen and Stoermer 2000; Haraway 2016; Latour and Porter 2017) This context of multiple breakdowns places industrial design at a crossroads. No longer is it sufficient for design to focus on incremental improvement or market-driven novelty. A more fundamental shift is required one that addresses the cognitive, ethical, and ecological dimensions of contemporary life.

Traditional industrial design emerged within paradigms of growth, utility, and control. It has largely served the logics of industrial capitalism, reinforcing systems of extraction, consumption, and anthropocentrism. However, as several authors argue, the challenges we face today demand a design that is not only sustainable but also systemic, plural, and regenerative (Manzini and Coad 2015; Escobar 2018) Design must move from object-making to world-making, from user-centred to life-centred practices. These shifts align with what systems theorists describe as an integrated view of living systems, cognition, and ethics (Capra and Luisi 2014).

The Anthropocene is not only a planetary crisis but also a collapse of meaning. As global systems become increasingly complex and unstable, individuals and communities struggle to interpret events, imagine futures, and act with coherence. In such conditions, the need for cultural and cognitive realignment becomes urgent. Beck notes that we live in a "world at risk" (Beck and World at Risk 2009) where inherited reference points fail to guide action. Weick describes this process of interpretation under uncertainty as sensemaking, a form of cognitive navigation essential in ambiguous contexts (Weick 1995). Applied to design, sensemaking becomes a strategy for restoring coherence between inner experience and external conditions.

Design, in this light, must evolve from solving problems to cultivating meaning. It shapes how we perceive, feel, and relate, not only to artefacts and systems, but to one another, to ourselves, and to the more-than-human world. It mediates the symbolic and material dimensions of life, acting as a translator between epistemological frameworks and experiential reality. This expanded role requires rethinking the ontological and ethical foundations of design.

This chapter introduces Bioneurodesign as a framework capable of responding to these layered demands. Initially conceptualised by Félix (2025), Bioneurodesign is grounded in cognitive neuroscience, ecological ethics, and human-centred design, and it aims to support wellbeing across personal, social, and planetary dimensions. Unlike conventional methodologies, Bioneurodesign does not treat cognition as background noise or ethics as an accessory. It foregrounds the relational entanglement between neural systems, affective atmospheres, and ecological realities. This perspective aligns with the concept of cognitive ecologies, which emphasises the interplay between mind, body, and environment in shaping human understanding and behaviour (Capra and Luisi 2014; Varela et al. 1991).

While neuroethics broadly concerns the ethical implications of neuroscience and cognitive technologies, Bioneurodesign calls for a more situated and relational model: a bioneuroethical approach. This orientation incorporates not only the ethical responsibility of design to the human brain, but also to the cultural and planetary systems that sustain life. As demonstrated in prior work on design innovation and education (Félix and Santos 2020; Félix and Silva 2021; Félix 2024), Bioneurodesign positions design as both a cognitive and regenerative practice.

This perspective is further informed by affect theory, posthumanism, and regenerative design (Massumi 2002; Braidotti 2013; Wahl 2016). It considers emotions not merely as reactions, but as constitutive forces in the formation of experience and consciousness (Damasio 1999; Chatterjee 2014). From this standpoint, design becomes neuroethical: it influences perception, attention, moral judgement, and the construction of meaning.

The chapter unfolds in three movements. First, it explores design as a neuroethical practice of sensemaking, with attention to the interplay between environment, cognition, and affect. Second, it advances the notion of multispecies ethics, aligning with recent scholarship on more-than-human design (Latour and Porter 2017; Tsing et al. 2017; Azouzi and Lucchio 2024), to expand design responsibility beyond the human. Third, it introduces a bioneuroethical manifesto, articulating principles for a new generation of designers working toward regenerative futures.

Ultimately, this chapter is an invitation to reimagine design not as a tool for optimisation, but as a cultural and cognitive force. It suggests that the future of design lies not in producing more, but in perceiving differently; not in control, but in care; not in abstraction, but in attention.

This chapter adopts a transdisciplinary and conceptual methodology, drawing on literature from neuroscience, design theory, posthuman ethics, and environmental philosophy. Rather than presenting empirical data, it constructs a theoretical framework supported by interdisciplinary dialogue and reflective analysis. The propositions emerge from a synthesis of scholarly sources, recent design debates, and observations drawn from practice-based contexts, including insights generated through the Design Commit conference ecosystem.

2 Sensemaking and Design: A Neuroethical Turn

The complexity of the Anthropocene demands more than functional or aesthetic responses from design. It calls for epistemological tools capable of navigating ambiguity, contradiction, and disruption. Among these tools, the concept of sensemaking has emerged as a valuable cognitive process through which individuals interpret shifting environments and construct meaning (Weick 1995; Kahneman 2011; Snowden 2005). In a world marked by systemic uncertainty and symbolic erosion, this process becomes central to design practice. It shifts the designer's role from problem-solver to cognitive mediator, someone who orchestrates intelligibility across cultural, emotional, and material registers.

Sense making, in this context, refers to the human capacity to synthesise multiple sensory, symbolic, and relational inputs into coherent mental frameworks that inform decision-making and action (Snowden and Boone 2007). Design, as a cultural and material practice, shapes these frameworks not passively, but actively. Objects, spaces, interfaces, and systems become vehicles through which individuals generate narratives, filter stimuli, and orient themselves. This perspective aligns with the growing literature on design as interpretation (Tonkinwise 2015), where the emphasis shifts from solution to signification.

Recent neurocognitive research reinforces this view by demonstrating that physical and symbolic environments directly modulate neural plasticity, attentional networks, and emotional regulation (Coburn et al. 2017; Sternberg 2009). In other words, design does not merely serve cognition it sculpts it. This insight lays the groundwork for a neuroethical redefinition of design responsibility, if design shapes the brain's responses, then it must be conceived as an ethical practice of mind-making.

Here, the section proposes a bioneuroethical turn in design, anchored in three interdependent dimensions.

2.1 Neuroplasticity and Environmental Imprinting

Neuroscience has demonstrated that the brain is not a static processor but a dynamic, plastic organ shaped by experience (Damasio 1999; Chatterjee 2014). Every encounter with a material form or spatial condition modifies neural pathways. For example, prolonged exposure to chaotic environments has been linked to heightened cortisol levels and impaired cognitive function, whereas restorative spaces promote attentional recovery and emotional regulation (Kaplan and Kaplan 1989; Joye 2007).

Designers thus participate in what might be termed *cognitive conditioning*, shaping the neural environments of their users over time. In education, workplace design, healthcare, and public space, these effects are not neutral, they influence mental health, learning capacity, and even moral judgement (Goleman 2013). A bioneuroethical approach calls on designers to acknowledge this responsibility and to prioritise environments that support cognitive resilience, adaptability, and wellbeing.

2.2 Symbolic Perception and Cultural Legibility

Beyond the sensory, design operates in the symbolic realm. Forms, materials, textures, and arrangements evoke meaning, not only visually, but somatically and emotionally. Drawing on phenomenology and cultural semiotics, Bioneurodesign recognises that symbolic perception is both embodied and culturally encoded (Berghofer 2019; Malafouris 2013).

When a user interacts with an object, their brain is engaged in an interpretive act, modulating meaning through memory, context, and affect. This is especially relevant in globalised contexts where cultural pluralism challenges universal design assumptions. The bioneuroethical lens insists on interpretative plurality and symbolic coherence, rejecting cognitive imposition in favour of resonance and recognition.

Recent work on design justice reinforces this imperative: "Design must be accountable to those most affected by design decisions" (Costanza-Chock 2020). This entails recognising that symbols carry power, and that aesthetic decisions are not neutral, they shape identity, accessibility, and belonging.

2.3 *Cognitive Empathy and Ethical Engagement*

The third dimension concerns the integration of cognitive empathy into the design process. Empathy, in this context, extends beyond affective identification to include perspective-taking, attuned attention, and neurodiverse responsiveness (Papoulias 2018; Floridi 2023). Rather than assuming what users want or feel, designers are invited to cultivate listening frameworks that allow other minds to emerge within the process.

Technological tools such as neurofeedback, emotional mapping, and biometric sensing have opened new possibilities for understanding user states. Yet these must be deployed ethically and reflexively. A bioneuroethical stance does not seek to optimise users, but to engage them in co-constructed meaning-making, fostering agency and attentional sovereignty.

2.4 *Design as Neuroethical Sensemaking*

Taken together, these three dimensions call for a redefinition of design as a practice of neuroethical sensemaking. This approach situates the designer not as a neutral facilitator, but as a cognitive and cultural agent, one who constructs conditions for coherence in fragmented times. In doing so, it expands the field of design beyond functionality or symbolism, towards the ethics of mental formation.

This aligns with emerging discussions on cognitive ecologies, which frame cognition as distributed across brain, body, artefact, and system (Varela et al. 1991; Clark 2008). In this view, designers act upon cognitive fields, shaping not only how individuals interact, but how they perceive reality. This is particularly urgent in a digital culture where attention is fragmented and environments are saturated with stimuli engineered for manipulation.

A neuroethical lens compels designers to reverse this trend, to design for attention, not distraction; for interpretability, not opacity; for depth, not overload. As others have argued, regenerative futures will require not only ecological restoration, but cognitive restoration, an ability to think, feel, and connect differently (Wahl 2016; Raworth 2017).

This section has outlined how Bioneurodesign, when expanded through a bioneuroethical lens, offers a powerful response to the challenges of the Anthropocene. It situates design as a practice of sensemaking, grounded in neuroscience, guided by ethics, and attuned to cognitive diversity. In doing so, it calls for a new generation of design practitioners who understand their work not only as problem-solving, but as the subtle and strategic shaping of minds, meanings, and futures.

3 Designing Beyond Certainty: Toward Multispecies Ethics

The Anthropocene has rendered visible the limits of human-centred thinking. From climate disruption to zoonotic spillovers and species extinction, the crises we face today are not isolated environmental failures, they are signs of an ontological rift, rooted in an illusion of human separateness. As Haraway argues, we must learn to "stay with the trouble" of shared entanglements and abandon the fantasy of mastery (Haraway 2016). For design, this calls for an urgent expansion of ethical scope: from anthropocentric optimisation to multispecies reciprocity.

3.1 From User to Kin: Reframing the Design Subject

Traditional design practices have long assumed a universal, abstract "user" whose needs are to be met through functional efficiency or aesthetic delight. Yet this model collapses in the face of ecological interdependence. We are not alone in shaping or inhabiting the world. Animals, plants, fungi, bacteria, and the vast invisible networks they compose, are co-actors in every material and spatial system we touch.

Bioneurodesign, when reinterpreted through a multispecies lens, begins by challenging the very notion of the user. Inspired by the work of (Tsing 2015; Barad 2007), it acknowledges that meaning and impact emerge relationally, not hierarchically. This reframing transforms design into a practice of cohabitation, where responsiveness to the needs, rhythms, and vulnerabilities of non-human others becomes integral to the creative process.

3.2 Designing for Symbiosis

In ecological terms, symbiosis denotes mutually beneficial relationships across species. It is not a metaphor, but a fact of survival. Human life depends on multispecies assemblages: microbiomes, pollinators, trees, mycorrhizal networks. Yet mainstream design continues to marginalise or disrupt these relationships through toxic materials, habitat fragmentation, and monocultural aesthetics.

A symbiotic design ethic reimagines products, environments, and systems as support structures for shared thriving. Examples include urban furniture that provides refuge for pollinators, biodegradable packaging that nourishes soil ecologies, or architectural surfaces that host microbial diversity rather than sterilise it. These interventions are not decorative, they embed ecological intelligence into material culture.

This ethic aligns with regenerative design principles (Escobar 2018) and with what Escobar calls "designs for the pluriverse" practices that foster coexistence rather than dominance (Wahl 2016). In Bioneurodesign, this means integrating ecological feedback into the design process, acknowledging the embodied effects of place on all sentient and semi-sentient beings.

3.3 Materials, Affordances and the More-Than-Human Sensorium

Materials carry affect, memory, and metabolic consequence. They signal invitation or exclusion, not only to humans, but to entire ecosystems. A wooden façade may absorb moisture and host moss; a synthetic polymer may repel birds or leach toxins into soil. Design decisions at the molecular level ripple outward into multispecies lifeworlds.

From a bioneuroethical perspective, materials are not neutral, they shape what Gibson termed affordances: possibilities for action within an environment (Gibson 2014). Extending this to a multispecies context, design must begin to consider how surfaces, textures, and forms invite or inhibit interaction across species boundaries.

Emerging disciplines such as microbiome architecture, biophilic design, and living systems engineering reinforce this principle. These fields demonstrate that attending to invisible life, from gut flora to fungal networks, is not only ecologically sound, but cognitively and emotionally beneficial for humans as well (Joye 2007; Gilbert and Stephens 2018). Bioneurodesign thus functions as a translator between neurobiological needs and ecological interdependencies.

3.4 Decentring the Human: Toward Epistemic Humility

At its core, multispecies design demands an epistemological shift, from anthropocentric control to situated, relational understanding. As Puig de la Bellacasa emphasises, care is a way of knowing: an attentiveness to that which does not speak in human terms, but still requires response (Puig de la Bellacasa 2017).

This ethic of care resists the extractive logic that underpins much of modern design. Instead of asking "What can this material or species do for us?", it asks "How can we co-exist without harm?" This decentring of the human does not diminish design's creative agency, it amplifies it through responsibility.

Bioneurodesign, rooted in systems thinking and affective neuroscience, provides a methodology for this shift. It offers tools for perceiving the entangled effects of our actions, for integrating ethical foresight into ideation, and for creating environments that foster empathy, not supremacy.

3.5 *Design as Ecological Listening*

To design beyond certainty is to embrace design as listening, a process attuned to signals from damaged landscapes, stressed species, and disoriented human communities. It requires slowness, reciprocity, and a redefinition of success, not in terms of scale or novelty, but of contextual responsiveness and long-term flourishing.

Projects grounded in multispecies justice, such as indigenous-led land restoration, animal-inclusive public infrastructure, and interspecies co-design workshops, show that alternative futures are not only possible, but already emerging. These practices remind us that design is not simply about what we make, but how we relate.

By integrating multispecies ethics into the Bioneurodesign framework, we move closer to a model of design that thinks with, rather than for, the world. It is not a utopian gesture, but a necessary evolution, one capable of meeting the complexity, fragility, and potential of planetary life.

4 Cognitive Ecologies: Design as Mediator of Inner and Outer Worlds

If the previous section repositioned design within a multispecies ethical framework, this chapter turns inward, to explore how design mediates not only relations between species, but between mind and environment. In the face of symbolic collapse, overstimulation, and disembodiment, the question is no longer whether design can affect cognition it is how it should. This leads us to the notion of cognitive ecologies, the idea that cognition does not reside solely within the brain, but emerges through interactions between neural systems, bodies, tools, spaces, and cultural codes (Barrett et al. 2015; Chatterjee et al. 2021).

Bioneurodesign embraces this relational model. Rather than treating the user as a fixed entity navigating static environments, it positions design as a co-constitutive force in the construction of experience. This shift challenges technocratic paradigms that reduce design to efficiency or performance. Instead, it invites a new agenda: to create conditions for emotional safety, mental clarity, and neurodiverse flourishing.

4.1 Designing with the Nervous System

Neuroaesthetic research has shown that form, light, colour, texture, and rhythm influence not just perception, but also autonomic regulation, emotional states, and attentional patterns (Clark 2008; Citton 2017). These findings are not limited to museums or product packaging—they extend to schools, hospitals, housing, and public space. In these contexts, poor design contributes to stress, cognitive fatigue, and disengagement. Conversely, attuned design can act as a form of care.

Bioneurodesign proposes that designers should engage with environments as if they were extended nervous systems, fields of sensory and symbolic input that can either overload or support the brain's capacity to adapt. Design choices such as spatial rhythm, acoustic dampening, or multisensory wayfinding are not merely technical; they are interventions in emotional and cognitive ecologies.

In education, this becomes particularly salient. Research in learning environments suggests that spatial layout, light exposure, and materiality directly impact memory retention, motivation, and emotional regulation (Joye 2007). Bioneurodesign advocates for adaptive learning spaces that reflect these insights, spaces that adjust to circadian rhythms, offer varied sensorial zones, and acknowledge the needs of neurodivergent learners.

4.2 Design for Neurodiversity and Emotional Safety

Cognitive ecologies are never universal. They are modulated by developmental, cultural, and neurological differences. For neurodivergent individuals, such as those with autism, ADHD, PTSD, or sensory processing disorders, design often fails to accommodate complexity. A one-size-fits-all model of user experience reinforces exclusion by privileging normative cognition.

A bioneuroethical approach demands that design embrace cognitive pluralism. This entails creating spaces and interfaces that do not merely tolerate difference but are informed by it. Examples include sensory-sensitive classrooms, mental health-responsive furniture, and digital environments that adapt not just to preferences, but to emotional thresholds.

This move intersects with recent scholarship on trauma-informed design (Kopec 2012), which calls for environments that prioritise predictability, choice, and refuge. In Bioneurodesign, such values are not secondary, they are foundational. They recognise that cognition is not abstract but lived, and that the affective dimension of experience is central to perception, learning, and identity formation.

4.3 The Embodied Mind Affect, Agency and Environmental Feedback

Design has always been performative. But Bioneurodesign argues that it is also affective and dialogical. It co-produces states of attention, perception, and action. The more a space resonates with the user's inner state, the more it fosters agency and regeneration. This has implications for how we approach public space, healthcare, and domestic environments.

Projects that integrate affective computing or responsive environments, such as lighting that adjusts to heart rate or soundscapes that shift with emotional state—represent early steps in this direction. However, Bioneurodesign cautions against bio-surveillance or datafication of affect. Instead, it proposes a model of responsive intimacy: environments that gently respond to shifts in mood, energy, or focus, not to control behaviour, but to mirror and support it.

This feedback loop reinforces a core principle of cognitive ecologies: that environment is not background, but co-author of thought. In therapeutic architecture, for example, the spatial experience can trigger memories, invite calm, or activate embodied metaphors that aid recovery (Sternberg 2009). Design becomes a silent companion in mental life.

4.4 Depth Over Speed: Reclaiming Cognitive Sovereignty

We live in a culture of acceleration, marked by attentional fragmentation and cognitive overload. Design, rather than mitigating this condition, often exacerbates it—through frictionless interfaces, pop-up alerts, and performative minimalism that hides complexity under false clarity.

Bioneurodesign counters this with a plea for slowness and depth. It proposes environments that cultivate sustained attention, contemplative engagement, and meaningful friction. This resonates with the ethics of attention articulated by philosophers such as Simone Weil and media theorists like Citton, who argue that the preservation of attention is a political and cultural act (Varela et al. 1991).

Designers, then, become custodians of cognitive sovereignty. By shaping how individuals relate to information, to space, and to themselves, they influence not just behaviour, but capacity for reflection, empathy, and ethical orientation.

This section reframed Bioneurodesign through the lens of cognitive ecologies, situating design as a bridge between inner states and external systems. It argued that design is not neutral, but deeply affective, and that spaces, interfaces, and artefacts shape how people think, feel, and become. By embracing neurodiversity, emotional safety, and affective responsiveness, Bioneurodesign offers a regenerative model for environments that support coherence, agency, and care in a cognitively saturated world.

5 Results: A Bioneuroethical Manifesto for Designers

As the preceding sections have shown, design in the Anthropocene must evolve beyond its disciplinary boundaries and cultural assumptions. It must grapple with a fractured world where ecosystems are collapsing, attention is fragmented, and meaning is unstable. Bioneurodesign, emerging at the intersection of neuroscience, ethics, and systems thinking, offers a framework for this evolution. Yet frameworks are not enough. What is needed now is a set of orienting principles, not prescriptive rules, but navigational tools to guide design action in conditions of complexity, urgency, and care.

This section presents a Bioneuroethical Manifesto for Designers. Each principle is grounded in the preceding theoretical contributions, but translated into actionable, generative commitments. These principles do not aim to exhaust the possibilities of design ethics, but to provoke reflection and transformation.

The manifesto is also informed by the discussions and collaborative energy of the *Design Commit 2024* international conference, which brought together designers, scholars, and activists to imagine post-anthropocentric, inclusive, and regenerative futures. Its upcoming 2026 edition, under the theme *Learning with Nature*, further reinforces the urgency of design that listens, heals, and co-evolves. This manifesto emerges from that same impulse.

5.1 *Design for Emotional Resilience*

Design has long prioritised clarity, efficiency, and performance. But in a time of collective trauma—pandemics, displacement, climate grief—emotional resilience becomes a vital design objective. Drawing from affective neuroscience and trauma-informed practice, this principle calls for spaces, systems, and objects that help individuals regulate stress, process uncertainty, and sustain hope.

Emotional resilience in design does not mean aestheticising calm or imposing tranquillity. It means enabling emotional literacy through choice, rhythm, tactility, and spatial generosity. Design for resilience is not neutral; it makes a claim: that emotions are central to cognition, identity, and justice (Damasio 1999; Chatterjee 2014).

This is particularly vital in education, healthcare, and public services, where design can support, or hinder, the recovery of affective agency.

5.2 *Design for Cultural Plurality*

Design is never ideologically neutral. It carries epistemologies ways of knowing and being. Historically, many design systems have been shaped by colonial, capitalist, and

Eurocentric worldviews that silence other cosmologies. This principle calls for design practices that engage and integrate diverse cultural perspectives, not as decoration, but as ontological contributions (Sousa Santos 2016; Mignolo and Walsh 2018).

To design for cultural plurality is to practise epistemic humility. It requires recognising that different communities perceive time, matter, agency, and value in divergent ways. From Andean relational ontologies to African philosophies of *ubuntu*, these perspectives enrich design's ethical and conceptual vocabulary.

This is not a call for relativism, but for co-presence. Design can serve as a mediator between local specificity and global solidarity. This principle resonates strongly with the agenda of *Design Commit 2026*, where pluriversal design is a central theme.

5.3 Design for Neuroethical Agency

As shown in Sect. 2, environments and interfaces shape perception, cognition, and emotion. This shaping power places designers in positions of neuroethical responsibility. This principle affirms the need for attentional sovereignty, cognitive justice, and meaningful consent in all designed systems.

Interfaces should not manipulate. Spaces should not overwhelm. Materials should not alienate. Design for neuroethical agency insists on transparency, choice, and the recognition of diverse cognitive patterns. It aligns with principles of design justice (Costanza-Chock 2020) and with the emerging field of the ethics of attention (Citton 2017).

This principle also critiques the extractive logic of surveillance capitalism. Neuroethical design resists the commodification of cognition. It refuses to design for addiction, distraction, or overexposure. Instead, it protects the user's right to focus, reflect, and resist.

5.4 Design for Multispecies Reciprocity

In Sect. 3, we argued that design must expand its ethical horizon beyond the human. This principle articulates that expansion in affirmative terms: design for reciprocity, not resource extraction; for care, not command.

This requires not only ecological literacy, but ecological humility. Designers must learn to read the signals of other species and adapt accordingly. From biophilic public space to microbial urbanism, this principle embraces the idea that habitats should serve many kinds of life.

Design for multispecies reciprocity also includes temporal sensitivity: aligning interventions with ecological rhythms and resisting the timeframes of acceleration and profit. It is not always design as action, it can be design as restraint.

5.5 Design for Sensemaking

Finally, this manifesto affirms design as a cultural act of sensemaking. In times of symbolic erosion, climate misinformation, and algorithmic opacity, design must help restore coherence, not by simplifying reality, but by making it navigable (Weick 1995; Snowden and Boone 2007).

This principle does not advocate for visual clarity alone. It calls for narrative legibility, ethical transparency, and material storytelling. It invites designers to think not only in terms of solving problems, but of shaping the stories through which people understand themselves and their world.

In this light, Bioneurodesign becomes a method of situated cognition—a way of engaging with the world that is sensitive to neural, cultural, and ecological dimensions. To design for sensemaking is to design with awareness of meaning, not just function.

These five principles are not exhaustive, nor fixed. They are living commitments, open to challenge and transformation. Together, they offer a starting point for rethinking design in the Anthropocene, not as a professional skillset, but as an ethical, cognitive, and planetary practice.

The Bioneuroethical Manifesto does not ask designers to be neuroscientists or philosophers. It asks them to be attentive, responsive, and responsible. It asks them to act not from certainty, but from care. In doing so, it offers a compass for navigating design's evolving role: as a bridge between worlds, a mirror of inner life, and a guide toward regenerative futures.

The following manifesto synthesises the conceptual trajectory of this chapter and constitutes its primary outcome, a set of principles designed to guide future design practices from a bioneuroethical perspective.

6 Conclusion: Toward an Ethics of Attention in Design

We are living in an era marked by overload of information, images, crises, and contradictions. Attention, once considered an abundant faculty, has become a contested resource. In this landscape of cognitive saturation, symbolic collapse, and ecological grief, design is not peripheral. It shapes the channels through which we perceive, decide, and relate. It configures the environments that train our habits of attention—and thus, our capacity to care.

Throughout this contribution, we have argued that Bioneurodesign, when reframed through a bioneuroethical lens, offers more than a methodology. It presents a paradigmatic repositioning of design as a cognitive, cultural, and ecological force. Each section contributed to this reimagining:

In the introductory section, we recognised the Anthropocene as a crisis not only of climate and systems, but of meaning and reference. In response, we repositioned design as a form of sensemaking, capable of mediating between disruption and coherence.

In the section on neuroethical design, grounded in neuroscience and systems thinking, design was shown to influence neuroplasticity, symbolic perception, and empathy, shaping not only what users experience, but how they understand and engage with the world.

The segment dedicated to multispecies design expanded the ethical terrain beyond the human. Drawing from multispecies studies, it argued that design must operate beyond the logic of human centrality, embracing symbiosis, responsiveness, and material affordances that support biodiversity and relational ethics.

The section exploring cognitive ecologies turned the focus inward, analysing how thought, emotion, and space are co-constructed. Here, Bioneurodesign offered tools for designing adaptive, inclusive, and emotionally intelligent environments that support well-being and attentional sovereignty.

Finally, the Bioneuroethical Manifesto crystallised key principles, emotional resilience, cultural plurality, neuroethical agency, multispecies reciprocity, and sense-making, into actionable commitments. These were not prescriptive rules, but ethical coordinates for a new generation of design.

Now, in closing, we propose that the task ahead is technical or creative. It is attentional.

Design as the Stewardship of Attention

To design in the Anthropocene is to design under conditions of distraction, despair, and disorientation. What is needed is not more novelty, but more depth. Not more speed, but more listening. Design must cultivate what Simone Weil (1952) once called *"the rarest and purest form of generosity"*: attention (Weil 1952).

This does not mean reverting to aesthetic minimalism or anti-technological nostalgia. Rather, it calls for intentional environments, interfaces that slow down, architectures that invite reflection, systems that honour the complexity of lived experience. Attention, in this sense, is not only about focus, it is about ethical attunement, the capacity to notice what is neglected, to hold what is fragile, to care for what is yet to come.

Such an ethics of attention is inseparable from the ecologies of the mind, from the systems of meaning and spaces of relation that design helps to sustain. Whether in education, healthcare, urban planning, or digital culture, the design of attention becomes the design of futures.

The Future of Bioneurodesign

This text does not claim to offer final answers. Rather, it gestures toward a horizon of ongoing inquiry and transformation. The work of Bioneurodesign is unfinished, by necessity. It must evolve through dialogue, prototyping, feedback, and lived experimentation. Conferences such as *Design Commit 2024* and the upcoming 2026 edition have begun to map this terrain, fostering spaces where theory and practice converge in service of planetary well-being.

At its core, Bioneurodesign asks a simple but radical question: *What kind of minds, relations, and worlds are we designing for?* In responding, we are invited to shift our metrics, from performance to coherence, from consumption to reciprocity, from distraction to care.

This closing reflection leaves an invitation to designers, educators, researchers, and decision-makers: to treat attention not as a commodity to capture, but as a capacity to cultivate. To reimagine design not as optimisation, but as orientation. Not as domination, but as a practice of tending, to minds, systems, and futures, with discernment, humility, and hope.

References

Azouzi S, Di Lucchio L (2024) Design after the Anthropocene. In: Gambardella C (eds) For nature/ with nature: new sustainable design scenarios. Springer Nature Switzerland, Cham, pp 13–32

Barad K (2007) (: Meeting the universe halfway: quantum physics and the entan-glement of matter and meaning. Duke University Press, Durham, NC

Barrett P, Davies F, Zhang Y, Barrett L (2015) The impact of classroom design on pupils' learning: final results of a holistic, multi-level analysis. Build Environ 89:118–133. https://doi.org/10.1016/j.buildenv.2015.02.013

Beck U (2009) World at risk. Presented at the

Berghofer P (2019) Husserl's noetics–towards a phenomenological epistemology. J Brit Soc Phenomenol 50. https://doi.org/10.1080/00071773.2018.1525798

Braidotti R (2013) The Posthuman. Polity Press, Cambridge

Capra F, Luisi PL (2014) The systems view of life: a unifying vision. Cambridge University Press, Cambridge

Chatterjee A (2014) The aesthetic brain: how we evolved to desire beauty and enjoy art. Oxford University Press, Oxford

Chatterjee A, Coburn A, Weinberger A (2021) The neuroaesthetics of architectural spaces. Cogn Process 22:115–120. https://doi.org/10.1007/s10339-021-01043-4

Citton Y (2017) The ecology of attention. Polity Press, Cambridge

Clark A (2008) Supersizing the mind: embodiment, action, and cognitive extension. Oxford University Press, Oxford

Coburn A, Vartanian O, Chatterjee A (2017) Buildings, beauty, and the brain: a neuroscience of architectural experience. J Cogn Neurosci 29:1521–1531. https://doi.org/10.1162/jocn_a_01146

Costanza-Chock S (2020) Design justice: community-led practices to build the worlds we need. https://doi.org/10.7551/mitpress/12255.001.0001

Crutzen PJ, Stoermer EF (2000) The anthropocene. IGBP Newsletter 17–18

Damasio A (1999) The feeling of what happens: body and emotion in the making of consciousness. Harcourt, New York

de Sousa Santos B (2016) Epistemologies of the south: justice against epistemicide. Routledge, London

Escobar A (2018) Designs for the pluriverse radical interdependence, autonomy, and the making of worlds. Duke University Press, Durham

Félix MJ (2024) Transdisciplinary innovation in design: pathways to a sustainable and inclusive future. In: Machado J, Soares F, Trojanowska J, Ivanov V, Antosz K, Cagá\vnová D, Manupati VK, Pereira A (eds) Innovations in industrial engineering III. Springer Nature Switzerland, Cham, pp 253–264

Félix MJ (2025) Designing the future: bioneuro design and the new frontier of human-centric innovation. In: Raposo D, Neves J, Silva R, Correia Castilho L, Dias R (eds) Advances in design, music and arts III. Springer Nature Switzerland, Cham, pp 281–294
Félix MJ, Santos G (2020) Development of products and processes through technology and innovative design principles: contribution for modernization of small and medium-sized enterprises. In: Farinha L, Raposo D (eds) Handbook of research on driving industrial competitiveness with innovative design principles. IGI Global, Hershey, PA, pp 14–33
Félix MJ, da Silva FM (2021) The 4th industrial revolution—the future of design research and doctoral programs. In: Springer Series in Design and Innovation
Floridi L (2023) The ethics of artificial intelligence: principles, challenges, and opportunities. Oxford University Press, Oxford
Gibson JJ (2014) The ecological approach to visual perception. Psychol Press
Gilbert JA, Stephens B (2018) Microbiology of the built environment. Nat Rev Microbiol 16:661–670. https://doi.org/10.1038/s41579-018-0065-5
Goleman D (2013) Focus: the hidden driver of excellence. HarperCollins Publishers, New York, NY, US
Haraway DJ (2016) Staying with the trouble: making kin in the chthulucene. Duke University Press, Durham
Joye Y (2007) Architectural lessons from environmental psychology: the case of biophilic architecture. Rev Gen Psychol 11:305–328. https://doi.org/10.1037/1089-2680.11.4.305
Kahneman D (2011) Thinking, fast and slow. Farrar, Straus and Giroux, New York
Kaplan R, Kaplan S (1989) The experience of nature: a psychological perspective. Cambridge University Press, New York, NY, US
Kopec D (2012) Environmental psychology for design. Fairchild Books, New York
Latour B, Porter C (2017) Facing Gaia: eight lectures on the new climatic regime. Poli-ty Press, Cambridge
Malafouris L (2013) How things shape the mind: a theory of material engagement. https://doi.org/10.7551/mitpress/9476.001.0001
Manzini E, Coad R (2015) Design, when everybody designs: an introduction to design for social innovation. The MIT Press. https://doi.org/10.7551/mitpress/9873.001.0001
Massumi B (2002) Parables for the virtual: movement, affect. Sensation. Duke University Press, Durham
Mignolo WD, Walsh CE (2018) On decoloniality. Duke University Press
Papoulias C (2018) Showing the unsayable: participatory visual approaches and the constitution of 'patient experience' in healthcare quality improvement. Health Care Anal 26:171–188. https://doi.org/10.1007/s10728-017-0349-3
Puig de la Bellacasa M (2017) Matters of care: speculative ethics in more than human worlds. University of Minnesota Press. Minneapolis
Raworth K (2017) Doughnut economics: seven ways to think like a 21st-century economist. Random House Business Books, London
Snowden D (2005) Multi-ontology sense making: a new simplicity in decision making. J Innov Health Inform 13:45–53. https://doi.org/10.14236/jhi.v13i1.578
Snowden D, Boone ME (2007) A leader's framework for decision making. Harv Bus Rev 85:68–76
Sternberg EM (2009) Healing spaces: the science of place and well-being. Harvard University Press, Cambridge
Tonkinwise C (2015) Design for transitions—from and to what? Des Philos Pap 13:85–92. https://doi.org/10.1080/14487136.2015.1085686
Tsing AL (2015) The mushroom at the end of the world: on the possibility of life in capitalist ruins. Princeton University Press, Princeton
Tsing AL, Swanson HA, Gan E, Bubandt N (2017) Arts of living on a damaged planet: ghosts and monsters of the Anthropocene. University of Minnesota, Minneapolis
Varela FJ, Thompson E, Rosch E (1991) The embodied mind: cognitive science and human experience. Cambridge

Wahl DC (2016) Designing regenerative cultures. Triarchy Press, Axminster
Weick KE (1995) Sense making in organizations. Sage Publications
Weil S (1952) Gravity and grace. Routledge & Kegan Paul, London

Non-place as a Spatial Strategy to Establish Temporal Interaction for More Creativity in Higher Education: Lisbon School of Architecture Case

José Manuel Silveira Dias

Abstract In the current context that favors the emergence of individualistic and atheistic behaviors, where there is a growing standardization of creativity, it is possible to counteract this trend through the practice of a strategy of temporal appropriation for greater creative autonomy. Non-place in the context of higher education results from a reflection on the appropriation of sensitive space for more robust learning in the face of the current temporal dispersion. The ways of staying, delaying, remembering, and raising awareness were explored to understand this evasive strategy in institutional space. In teaching experience of Lisbon School of Architecture in the last five years, there have been illustrative situations that have provoked this reflection, created by the teachers to promote more creative interactions in the teaching time and by the students in the extraordinary appropriation of the teaching space in the learning time. Field observations, exploratory photography, and critique of literature supported by Spatial and Ephemeral design theory were conducted. Behaviors have been observed showing that they oppose immersion in a virtual world based on the increasing acceleration regime driven by social and informational networks benefiting the non-factual regime. Creative form of this occupation was identified in contrast to the pre-established space-time typologies. The aim was to theoretically consolidate this evasive occupation as a catalyst for autonomous creative practice in space. As a viral strategy to slow down the simulation of the virtual world that makes what is not appeared to be, non-place corresponds to a dissimulated three-dimensionality where space is without appearing to be.

Keywords Non-place · Spatial strategy · Temporal interaction

J. M. Silveira Dias (✉)
CIAUD, Research Centre for Architecture, Urbanism and Design, Lisbon School of Architecture, Universidade de Lisboa, Lisbon, Portugal
e-mail: silveiradias@fa.ulisboa.pt

D. Raposo et al. (eds.), *Perspectives on Design IV*, Springer Series in Design and Innovation 59, https://doi.org/10.1007/978-3-032-00733-9_20

1 Non-place

Since the last decade of the twentieth century, an increasingly accentuated dramaturgy of urban life has been identified, facilitated by all the paraphernalia of visual recording devices accessible to all. Surrounded by cameras, we are invited to assume the role of actors and directors: the video surveillance cameras that follow us everywhere; the cameras in stores that reproduce our image on countless screens; webcams and mobile phone applications that encourage the portrayal of everyday life for social and information networks. We have acquired the skills to be in front of the cameras as well as behind them. At the same time, we are consumers and producers of creative images. We stage, transform costumes, frame, simulate behaviors, and adapt narratives, removing responsibility to divine power and social structures based on our biological and socio-cultural condition (Campos et al. 2011).

As Stoichita would say, when we imitate them, we are simulacra of the images we consume and simultaneously develop our creativity through daily staging (Stoichita 2011).

While the television generation had a univocal relationship as a passive receiver of information, consuming only images, the Internet generation establishes a univocal relationship with electronic devices. This generation interferes with the content and participates in the production of information. It is not only a consumer, but also an image producer, creating visual records of its daily life through its mobile phone, webcam, digital camera, video, and social networks. Mechanical digital communication devices are recognized as technologies of memory, representation, and storytelling. Visual and audiovisual technologies and grammars become instruments of experimentation in the territory of the image, as a pictorial artifact or visual narrative, but also as forms of recording everyday life that use dramaturgy and performance (Goldberg 2012). The generation that produces images does not only produce technological images, but also interprets and produces complex and innovative visual grammars that are realized in real time, globally and medially as a response to the way we experience the world around us, thus giving a referential place and identity to the space that conditions it, which we identify as non-place because it reveals the cunning form of this occupation in relation to the pre-established typologies of the socio-cultural system.

Following Augé, the non-place does not intend to be the negation of place as a referential space with characteristics contrary to recognized places, and even less to oppose the material place to the immaterial place (Augé 2012). The non-place will be another process in which it is intended to carry out the synthesis of material and mental experience as an inducing space for phenomenological cognition, fully perceptible. In it, the meaning of the work or the place of lived experience is incorporated.

The non-place reveals itself in its modeling or transformation, moving away from the conventional place and assuming an aesthetic autonomy whose origin is linked to its producers and users. It is identified as a space of evasive expansion that results from the transformative appropriation of the everyday (Fig. 1).

Fig. 1 Time for remain in the shade (author, 2019)

2 Time for Remain

As Baudrillard would say, the concept of ambience or atmosphere arises when people begin to live less in proximity to their peers and become more absorbed in the over-abundance of the virtual world, losing the ability to communicate and pay attention to others (Baudrillard 2011). Baudrillard had already mentioned in the "System of Objects" that both phenomena are promoted by the signification of the objects that surround us, since objects do not appear in isolation, without the context of the objects that give them meaning, of the scenarios that present and distinguish them (Baudrillard 2000).

Later, Zumthor would explain that the atmosphere is like a reciprocal game, of give and take, between us and the surrounding space, which is designed to promote human interaction. As the layout of the built space, it communicates with those it relates to, exerts perceptual influence, and infects users (Zumthor 2006).

Bachelard, in the "Poetics of Space", defends the place as a concrete manifestation of space, where uses and habits build the image of that place. In constant transformation, it infects its identity and value to those who remain in it (Bachelard 1978). Space results from the organization and action in the environment (physical and mental) that creates conditions for human activity. It determines the place that encourages human action through elements such as color, smell, sound, light, form, which modify and qualify the perceived space, interfering in the relationship that is established with the constructed that delimits it. The perception of space through movement, surfaces and textures of materials also defines this place for possible translations that modify and qualify space. The stimuli related to the organization of flows and permanence in the environments that space provides and that allow the development of the sensation that human being experiences about his existence are identified there. Configured by space, the perceptual atmosphere reflects lifestyles in the appropriation of well-being in space. As a practice of humanization, this arrangement results from the search for balance between the elements that make up the space, with the aim of creating conditions to establish itself as a place of

Fig. 2 Time for delay in the tranquility of well-being (author, 2019, 2020)

reference. It emerges as an interface for the critical reflection of the production of the subject in the process of social interaction, a comprehensive territory of different cultures and values, where tradition is preserved to create conditions for innovation.

In fact, thinking about the non-place means thinking about one's life in society, trying to understand the production process and the productivity of the individual, in a symbolic universe and in the current globalization. In a context of dispersion, as a conquered space of permanence, it is conceived as a temporary reference place to respond to contemporary needs for well-being, optimizing existing resources through a co-responsible action in its spontaneous creation. To think about the non-place is to reflect on the evasive catalytic dynamics to establish a creative practice resulting from interaction in space (Silveira Dias 2017; Silveira Dias and Neto 2015) (Fig. 2).

3 Time for Delay

Acceleration presents itself as a strategy to respond to the problem of limited life span that haunts us. We believe that increasing speed increases the number of options, as a product of quantifying the opportunities that can be seized through this acceleration to achieve such an ideal of abundance. Acceleration is only an incremental quantifier to compensate for the lack of permanence of things. The paradox lies in the desire to achieve immortality in this way, since human being can remain in things, since things remain in the references of the world that last. One feels the need for a synthesis resulting from the narrative developed by the intensity of events. A narrative generated by the consummation and not by the enumeration of the events that make up everyday life, giving the possibility of concluding with meaning. The lack of narrative tension makes it impossible to materialize or maintain a long attention to things. When time is discontinuous, it does not articulate events among themselves, and no spatial relationship is generated for delay. The anxiety of always starting over, because of the inability to finish, of always choosing new versions or modalities, causes the sensation of acceleration that results from the absence of the experience of duration. Time has duration when it acquires dimension in depth and length, and only then does it conquer its space. Time becomes virtual, however, when it loses narrative tension, when the historical line breaks up into random and aimless points. Events only follow one another, and the informational data are isolated, and since they have no narrative extension, they are not linked to temporal space. A time of points has no history because it imposes a trajectory on the succession of events, articulating a linear narrative that unifies points. Through acceleration, this time of points tends to suppress or reduce intervals in favor of simultaneity, and everything passes in real and immediate time.

Electronic memory allows repetition, eliminates the time interval responsible for forgetting. The past is always available. Time itself is liberated from space. The instantaneity of e-mail removes the way as a spatial area. Distance is eliminated, everything is available, here, and now. As Byung-Chul Han would say, what cannot be present does not exist (Han 2016).

Possibilities are constantly emerging, resulting in discontinuous time, and the ability to decide is lost because new possibilities are constantly emerging. Thus, the network space is defined by its multiple possibilities of connection. There is no continuity of stages and transience, only random events, or circumstances. As a result, no option gains absolute predominance over the others. According to the same author, the loss of the linear condition of the world develops new forms of perception in the face of non-causal behavior. The production of digital technologies and products is both a product and a result of this temporal shortening. Obsolescence increases due to the drive for novelty and reduces cycles of renewal. There is no place for effective work, goals are not achieved, but a multiplicity of versions and modalities are acquired. Everything is present and simultaneous, with the possibility of being, but only at this moment.

The temporal framework of the present approaches the instant, and thus the sense of the time of delay, of the temporal extension of duration, of the space that predisposes to contemplation and allows one to speculate on the relations of things, is lost. In contrast to this panorama, the non-place emerges as the space that is created when things are correlated by any kind of affinity or proximity. To do this, it is necessary to discover and understand the ways in which things are connected and correspond, how they are articulated and how they communicate with each other in space (Fig. 3).

4 Time for Remember

As Pallasmaa would say, sensory experiences are absorbed and integrated into the body and the way of being human (Pallasmaa 2012). The human being is in constant interaction with the environment, and therefore the environment and the human being are constantly and mutually redefining each other. The perception of the human body and the image of the world become a continuous existential experience.

Fig. 3 Time to remember creative events (author, 2024, 2021)

Regarding the body, it does not exist separately from the space it occupies, and as such, it is not possible for space to detach itself from the unconscious image of our perceptual essence, that is, as a translation of the stimuli that surround us, reflected in the patterns of behavior and in the factors selected by the active senses of the human being (Sapolsky 2018). Human perception depends on subjective factors, such as lived experiences, the cultural values of the social group to which one belongs, and the selection of meaningful reference codes for the interpretation of reality.

In the recognition of past experiences, memory is evoked by the appeal to recall, giving meaning to the sensory chaos, giving order to the scattered data. It is only after lived experience that suggestion, conjecture, and analogy can arise as causes of recognition, and it is precisely in the experience of the present, as it acquires form and meaning, that memory is constructed. Any sensitive data captured by the senses conceals the meaning of the original data (Stoichita 2011). This illusion masquerades as an authentic perception, in which the meaning emerges from the sensible world, imitating this experience to signify what it really is. At each moment, the past experience that lies on the horizon of consciousness is reopened for recognition as an act of memory, and at the same time it immediately provides a present meaning and significance for understanding. Merleau-Ponty argues that remembering is not about bringing to consciousness an image of the past that persists per se, but about going into the past and gradually developing new perspectives that fit together until the experiences to which they relate are relived in space (Merleau-Ponty 1999). The non-place is a space that is always available to consciousness, which in all its perceptions includes an atmosphere of "montages" that give it temporality, such is the presence of the trajectory of the past configured in memory. To understand in space is not to experience a multiplicity of impressions that bring memories incapable of completing a narrative, but it is as a non-place that makes an immanent meaning emerge from a set of sensible data, without which memories would not be possible in the construction of the memory where they are stored, because we allow ourselves to take the time of the things of the world (Fig. 4).

Fig. 4 Time for knowledge awareness (author, 2022, 2024)

5 Time for Awareness

The sense of shelter has always been a primary human need. The need for shelter to protect corresponds to the sense of belonging and domicile that is inherent to human behavior as a place of reference and action in the realm of experienced space (Bachelard 1978).

The shelter is the home, the place of the authentic, where unique references and memories are kept. Inside the shelter there is a reencounter of the self, the place of subjectivity, and it is from there that one always comes out to the outside, a place of objectivity.

Space qualifies time from things that have no place. Things can be inside or outside, depending only on the position of the subject in relation to the object. Hence the term place, the space in which the subject finds itself when it becomes conscious of itself. The place of subjectivity defines the imaginary space that proposes new forms of interaction based on a "solitary contractuality", as a place of evocation through words that report images that build the imaginary of each one. This space reconciles functionality and the status quo through imagination. It exists only for the words that evoke it, creating the image that produces the myth and that becomes the new reality through simulation (Stoichita 2011).

In fact, it determines the place for the knowledge of our existence and of what surrounds us. As Damásio would say, consciousness is a mental state lived in its own perspective, exclusive to each organism in which the mind works, knowing that said existence occupies a certain situation surrounded by objects and events (Damásio 2010). Although this experience is exclusive, it does not mean that one cannot adopt an objective perspective in the sense of enriching the material self-capable of bringing knowledge to the mind, that is, of functioning as a knower. Conscious states of mind always have content, they are always about something. They reveal the distinct qualitative properties in relation to the different contents that are being apprehended. Necessarily, conscious mental states deal with knowledge based on different sensory material (it is qualitatively different to see or hear, touch or taste, and even smell) and manifest different qualitative properties for the different sensory streams, and where conscious mental states are felt in space.

The human being is regulated by the period of his life, which is interwoven with the dramatic sense of his "announced death" as soon as he is born. Therefore, it is necessary to praise the present moment in the succession of the infinite moment of everyday life. It becomes visible in the experience of the moment whose brief existence intensifies it. The constant mutability of things allows us to explore the relationships between spaces, environments, and people, in an experience of correlation of the participants. As such, the space for experimentation requires new territories and visionary technologies based on innovative strategies. As a place for the manipulation of emotions, it triggers sensations resulting from formal, material, and sensory experiments, developing the cognitive process: on the one hand, it models space in time, and on the other hand, it models perception through the experience apprehended in everyday time.

The Non-place does not have to be configured in the street or in the square, it can be in domains beyond the geographical situation or the architectural materialization. It can emerge from the consciousness of intimacy and everyday life, from cyberspace and the informal communication network, through verbal means even more than through social networks and information. By analogy with Byung-Chul Han on random data produced without narrative in the "cloud", non-place can also be conveyed through the dissemination of communication that promotes a viral culture, taking advantage of gaps that expand with an opportunistic character, infiltrating the institutional means of production for its manifestation (Han 2015). In our perspective, as a space of evasive awareness of the institutional space for a greater durability of temporal interaction (Fig. 5).

Fig. 5 Time for performance interaction in the institutional space (author, 2024)

6 Temporal Interaction

Every human being is constantly experimenting with the built space, whether at home, at work, in entertainment, in education, where one can see how the respective scenarios influence the way one thinks and acts. Therefore, we sometimes seek out certain places to experience certain influences: a church for sublimation or a theme park for excitement. Emotionally, although one feels and reacts to the concept of a built space, and these sensations influence one's performance in the space, one usually does not have time to reflect on, analyze, and understand the daily reactions to the place.

Understanding how places work and contributing to the creation of qualified space is urgent in the global situation. Growing immigration, population concentration in urban centers due to the quality of urban life, visible climate change and changes in energy balances are factors that require us to rethink the way we design the space where we live. Through new forms of knowledge expression, spatial design can be guided not only to ensure survival and mental health, but also to develop scenarios and environments that enhance the connection between diverse ideas and mental and physiological states.

The task of observing the intrinsic relationships between lived experiences and the places that contain them must be carried out, participating in, and contributing to scientific knowledge, and using artificial intelligence to optimize such relationships. Technologies used to study the human response to experienced space, from location applications to integrated biometric sensors that analyze passersby, can also be applied to space. Such technologies can redefine the entire experienced space and the meaning of the elements that configure that environment, to reformulate how humans can affect and be affected by what surrounds them, and to redesign the environment in the experienced space according to the behavioral results evaluated and the value of that information. As Ellard reminds us, the understanding of how space can affect people, supported by new technologies guided by design methods, must collaborate with the human sciences (Ellard 2019). In this way, it can facilitate the understanding of how feelings and behaviors can be influenced during experimentation, but also provide the deliberate co-optation of natural responses to the lived space, such as feelings, intentions, decisions, and actions, modifying and qualifying human behavior (Sapolsky 2018).

The duration of human exposure to grandiose built works, such as the dome of the Pantheon in Rome, or natural works, such as Niagara Falls, exerts an overwhelming influence, affecting how one feels about oneself, how one treats others, and how one perceives time. The configuration of the space of grandeur not only affects the way one feels, but also conditions attitudes and behaviors, exalting the sublime of the divine or transcendence to the ordinary. Otherwise, the effect of brand touchpoints in a shopping mall contributes to the emotional excitement motivated by seduction, promoted in the space with the aim of stimulating consumption (Silveira Dias 2019). In another way, in an industrial roundabout crowded with similar warehouses, one can feel the slow passage of time that promotes the boredom that it is, the opposite

Fig. 6 Strategies for the appropriation of temporal space (author, 2020–2024)

of that caused by the hustle and bustle of an urban market full of colorful products, delicious aromas, sound cries that align all the senses for the celebration.

The contrast between reactions to such spaces can easily be read in the human body, in body language, in the movements of the eyes, hands, head, and brain activity. As Damásio argues, the nervous system and the mind are shaped by what is experienced and perceived in these spaces, that is, to establish temporal interaction in space (Damásio 2010). As we know, there is no space without time (Fig. 6).

7 Spatial Strategy

In fact, throughout human history, space has been projected to change perceptions, influence thoughts and feelings, and organize human activity according to social, political, and economic behaviors. As Sapolsky argues, the space we experience always is designed to change the way we feel, to encourage us to reevaluate our relationship with others, with nature, with emotions, and with technology itself, and to influence our behavior and performance during and after its experimentation (Sapolsky 2018). By applying scientific principles resulting from the interdisciplinarity of the fields of architecture, design, and social sciences, we can better understand how built systems, whether built or scenographic, function. In addition, we can develop these real and/or virtual interfaces for social participation and make predictions about how changes in these spatial systems at any scale, from a room to a city, can affect human behavior (Silveira Dias 2019).

It is also worth noting the potential mismatch between the architect's vision and the needs and desires of the end user, as the space designed for experimentation must be useful to its users throughout the life of that space. In the public domain, the architect is responsible for framing the space created within the landscape into which it is inserted. Psychological analysis and the testing of possibilities are methods to be applied to fulfill this responsibility, but above all, conditions must be created to involve the users themselves. The sense of participation and collaboration of those who live in the space of the problem will certainly be the best way to explore possible solutions. As co-design and social design explore, any informed citizen can interact, express an opinion, and contribute with their experience to the discussion of how to develop the space to solve any problem that is experienced under methodological guidance (Silveira Dias 2023).

Using the connectivity provided by the Internet and mobile technology to counteract the hyper communication that these media foster, recognizing the influence that the networked world has on the ability to pay attention and remember, recognizing how instantaneous and global connectivity changes the nature of social relations and politics, are phenomena that can guide and discipline the abundance of information without narrative that prevails in the media and social networks. The widespread availability of technologies that allow the collection of location-based data on responses to the environment, including the responses of the body's own physiology, can also channel the contribution of citizens to the construction of more quality space. Digital applications already offer the possibility of adding geographic location to opinions, photos, travel patterns, heart rate, body temperature, and even arousal levels. Some applications even allow users to view their own data and compare individual statistics with those of other users. An "open data" policy that encourages municipalities, states, and countries to release data on activity, traffic, and economic patterns is also a useful tool for democratizing atmospheric design (Ellard 2019). Access to this information, which is fundamental to understanding how places work, should be made easier for anyone interested in learning how to use, interpret, and record it.

Design methodologies can guide both the collection and analysis of information in search of patterns that can contribute to the development of knowledge about the space to be designed. The collaboration with competent experts in the fields that are important for the creation of the space is also fundamental, since it is necessary to find an agreed language, under technical guidance, to develop arguments for discussion with the responsible institutions, to change what they dictate for the built space, going beyond the conventional and generic, parameterized. Such efforts of collective and popular collaboration transform citizens into active participants in the processes of co-responsible projection of collective space.

An informed public can better interpret the way their senses feel and understand how the space that affects them must be designed to create better conditions. The application of the strategy of non-place serves to recognize the ideal conditions of permanence, delay, memory, self-awareness, and thus to stimulate a diversified creative interaction under the methodological orientation of Design.

In this way, Martin and Hanington recall that methodologies selectively agglutinate methods and techniques, affirming their performance in the way of: guiding communities and designers to be able to establish knowledge; creating scientific credibility through critical reflection in an assertive way with the aim of situating and articulating new knowledge through innovation (Martin and Hanington 2012).

8 Afterword

Through this critical reflection, it was identified through exploratory ways of occupying space with an evasive character as a counterpoint to the typologies of space-time pre-established in the teaching system of FA-ULisboa. Behavioral results, recorded through photography, field observation and literary criticism, were observed, arguing how non-place, as a strategy for appropriating space, can serve to counteract in some way the immersion in an increasingly virtual world based on the regime of increasing acceleration driven by social and information networks. In the current regime, where the simultaneity and media coverage choreographed by globalization promote a denialist ideology in the face of sensible and factual reality, it is verified that the relationship between body, mind and space can be preserved and enhanced through the strategy of non-place. In the current context that favors the emergence of individualistic and unethical behaviors, where the uniformity of the creative diversity of the agents involved (teachers and students) is verified, it is possible to counteract this trend with the strategy of non-place as a possible methodological practice of temporal appropriation in space.

The phenomenology of the non-place as a strategy for the appropriation of space without an assigned place can typologically support its application as a viral space that aims to counteract the growing temporal dispersion. Because it is subject to the action of our senses or because it impresses us in some way (physically or mentally), either by its circumstantial or temporary state, it surprises by its extraordinary character, as something first strange and then ingrained, as Fernando Pessoa claimed with the advertising slogan of the Coca-Cola campaign in 1927.

Without the time necessary for immersion in meaningful things, time is not established in space because: (i) without permanence in space, the tranquility of well-being and the healthy enjoyment of comfort are not allowed; (ii) without conditions for the contemplative delay of things, a referential world is not recognized; (iii) there are no conditions for the transmission of data (facts, information, dogmas, experiences), so there can be no innovation without tradition; (iv) there is no temporal space for an awareness of self and other in a world of all.

The spontaneous strategic action of the non-place allows us to achieve: the tangibility of the things around us as raw material available to any creative mind; the place of the autonomy of creativity in counterpoint or complementarity to the institutional typological space as a parallel laboratory of experimentation in the place of teaching itself.

In short, the non-place as a strategy of viral space, which appears and disappears without warning, adopts the tactics of the digital world of dissemination without space in time. The difference is that the non-place requires concealed three-dimensional space, the space that exists without the appearance and not the simulation of the virtual world that makes it appear, that which is not.

Acknowledgements This paper was written in the context of the Embryo-project SPIRAL (Sustainable Project into Redesigning Atmospheres and Landscapes) which is part of the research group REDes (Research and Education in Design) of CIAUD (Research Centre for Architecture, Urbanism and Design). This Embryo-project is financed by national funds through FCT—Fundação para a Ciência e Tecnologia, I.P., under the Strategic Project with references UIDB/04008/2020 and UIDP/04008/2020.

References

Augé M (2012) Não-Lugares. Introdução a uma Antropologia da Sobremodernidade. Letra Livre, Lisboa

Bachelard G (1978) A Poética do Espaço. Abril Cultural, São Paulo

Baudrillard J (2000) O Sistema dos objetos. Perspectiva, São Paulo

Baudrillard J (2011) Sociedade de Consumo. Edições 70, Lisboa

Campos et al (2011) Uma Cidade de Imagens, Produção e Consumos Visuais em Meio Urbano. Editora Mundos Sociais, Lisboa

Damásio A (2010) O Livro da Consciência. A Construção do Cérebro Consciente. Círculo de Leitores, Lisboa

Ellard C (2019) A Alma dos Lugares. Como a paisagem e o ambiente alteram o nosso comportamento e as nossas decisões. Contraponto, Lisboa

Goldberg R (2012) A Arte da Performance, 2ª edn. Orfeu Negro, Lisboa

Han B (2015) Psicopolítica. Relógio d'Água, Lisboa

Han B (2016) O Aroma do Tempo. Um Ensaio Filosófico sobre a Arte da Demora. Relógio D'Água, Lisboa

Martin B, Hanington B (2012) Universal methods of design. 100 ways to research complex problems, develop innovative ideas, and design effective solutions. Rockport Publishers, Massachusetts

Merleau-Ponty M (1999) Fenomenologia da Perceção, 2ª edn. Martins Fontes, São Paulo

Pallasmaa J (2012) The eyes of the skin: architecture and the senses. Wiley, New Jersey

Sapolsky R (2018) Comportamento. A biologia humana no nosso melhor e pior. Círculo de Leitores, Lisboa

Silveira Dias J (2017) Non-object; designing a conceptual model for the design process. In: Kong M, do Rosário Monteiro M, Neto MJP (eds) Progress(es)—theories and practices. CRC Press, Boca Raton, pp 213–218

Silveira Dias J (2019) POPPING UP. Desenhando um modelo conceptual para o processo em Design ou A efemeridade em busca do seu desígnio. Tese de Doutoramento em Design. Faculdade de Arquitetura da Universidade de Lisboa, Lisboa

Silveira Dias J (2023) SPACE BETWEEN US - Metodologia para o Design de Ambientes. O espaço sentido no lugar do tempo vivido – Relatório de Pós-Doutoramento. Faculdade de Arquitetura da Universidade de Lisboa, Lisboa

Silveira Dias J, Neto MJ (2015) The catalytic effect of the pop up in the otherness process. The aesthetics for the development of the ethical sense. Int J Stud Humanit Social Sciences (SHSS) 94–101

Stoichita V (2011) O Efeito Pigmalião. Para uma Antropologia Histórica dos Simulacros. Imago, Rio de Janeiro

Zumthor P (2006) Atmósferas: entornos arquitectónicos, las cosas a mi alrededor. Gustavo Gili, Barcelona

Mirrors and Projections—Making New Art/Design from Old Technology

Iwona Abrams

Abstract The 'Mirrors and Projections' is a practice-based research project exploring redundant and imperfect analogue projection technology as a relevant mode of artistic expression. From a media-archaeology perspective, the paper demonstrates how analogue technology can be employed not merely for nostalgic reasons, but because it is important to experiment with apparatus previously known as 'high-tech' and that became nearly obsolete as a result of rapid digital evolution. A significant part of contemporary artistic and design production moves towards practices increasingly hybrid and immaterial. In response to this technological acceleration there has been a surge of artists and designers seeking to explore ephemeral materials and production processes. These artmaking practices often incorporate explicit references to machines from earlier phases in the development of our techno culture. With the current obsession for ever more powerful gadgets, this paper is distinct in exploring the narratives of blurred edges of redundant technology. The paper, through showcasing works produced using an overhead projector (OHP), celebrates not only the heritage of this 'extinct' machine but also discuss technological evolution by asking: What is the value of experimentation with obsolescent apparatus in the artwork production and how these experiments expand the field of heterogeneous art/design practices?

Keywords Analogue · Projector · Human-machine · Experiences · Magic

1 Introduction

Throughout history our societies often have been defined by the technologies they employ as models of progressive advancement ('Stone Age', 'Industrial Age', 'Personal Computer (PC) Age', 'Ai Era', etc.). Since the digital revolution, contemporary art and design has seen an increased democratisation of the image production that has been led by technological developments and artists' adoption of hybrid

I. Abrams (✉)
University of Westminster, London, UK
e-mail: I.Abrams@westminster.ac.uk

D. Raposo et al. (eds.), *Perspectives on Design IV*, Springer Series in Design and Innovation 59, https://doi.org/10.1007/978-3-032-00733-9_21

design tools. Consequently, online media are saturated with highly polished images, often divorced from authentic experiences which might explain a rise of interest in the use of analogue technology.

The aim of the 'Mirrors and Projections' as a practice-based research project is to demonstrate, in contrast to the digital and immaterial modes of image production, how analogue technology can be employed not merely for nostalgic reasons, but because it is important to experiment with apparatus previously known as 'high-tech' that became nearly obsolete as a result of rapid digital evolution and to suggest new reading of the role of mechanical tools in art production. With reference to examples of artworks produced using an overhead projector (OHP) encompassing personal reflections, the paper explores new methods of creating images from old media. Zielinski (2006a) in his seminal book 'Deep Time of the Media', argues that 'Nothing endures in the culture of technology; however, we do have the ability to influence how long ideas and concepts retain their radiance and luminescence'; and challenges us (Ibid., p. 3): 'do not seek the old in the new, but find something new in the old'.

The paper's case studies celebrate not only the heritage of this 'extinct' machine but also contribute to the discussion about technological evolution. By exploring OHP (Fig. 1) as a creative tool in artmaking, the aim is to highlight the value and potential of this dated technology in creating a unique artistic experience in both, an art production and audience participation processes.

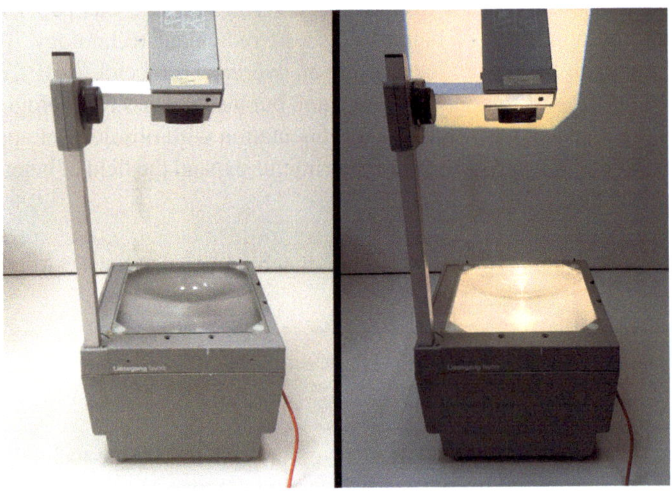

Fig. 1 Overhead projector (OHP), Liesegang favorit 400

1.1 *Overhead Projectors—Context and Historical Background*

Overhead Projectors (OHP) were designed to use light for live projections of magnified images on a screen, to share text or pictorial information with an audience. The OHPs were mostly used in education and business as an equivalent to contemporary digital projectors to aid a delivery of presentations, lectures or training sessions. The images were either printed in advance or drawn directly onto transparent sheets of thin and flexible polyester acetates. The transparencies were placed on the projector's glass flatbed with the light source beneath, while a mirror and lens fixed above projected magnified images. The golden era for the overhead projector span over the period from the sixties until the eighties of the twentieth century.

Going back to the origins of technology behind an overhead projector is a journey deep inside history of knowledge. There is a record, that as early as in ca. 1420 there were experiments carried out to project figures by Giovanni da Fontana, the Venetian rector of the Padua art academy (Ibid., p. 135).

The technical principals of contemporary overhead projectors point to similarities with early instruments perfected and publicised by Athanasius Kircher, German Jesuit scholar. In his 1645, nearly one thousand page book 'Ars Magna Lucis et Umbrae' ('The Great Art of Light and Shadow'), Kircher laid out his ideas, models and building instructions, including the first complex graphic representations of apparatus he called laterna magica, a magic lantern.

The working method of the lantern involved focusing lens and placement of the transparent strip of glass with the image or text onto a concave mirror reflecting sunlight. Although it is claimed that Kircher himself used this device for projecting bright images in dark rooms in lectures and theatrical productions, he cannot be credited with the invention of the magic lantern, instead Kircher's great achievement was the publishing of a technical artifact disseminating an impressive compilation and analysis of existing at the time knowledge on optics.

One could say that the seventeenth century thinkers were fixated about light, inventing optical devices and concepts of mathematical and astronomical instruments. In this period of the Scientific Revolution seminal works of Galileo Galilei and Johannes Kepler were published. Kepler's 1604 'Astronomiae Pars Optica' is still considered as the foundation of modern optics (Casper 1993). In his 1611 'Dioptrice', Kepler introduced the concepts of real versus virtual images, upright versus inverted images, explained through the theory of double-convex converging lenses and double-concave diverging lenses (Ibid., pp. 198–202).

In addition to the broad intellectual and scientific underpinnings of the invention of a magic lantern, there was a lively interest in its media potential and turning it into a marketable and economically viable media apparatus for projecting images. Further development of a magic lantern proved to be successful and there is a record that by 1664, the Dane Thomas Wallgenstein worked in Europe, with a portable version of the apparatus (Zielinski 2006b).

2 Methodology

My own fascination with the OHP begun a long time after its heyday, while teaching in the Drawing Research Centre at the Royal College of Art. In its recent, twenty first century history, the RCA's drawing studio, in addition to various props, was equipped with an analogue slide and overhead projectors formerly used for presentation of examples of artworks. Direct access to these machines allowed me to test their potential and as I had never encountered overhead projectors in the course of my education, to develop spontaneous working methods with an open creative mindset. In my artistic practice, I have been working mostly with Liesegang favorit 400, a classic overhead projector from a personal collection of vintage devices.

In the course of my ongoing research into unique production processes and practical experiments with an overhead projector, two main strands of enquiry emerged:

- Unique Imaging
- Unique Experiences and Magic.

2.1 Unique Imaging: What Is the Value of Experimentation with Obsolescent Apparatus, Such as Overhead Projector in the Artwork Production?

Parallel to researching history and science of instruments related to a modern overhead projector, I began to develop a deep interest in experimenting with an OHP as an artistic tool. My research focused on the technical aspects of designing a method for producing imagery through analogue projections.

The initial research was focused on identifying the properties and limitations inherent to OHP's technical function. The aim was to create a range of experiments ensuring adequate empirical data gathering leading to perfecting varied methods of projections. The conditions to facilitate chance were established in advance through a gamut of repetitions, by varying:

- Transparency of the tested materials (different stocks of films/acetates),
- Layering of materials,
- Negative-positive reversal,
- Left-right reversal,
- Grouping of elements/objects,
- Use of halftone, line and pattern,
- Colour dyes applications,
- Parameters of the individual elements,
- Placement and scale of projections onto objects,
- Sequencing and repetition.

As a practitioner with specialism in photography and print, over the years I acquired a substantial archive of exposed films for screen-printing (Fig. 2) and photographic negative-positive films for analogue photography. Historically, the method for translating photographs into print was achieved through halftone and I experimented extensively with the techniques for introducing dot/raster into the image, in both digital and analogue darkroom processes. The aesthetic attraction of the positive-negative and halftone qualities of the projected OHP images can be seen in signalling the properties of older art production processes. There is cultural currency in adopting manifestations of the visual language of the mid-twentieth century and associated with it notion of nostalgia. As Coldwell (2008) argues: 'Artists go forward and backwards in history to find ways of working and to find models that help them understand contemporary issues.'

There is a phenomenon of 'faux vintage' style in contemporary digital photography characterised by mimicking analogue photography through grains and dots of halftones, and early digital imagery through pixelation or blurriness. These vintage styles, often seen in online image content are designed stylisations, that could be perceived as inauthentic and merely derivative styles. In contrast, working directly with old machines such as OHP grants this image production method originality, and rejects dominant position of high-definition imagery, which is so representative of the aesthetic mores of our time. The attraction of seeking the new in the old is summarised by Faure Walker (2006):

'Why, when you have tasted the digital, would you want to go back to old world ways? Well perhaps you preferred the real thing after all and you wanted to figure

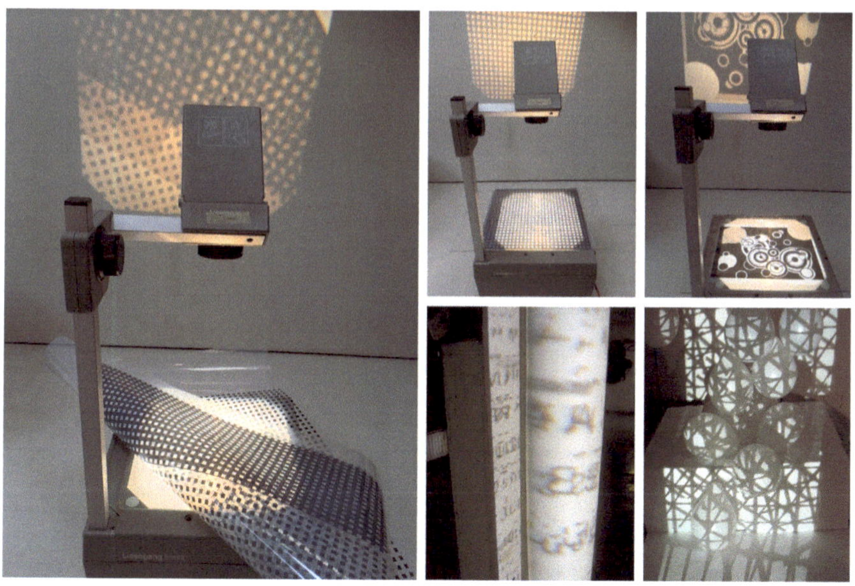

Fig. 2. OHP projections using varied screen-print films, studio experiments

out how to incorporate the new and the old, a combination that somehow makes something interesting out of the difference.'

Following tests and experiments conducted in my studio, gradually a system was developed to create three-dimensional installations designed for producing photographic documentations and drawings of the projections. The method involved projecting images, shapes, typography, ready-made objects and natural materials into the drawing studio's space and onto a range of objects to achieve distortions, contrasting light and shadow effects, etc (Fig. 3). The main principle was to create images using light and shadow. Further into this process I was able to share my findings and methods with the MA students at the Royal College of Art within a regular programme of drawing sessions.

The validity and merits of introducing students to working with the analogue overhead projector can be measured in their successful entries in national and international competitions with projects inspired by exposure to the creative potential of this redundant technology, including visualisations of music by Philip Glass ('Einstein on the Beach') for a typographic project and translation of projected 2D imagery of melting ice into 3D jewellery objects.

In my own artistic practice, including one of my ongoing projects involving ice, an OHP is used to explore an idiosyncratic documentation method of 'live' projections of melting ice to develop a new perceptual model of advancing existing approaches in representing temporal qualities of ice. The intention of these 'real time projections' is to incorporate activities and processes of performance as a method of making ice 'present' (Fig. 4). The morphology of ice is observed and recorded through an overhead projector's magnification method by placing slabs of ice directly onto an OHP's glass flatbed. The technical challenge of the potential for flooding the projector is

Fig. 3. OHP projections, drawing studio experiments, Royal College of Art

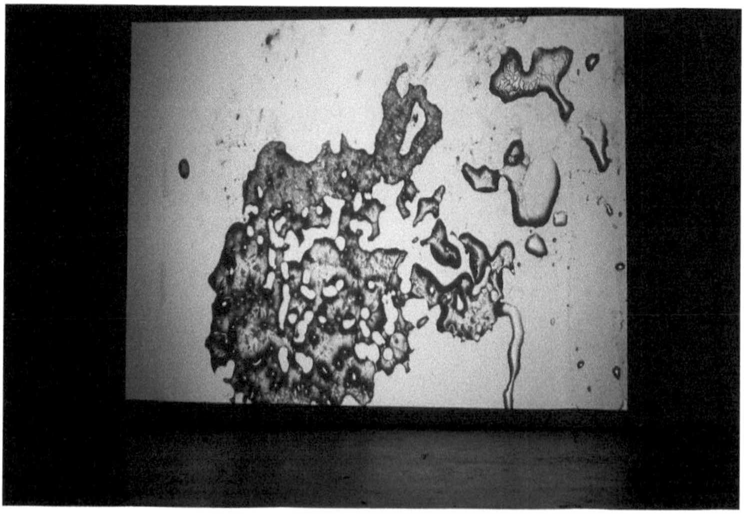

Fig. 4. OHP projection of melting ice, Ambika P3, University of Westminster

resolved by designing a special Perspex tray. Photography and drawing are used as documentation and mapping tools to record changes of ice in time. These experiments lead to another level of understanding of ice, relating to its dynamic properties. Produced works contribute to advancing existing approaches in the representation of natural environmental processes through devising new temporal methods of documentation and recording which incorporate temporal operations.

2.2 *Unique Experiences and Magic: How Does Experimentation with an Overhead Projector Expand Our Understanding of Human-Machine Relationship in Art/Design Practices?*

The second strand of my enquiry explored overhead projectors' potential as an 'active' and direct technology that is capable of generating and transmitting expression of unique and subjective experiences. With contemporary technology becoming increasingly invisible, the analogue technology presents opportunities for different experiences relating to human-machine relationships.

In my group work with the students (Fig. 5) and gallery audiences, the focus of the OHP projections was on the act of making, which often involved direct interpersonal communication and prioritising a process over a self-contained final product. Thus, OHP could be perceived as a tool for building connections. The magnifying lens of the OHP works like a microscope bringing us closer to details we could not otherwise see with our bare eyes. These magnifications also add a sense mystery

Fig. 5. OHP Ice Projection workshop, Royal College of Art

through revealing an unexpected imagery and playing with our sense of micro and macro perspectives.

Lem (2013) states that: 'Successful technologies used to have myth or superstition as their "theoretical foundation…"'. Live OHP projections using light-shadow immersive experiences proved to have a capacity for magic through the physical and almost ceremonial process of direct acting in time and place. The method, when refined through repetition, led to devising sequences of drawing 'performances', demonstrating morphic language of magnified forms and matter, at once mythic and ordinary.

When considering the notion of magic in technology, we can draw parallels between twentieth century overhead projectors and the development of diverse devices for production of special effects employed in the mid seventieth century Baroque. What was interesting about the Jesuit Order's scholars, such as Kircher who pioneered the construction of the popular magic lantern, was the intention of these effects to surprise and captivate the audience. The magic lanterns were used for creating a sense of mystery and staging illusions. Screenings took place in dark rooms with the projection equipment hidden from the spectators. The Grim Reaper and the female figure in flames were recurring subjects of the projections. Technology here was synonymous with magic and served as powerful instrument for religious conversions, as described by Austrian historian Fülöp-Miller (1930), in 'The Power and Secret of the Jesuits':

'Many Jesuit theatres had traps for scenes in which apparitions appeared, or disappearances were staged; further... enabled the Jesuit directors... to let gods appear, ghosts to materialise, and other difficult scenes with a high level of technical perfection.'

For the Baroque magic lantern users, the aim was to hide the technology behind the spectacle of the projections, just as the interface in the high-tech devices used in our daily lives. By contrast the analogue machines project an aura of simplicity and transparency through their visible mechanics, and this is where their power lies in stimulating our imagination. Devising these light and shadow immersive drawing sessions led me to appreciating the overhead projector's great potential as both an inspiring medium and means for visual expression, as well as a resourceful support for visual communication and abstract thinking.

Evolution in the natural world progresses very slowly and is counted in billions of years. By comparison, the changes in technology have been developing in a short space of time. The overhead projector is a great example demonstrating that the history of the development of technical instruments despite of spanning over centuries, very often is not a product of a predictable advancement from primitive to complex apparatus. It points at similarities not only in the technical principals but also evolving human-machine relationships.

3 Conclusions

A central concern with media constitutes a large part of the self-definition of contemporary art. The 'Mirrors and Projections' project contributes to this context, but is distinct, through experimentation not only with outdated technology but also by placing the overhead projector (OHP), a practical analogue tool designed for educational and business purposes, into the unfamiliar context of the art production processes.

The experiments with the OHP contribute to a rethinking of the relationship between artistic/design production and the use of tools/instruments and document the shift in emphasis away from contained artwork, to an idea of 'art/design' as a series of connected processes with the outcome being one of many components of the work, i.e., through an emphasis on time and process.

It goes without saying that the acceleration of contemporary technological advancement is so omnipresent that one does not need to be an expert to notice it. With the growing sophistication of digital equipment, comes the emphasis on dissolving the boundary between the human and machine. As the use of digital media machines has been so successful in contemporary lives, their interfaces have become increasingly invisible and so friendly that we as users often are unaware of dealing with pre-structured, calculated constructions of these machines. While many of these digital devices have a built-in obsolescence which will not allow future generations to access them, my old Liesegang favorit 400, has a potential to outlast me, unless it cannot find another creative user. With the advances in VR, fantasies of building and inhabiting parallel Worlds, how do we make best use of machines enriched by slower yet more direct and material ways of being and communicating?

This is where the attraction of the analogue technology, such as an overhead projector resonates with us, to coax new visual languages, meanings and magic from this world of old and crude but reliable machines.

References

Casper, M. (1993) Kepler. Dover Publication Inc., New York, pp 142–146
Coldwell P (2008) Between digital & physical. J New Media Caucus 3
Faure Walker J (2006) Painting the digital river: how an artist learned to love the computer. Prentice Hall, USA, p 17
Fülöp-Miller R (1930) The power and secret of the Jesuits. Putnam's, New York, p 514
Ibid., p 3
Ibid., p 135
Ibid., pp 198–202
Lem S (2013) Summa technologiae. University of Minnesota Press, Minneapolis and London, p 11
Zielinski S (2006a) Deep time of the media. The MIT Press, Cambridge, Massachusetts and London, p 2
Zielinski S (2006b) Deep time of the media. The MIT Press, Cambridge, Massachusetts and London, p 135

MIX
Papier aus verantwortungsvollen Quellen
Paper from responsible sources
FSC® C105338

If you have any concerns about our products,
you can contact us on
ProductSafety@springernature.com

In case Publisher is established outside the EU,
the EU authorized representative is:
Springer Nature Customer Service Center GmbH
Europaplatz 3, 69115 Heidelberg, Germany

Printed by Libri Plureos GmbH
in Hamburg, Germany